Canto is a paperback imp
which offers a broad range c
both classic and more rec
representing some of the b...
and most enjoyable of Cambridge
publishing.

ON
GROWTH AND FORM

ON
GROWTH AND
FORM

BY

D'ARCY WENTWORTH THOMPSON

AN ABRIDGED EDITION
EDITED BY
JOHN TYLER BONNER

CAMBRIDGE
UNIVERSITY PRESS

PUBLISHED BY THE PRESS SYNDICATE OF THE UNIVERSITY OF CAMBRIDGE
The Pitt Building, Trumpington Street, Cambridge, United Kingdom

CAMBRIDGE UNIVERSITY PRESS
The Edinburgh Building, Cambridge CB2 2RU, UK
40 West 20th Street, New York, NY 10011–4211, USA
477 Williamstown Road, Port Melbourne, VIC 3207, Australia
Ruiz de Alarcón 13, 28014 Madrid, Spain
Dock House, The Waterfront, Cape Town 8001, South Africa

http://www.cambridge.org

First published in this abridged edition 1961
Reprinted 1966
First paperback edition 1966
Reprinted 1969 1971 1975 1977 1981 1983 1984 1988 1990
Canto edition 1992
Sixth printing 2004

Printed in the United Kingdom at the University Press, Cambridge

Library of Congress Cataloguing in Publication data
Thompson, D'Arcy Wentworth
On growth and form. – Abridged ed./Edited
by John Tyler Bonner
1. Anatomy
I. Title II. Bonner, John Tyler
574.4 QH351

ISBN 0 521 43776 8 paperback

Cover illustration: Nautilus shell.
Photo: Sinclair Stammers/Science Photo Library

CONTENTS

FOREWORD

This Was a Man

STEPHEN JAY GOULD

Museum of Comparative Zoology
Harvard University

From Falstaff to the *Ring of the Nibelungen*, great constructions and great works of art have paid a price for amplitude beyond usual standards. D'Arcy Wentworth Thompson (1860–1948), Professor of Zoology at Scotland's University of St. Andrews, and perhaps the greatest polymath of our century, was scarcely *homo unius libri* (a man of one book). He composed two volumes of commentaries on all birds and fishes mentioned in classic Greek texts; he prepared the standard translation of Aristotle's *Historia animalium*; he labored for years over statistics for the Fishery Board of Scotland; and he wrote the section on pycnogonids (a small but fascinating group of arthropods) for the *Cambridge Natural History* series. But his enduring (indeed evergrowing) fame rests upon a glorious (and very long) book that served more as the active project of a lifetime than a stage of ontogeny—*On Growth and Form* (first edition of 793 pages in 1917, second edition enlarged to 1116 pages in 1942).

Much as it must pain any scholar and publisher of integrity to abridge such a work (for such an act does resemble the dissection of a body), one must not, as Jesus told us, light a candle and then place it invisibly under a bushel (Matthew 5:14–17). *On Growth and Form* is one of the great lights of science (and of English prose); it must be available at an affordable price and a totable heft: "Let your light so shine before men, that they may see your good works."

D'Arcy Thompson, according to a legend that could have been true, was offered his choice of professorships in three apparently disparate disciplines: classics, mathematics, and zoology. The greatness of *On Growth and Form* lies in its genuine integration (not just ostentatious show) of these three foci.

1. Classics and the humanities. *On Growth and Form* is the greatest work of prose in twentieth-century science. Consider the judgment of two preeminent scientific humanists: P. B. Medawar

called it "beyond comparison the finest work of literature in all the annals of science that have been recorded in the English tongue." G. E. Hutchinson regarded it as "one of the very few books on a scientific matter written in this century which will, one may be confident, last as long as our too fragile culture." D'Arcy Thompson's prose is like a Wagnerian opera: it flows on and on in waves of sumptuous sound, with occasional cadences at climactic moments. I can, for example, quote by heart the lovely last line of chapter 2, "On Magnitude," as the author descends from the ordinary gravitational world of our own species, through the realm of surface forces inhabited by insects, to the utterly unfamiliar domain of the bacterium: "We have come to the edge of a world of which we have no experience, and where all our preconceptions must be recast."

But D'Arcy Thompson can also become a victim of his own erudition, not so lightly worn. Few people today have his literary and linguistic background; even fewer will grasp the classical allusions (not to mention the untranslated lines of Greek and Latin) that are not mere adornments, but intrinsic parts of the text. The very beginning of the book can be quite daunting, with its Kantian statement in German (translated at least), followed by an unannotated Latin line from Roger Bacon and a similarly untranslated Italian footnote from Leonardo—all before the first paragraph even reaches its halfway point!

2. Mathematics. D'Arcy Thompson states his primary purpose in the epilogue: "to show that a certain mathematical aspect of morphology . . . [is] helpful, nay essential, to [the] proper study and comprehension of Growth and Form." (The 2-page epilogue then continues with two untranslated quotes in Latin, one in Greek and two in French, including the closing line.) The application of mathematics to organic form may sound like a highly modern enterprise, but D'Arcy Thompson's examples contain blessedly little of the apparatus usually employed in such efforts: theoretical modelling with differential equations or empirical treatment with sophisticated statistics. D'Arcy Thompson knew these fields, but he approached *Growth and Form* as a classical scholar, in particular as a Pythagorean geometer boosted with a knowledge of Newtonian mechanics. This book dwells in the Miraldi angle, the Fibonacci series, the logarithmic spiral and the golden ratio.

3. Zoology. D'Arcy Thompson was an invertebrate zoologist by

primary choice. He used his classical and mathematical training to full and integrative advantage in *On Growth and Form*, but his great treatise is still, primarily, a biological theory. The theory is easily lost amidst D'Arcy Thompson's elaborations and elegant examples—and also readily missed because the theory is so iconoclastic that we can't quite believe he is really saying such a thing. I have been astonished at how many people can read the book, enjoy the sweep of the prose and all the ingenious examples, and then thoroughly miss the coordinating theory!

This hybrid theory of Pythagoras and Newton argues that physical forces shape organisms directly (with "internal" and genetic forces responsible only for producing raw material, admittedly in gradients and programmed timings, for construction under principles of physics)—and that the ideal geometries beloved by classical Athens pervade organic form because natural law favors such simplicity as an optimal representation of forces. I do not know D'Arcy Thompson's religious views, but we may certainly say that his God was a geometer, just as Einstein's did not play dice with the universe. D'Arcy Thompson tells us in the epilogue:

For the harmony of the world is made manifest in Form and Number, and the heart and soul and all the poetry of Natural Philosophy are embodied in the concept of mathematical beauty . . . A greater than Milton had magnified the theme and glorified Him 'that sitteth upon the circle of the earth', saying: He hath measured the waters in the hollow of his hand, and meted out heaven with the span.

As usual, D'Arcy Thompson does not give us the explicit reference (and doesn't even include quotation marks to alert us), but the source of the last phrase is Isaiah 40:12, and the actor is Jehovah himself.

The sequence of chapters, and the logic of the entire book, focus upon this theme. We first have an introduction on mathematical approaches to form, followed by the crucial chapter "On Magnitude," with its central argument that surface/volume ratios must decline as organisms grow in size, and that large and small therefore inhabit different realms of forces. If physical forces shape organisms directly, then small creatures should be the product of surface forces, large creatures of gravitational (volumetric) forces. The next several chapters move up this continuum: small creatures are shaped by surface forces (protozoans as Plateau's surfaces of revolution, for example); intermediate sizes express a balance (jellyfishes like viscous drops, held up by surface tension, but oozing down by gravity); large creatures are ruled by gravity (the famous diagram comparing forces on a leg bone with the loading of a crane).

D'Arcy Thompson does not try to extend the argument to all nuances of form in complex creatures: the disparity between a hippo and an oak tree does not simply lie in differing external forces acting upon their growth. Two strategies for bringing complex forms into the general theory structure the remaining chapters of *Growth and Form*. (1) Parts or wholes, even when not shaped directly by physical forces, take optimal forms of ideal geometry as solutions to problems of morphology (the equiangular spirals of mollusks, ram horns and paths of moths flying to light as the only way to coil and maintain the same shape as size increases; alternating spirals of the Fibonacci series in phyllotaxis as a consequence of optimal space filling in systems adding new units at a pole). (2) Even if complex prototypes must be accepted as genetic givens, at least their transformations to related forms may be expressed as simple physical deformations of entire systems— the theory of transformed coordinates. (This theory, subject of the last and most widely influential chapter, is often misunderstood as a coda unrelated to the rest of the book. But when we realize that simplicity in geometry of the transformation grid, not complexity of the underlying prototype, expressed D'Arcy Thompson's main concern, then we can grasp the stunning unit of the entire book.)

We may view *Growth and Form* as a coherent triumph today, but D'Arcy Thompson suffered for his iconoclasm during life. He did not intend to spend his career at a small and isolated Scottish University; he had applied for key jobs at Oxford, Cambridge, and London, and had lost out every time. He received many accolades late in life, including an honorary degree from Oxford in 1945, but marginality was his fate throughout the central years of his professional life.

I wrote one of my earliest articles about D'Arcy Thompson in 1971, when I was still in my twenties—and I misread him (or grasped him only partially) as a consequence of my misplaced romanticism. I saw him as a noble Canute holding back the tides of modernistic philistinism—a Greek geometer more than two thousand years from his time of flourishing. This timelessness is, indeed, part of the man—as when, in explaining the tests of foraminifera (single celled protozoans), he rejects the contingency of genealogy, and speaks of geometric permanence, unaltered by mere time.

But D'Arcy Thompson was also a man of his own age, and I missed this theme. His doubts about Darwinism (combined with confidence about the fact of evolution) reflect a common view of 1917 (date of the first edition), though they had become passé by the second edition of 1942. His two major criticisms are exactly those highlighted by his more conventional contemporary, William Bateson, though D'Arcy Thompson gave a personal twist to each standard doubt: (1) Do not make up speculative stories about natural selection just because a gradual transition can be observed (for D'Arcy Thompson, such transitions might only reflect a changing set of external forces acting on unaltered biological material). (2) Some changes must be saltational rather than gradual (just as some geometries can only transform into others through a discontinuity).

Will the real D'Arcy Thompson stand up? Classicist? Prose stylist? Compiler of elegant examples? Iconoclastic morphologist? Contemporary critic of Darwinism? Greek geometer? He was all this and much more. Shakespeare may as well have the last word: ". . . the elements so mixed in him that Nature might stand up and say to all the world, 'this was a man!'"

1992

THE EDITOR'S INTRODUCTION

There are two justifications for a new edition of D'Arcy Wentworth Thompson's *On Growth and Form*: One is that a shorter version might make the work more available, at least to the general reader, and the other is that the 1942 edition contains many passages that are now out-of-date. If the book is to retain its importance, which it has maintained since its first publication in 1917, a mild freshening in the form of commentary does not seem out of place.

Of its importance there is no doubt, but I must agree with Medawar* when he says that its considerable influence 'has been intangible and indirect'. I will shortly mention some of the facets which make the book so distinctive and unique, and all of these have contributed to its success. But I believe that the cardinal point is that D'Arcy Thompson was consistently able to examine subjects of significance in biology from a fresh point of view, and the mere fact that there was another point of view (sometimes one first imagined in early antiquity) comes as a shock, and therefore a stimulus, to those who so easily fall into the scientific fads and fashions of our day, and make little effort to look beyond the horizon of the 'current views'.

The most conspicuous attitude in the book is the analysis of biological processes from their mathematical and physical aspects. I will not enter into the details of this approach here, for he discusses it repeatedly in the text, especially in the Introduction, and others† have given it careful examination. Except for the details of some specific analyses, the general approach, in this day of biophysics, mathematical biology, cybernetics, etc., is hardly novel, although it must be remembered that it was very much so in 1917 at the time of the first edition.

There are two particular aspects of his physics and mathematics

---

* All those who are interested in both the man as well as his work will find Miss Ruth D'Arcy Thompson's biography of her father required reading (*D'Arcy Wentworth Thompson*, Oxford University Press, 1958) which includes as a postscript a penetrating critique by P. B. Medawar of *On Growth and Form*. I must confess that both Medawar's essay and the equally successful one of G. E. Hutchinson (In Memoriam, D'Arcy Wentworth Thompson, *Amer. Sci.* 36 (1948), 577), have made this Introduction both an easy and a difficult task for me; easy because they have told me what to say, and difficult because it is hard not to use their well-chosen words, a temptation, as will be evident, that I have not resisted.

† G. K. Plochman, *Phil. Sci.* 20 (1953), 139; F. Mayer, *Anat. Rec.* 85 (1943), 111.

that deserve comment. One is that since he was clearly concerned with the explanation of biological growth and form in physico-mathematical terms the reader must often be prepared for disappointment if he expects to find immediate causes. Most experimental scientists are only mentally satisfied if they can understand a particular form by the configuration of its immediate precedents, and the precedents in turn are analysed in the same way so that an epigenetic chain is exposed; this is the basis, for instance, of 'causal' embryology. D'Arcy Thompson, on the other hand, was quite satisfied with a mathematical description or a physical analogy. No doubt this condition of mind is closely tied in with the fact that he himself was in no way an experimenter. He even refused, in his 1942 edition, to recognise experiments that bore on the facts of his 1917 edition. But this singleness of purpose on his part has the merit and the power that it constantly suggests experiments or new experimental approaches and, therefore, his particular blindness has been an advantage to all of us and perhaps because of it he should be praised rather than criticised.

The other aspect of his approach is that it is missing still another piece. Chemistry, except for those physical aspects of chemistry, is almost wholly absent. In biology, biochemistry has become the most potent source of new knowledge, the most rapidly growing forest of new information, yet the idea that form and its change is related to the reactive properties of chemical substances is almost completely ignored. It is true that if he had made this a central part of his story, the progress in that field has been so staggering that the book would probably have been out-of-date many times over and its influence long past. Again his weakness in an area which has been the main current of biological advances has been an advantage and made his work endure.

Another significant fact which contributes to the importance and uniqueness of *On Growth and Form* is that it is good literature as well as good science; it is a discourse on science as though it were a humanity. Medawar considers it 'beyond comparison the finest work of literature in all the annals of science that have been recorded in the English tongue'. And I do not think this very high praise unjustified.

The reasons may partly be found in D'Arcy Thompson's background. His father was a classicist and took a close hand in his son's education. It meant that in later life he was able, besides his science, to translate Aristotle's *Historia Animalium* and write a *Glossary of*

Greek Birds and a *Glossary of Greek Fishes* in which he gathered all the birds and fishes mentioned in Greek literature and from his zoological knowledge illuminated the references and fused science with the classics. He was, and again I quote Medawar,

an aristocrat of learning whose intellectual endowments are not likely ever again to be combined within one man. He was a classicist of sufficient distinction to have become President of the Classical Associations of England and Wales and of Scotland; a mathematician good enough to have had an entirely mathematical paper accepted for publication by the Royal Society; and a naturalist who held important chairs for sixty-four years, that is, for all but the length of time into which we must nowadays squeeze the whole of our lives from birth until professional retirement. He was a famous conversationalist and lecturer (the two are often thought to go together, but seldom do), and the author of a work which, considered as literature, is the equal of anything of Pater's or Logan Pearsall Smith's in its complete mastery of the *bel canto* style. Add to all this that he was over six feet tall, with the build and carriage of a Viking and with the pride of bearing that comes from good looks known to be possessed.

D'Arcy Thompson (he was always called that, or D'Arcy) had not merely the makings but the actual accomplishments of three scholars. All three were eminent, even if, judged by the standards which he himself would have applied to them, none could strictly be called great. If the three scholars had merely been added together in D'Arcy Thompson, each working independently of the others, then I think we should find it hard to repudiate the idea that he was an amateur, though a patrician among amateurs; we should say, perhaps, that great as were his accomplishments, he lacked that deep sense of engagement that marks the professional scholar of the present day. But they were not merely added together; they were integrally—Clifford Dobell said chemically—combined. I am trying to say that he was not one of those who have made two or more separate and somewhat incongruous reputations, like a composer-chemist or politician-novelist, or like the one man who has both ridden in the Grand National and become an F.R.S.; but that he was a man who comprehended many things with an undivided mind. In the range and quality of his learning, the uses to which he put it, and the style in which he made it known, I see not an amateur, but, in the proper sense of that term, a natural philosopher.

The style, of which Medawar speaks, is certainly a conspicuous feature of the book. I know from experience that in their first encounter undergraduates find it too rich, too 'fancy'; it lacks the conventional condensation and emasculation of modern scientific, machine-made, basic prose. But I also note that some of the students learn to admire it and see that there may not always necessarily be a cleft between their courses in science and their courses in letters, and that conventions of science, no matter how well established, are sometimes profitably disregarded.

The historical sense is also a strong part of D'Arcy Thompson's writing. A distinction often made between science and the humanities is that in science, a cumulative discipline, one need study only the latest paper on any subject to obtain all the background necessary for further investigations. Relatively few papers in a journal of bacteriology, for example, mention Pasteur or van Leeuwenhoek, even though every paper in the journal might be traced ultimately to the work of those great men. In the humanities, on the other hand, contributions are important without reference to time or sequence, and we still read and profit from Shakespeare or Gibbon. History in particular is a sequence of facts, ideas and interpretations, and in this sense *On Growth and Form* is a history of science along with the scientific end-product. As Hutchinson says

On Growth and Form is a great book because it shows us science as a traditional activity and tells us that the tradition is one of daring and of imagination.... To foster the values of civilized men takes many talents, and few possess a great share of them. D'Arcy Wentworth Thompson had more than most of us and he used them superbly. What he wrote brings home to the scientific mind, perhaps better than the work of any other writer, what it means to be civilized; the splendour of the panorama that he has presented gives value to what we try to do, and in giving us this he achieved a reward given to few men.

But with all this impressive armoury of talents he was, in the world of science, a lonely figure. He himself was well aware of this and was even proud that he did not run with the pack. I have already showed how this quality contributed in various ways to the making of *On Growth and Form*, but I have not yet said anything about his views on evolution which are constantly creeping to the surface throughout the book.

They are ideas that were heretic in 1917 and it must be admitted that, for partly different reasons, they remain so today. Yet despite the fact that they are not wholly palatable, there is no doubt that they stimulate and jar any unthinking complacency. He does not deny natural selection, but suggests that it operates only to eliminate the unfit and does not serve as a progressive force in evolution. The idea that every structure is an inherited adaptation arising through selection he considers particularly unjustified and unreasonable; in many cases he would consider the structure to have arisen by direct physical forces, molecular in very small structures and mechanical

xvii

in larger ones (that is, 'direct adaptations'). Heredity and the activity of genes in development are wholly missing in the book except for a few passing references which seem to imply that they do not fit into his scheme of things. (Again we have an instance here, as we did in the case of biochemistry, of a complete omission of one of the most important trends in biological thought in the past fifty years.) Phylogeny, the study of animal ancestry and relationships, which was the central concern of the comparative anatomists at the turn of the century, is pushed aside and replaced by the idea that the functional aspect of form is far more important than blood relationships and family trees.

I do not think the modern reader should be disturbed by these deviations from the accepted dogma of today, for to dismiss D'Arcy Thompson on these grounds means a blind loss of many riches. And furthermore the particular stress on the formative power of physical forces is bound to make any evolutionary scheme more rounded, for these forces are real and cannot be ignored. This is no place for a complete discussion of modern evolutionary theory, but let me briefly show how the heterodoxies of D'Arcy Thompson can be easily transformed into orthodoxy.

Nowadays we think of variation as having a genetic foundation and that evolutionary progress can be reasonably accounted for on the basis of gene change (mutation) and selection. Furthermore, during development, during the period of growth and the achievement of final form, the particular gene complement of an organism is directly, through spatially organised chemical reactions, responsible for the 'growth and form'. The fact that often the form is mechanically efficient is explained in two ways. If any given gene combination produces a structure of good mechanical design, it may, because of this, have a high adaptive value and remain in the population through favourable selection. In other instances there is clearly a direct effect of the environment which causes, by mechanical or physical forces, the form of a living structure. In the case of these direct adaptations we may assume that such responsiveness to the environment is adaptively advantageous and, therefore, the gene complement that favours responsiveness or reactivity to these environmental conditions is favoured and maintained by selection. Then there are those curious cases, such as the callosities of an ostrich, where there is both a direct adaptation in the form of increased thickening by abrasion, and a complete inheritance of the callosities existing in the embryo before hatching, before the mechanical forces can operate. There is no space here for a discussion of this interesting phenomenon, but it is briefly examined in Chapter VIII where it is shown that no doubt it can be reconciled with our current concepts of development, genetics and evolution.

Lastly, in this general introduction, I should like to explain the basis of the deletions I have made in the text; perhaps it would be more correct to say 'justify', for it is a grave responsibility tampering with a classic. The fact that some editing might be possible has been on my mind for a long time because it is generally recognised that the 1917 edition of 793 pages is in many ways a better book than the 1942 edition of 1116 pages. The argument is that the basic ideas are all there in the 1917 edition, and the great increase in bulk of the second edition is largely due to additional examples interpolated here and there. To put it another way, the 1917 edition is superior because his ideas have been stretched on a canvas of a more appropriate size. If this is so then it is just possible that a further shrinking would make his points all the more forceful, especially if the out-dated or erroneous passages are removed.

But in this reducing I have taken great pains never to condense, rewrite, or digest his material; the words are all his own.* Sometimes, as I have just intimated, my desire to keep the text modern has been my guide; sometimes it is simply a desire to keep the interest of the reader. Frequently I found the 1917 edition helpful in suggesting deletions (although in one case an eliminated section of the 1917 edition was restored). Inevitably in this process some material of importance is bound to be lost; it can only be hoped that the loss has not been too great.

While there are minor deletions in all the chapters, there are some major ones that should be mentioned here. The largest omission is the complete chapter on 'The rate of growth'. This large chapter of over 200 pages is, as numerous critics† have pointed out, weak for a number of reasons. Its organisation is loose and repetitive, the emphasis on growth in man is not always helpful, and the author's lack of appreciation of the multiplicative aspects of growth led him into difficulties, especially into failing to appreciate the importance of differential or relative growth (allometry) to which J. S. Huxley and G. Teissier have made such notable contributions. Perhaps it is simplest to say that those interested in the rate of growth would do

* The only exception is that on a few occasions I have had to alter a connecting sentence to close the gap after a portion has been removed.

† E. W. Sinnott, *Quart. Rev. Biol.* **18** (1943), 65; F. Mayer, *Anat. Rec.* **85** (1943), 111; P. B. Medawar, *op. cit.*

well to consult elsewhere and examine, for instance, S. Brody* or J. S. Huxley.†

The chapter 'On the internal form and structure of the cell' has also been eliminated, largely because it is so completely out-of-date. Cytology, with the current rapid progress in electron microscopy, is becoming a wholly new science. The problem of mitosis has hardly been solved, but we now have a vast array of new information that has been marshalled in some excellent recent reviews, such as those of F. Shrader,‡ A. Hughes§ and M. M. Swann.¶ D'Arcy Thompson's chapter has lost its usefulness.

In the short chapter called 'A note on absorption', which has also been eliminated, there is an attempt to show that the phenomenon of absorption, which was brought to light so clearly by Willard Gibbs, is possibly playing a role in cell form. Today we would continue to accept the notion that absorption, along with all other physical forces, must operate within cells and affect them, but somehow D'Arcy Thompson's brief analysis of this one aspect does not seem especially helpful to us at the moment.

Another short chapter that is gone is 'A parenthetic note on geodesics', where again this diversion into geometry, and the fact that many organisms (such as the chloroplasts of *Spirogyra*) can be described as geodesics, seems of minor importance to the main stream of thought.

The decision to remove the chapter 'On leaf arrangement, or phyllotaxis' was a more difficult one to make. The main reason was that D'Arcy Thompson has really contributed no new information to this old subject, although his chapter is an excellent summary of the old views and the games with numbers. Also he makes no mention of the recent revival of interest in the problem, along with some intriguing experimental work by such authors as Snow and Snow, Wardlaw and others.‖

The last major section that has been eliminated is 'on the shapes of eggs, and of certain other hollow structures'. The whole first portion is a delightful discussion of the shape of the eggs of birds. While there are many facets to it, and some of lasting interest, the

---

* *Bioenergetics and Growth* (Reinhold, New York, 1945).
† *Problems of Relative Growth* (Methuen, London, 1932).
‡ *Mitosis*, 2nd ed. (Columbia University Press, 1953).
§ *The Mitotic Cycle* (Academic Press, New York, 1952).
¶ *Cancer Res.* **17** (1957), 727; **18** (1958), 18.
‖ R. Snow gives a short but useful review of the problems of phyllotaxis in *Endeavour*, **14** (1955), 190.

main point rests on the argument that the peristaltic waves in the oviduct shape the egg blunt end forward, for it is a well-known fact that when the egg is laid the blunt end does indeed emerge first. But the sad truth, as Bradfield* has shown recently in some radiographic studies, is that the egg passes down the oviduct pointed end first and is flipped around just before laying. So all the compelling argument is precisely the reverse of the facts. To make matters worse, this was suspected for a long time and was even discussed by Needham in 1931.† There is also a section in this chapter on the shape of sea-urchin tests which is only moderately convincing, and finally there are a few excellent pages on the form or angle of branching of blood vessels, most of which has been preserved and appended to Chapter IV.

A discussion of what has been removed can only be confusing and unsatisfactory, and now we can turn to more positive matters and give the heart of the book which has survived, and even shines, despite the erosion of years and the whims of an editor.

* *J. Exp. Biol.* **28** (1951), 125.
† *Chemical Embryology* (Cambridge University Press, 1931), p. 233.

1961

Addendum

I wrote the above introduction to this book over thirty years ago and in those thirty years there has been quite extraordinary progress in the field of biology. The most spectacular change has been the rise and the continuous blossoming of molecular biology. This has had a huge impact on developmental biology which, as a result, has become part of cell biology. More closely related to D'Arcy Thompson's ideas are the significant advances in the use of mathematics to analyse the mechanics of development. This approach received its first big push from the distinguished mathematician Alan Turing, who showed one could develop useful models which combine the activity of chemical reactions inside the embryo with physical forces such as diffusion. As was true thirty years ago, all these new accomplishments of biology do not supplant D'Arcy Thompson's message: they complement it.

There is one specific way in which his views fit in with modern evolutionary and developmental biology. Ever since the rise of genetics (and greatly enforced by the more recent molecular genetics) there has been a tendency to consider that all the growth and form of an organism can be explained in terms of gene instructions. However, today there is an increasing awareness of what seems rather obvious: that the genes are not acting alone, but are governed by physico-chemical rules which severely limit what shapes, what morphologies are possible. In essence this is what D'Arcy Thompson says; it is simply put in different words. The genes, which are responsible for producing proteins, are, in turn, responsible for producing structure or form, but there are severe limits to their role. Those limits are due to the nature of the very structures for which they provide the assembly instructions. There are physical limits to what can be created in the way of the form. The shape of the structure to be built during developments is, to a considerable degree, imposed by the materials used and the way they are put together. Think of it in terms of a simple analogy. An architect and his builder cannot create any shape house they please; they are confined by the limits of the physical properties of steel, bricks, and wood. So it is with living organisms: the laws of physics set the ultimate rules and there can be no exceptions.

In the recent literature these physical rules which limit morphology are called 'constraints'. There are those, the so-called 'structuralists', who believe that the physical forces which set those constraints are themselves entirely responsible for guiding evolution and development. This, however, seems to me an extreme position, one often motivated by a desire to emphasize their belief that something is basically wrong with Darwinism. My view is more conservative. Neither the role of natural selection nor the laws of physics can be ignored; together they are responsible for development and evolution.

What is different today as compared to thirty years ago is our appreciation of the importance of constraints and the physical laws that are responsible for them. If one thinks of D'Arcy Thompson's principal message in this way, if anything, it glows even more brilliantly in the 1990s than it did in the 1960s.

1992

TYPOGRAPHICAL NOTE

In this book the editor's commentaries are set in type smaller than that used for the author's text. The editor's footnotes are indicated by symbols (*, †, ‡ etc.) and are separated by a horizontal line from those of the author, which are indicated by raised numerals ([1], [2], [3] etc.).

CHAPTER I

INTRODUCTORY

Of the chemistry of his day and generation, Kant declared that it was a science, but not Science—*eine Wissenschaft, aber nicht Wissenschaft*—for that the criterion of true science lay in its relation to mathematics. This was an old story: for Roger Bacon had called mathematics *porta et clavis scientiarum*, and Leonardo da Vinci had said much the same. Once again, a hundred years after Kant, Du Bois Reymond, profound student of the many sciences on which physiology is based, recalled the old saying, and declared that chemistry would only reach the rank of science, in the high and strict sense, when it should be found possible to explain chemical reactions in the light of their causal relations to the velocities, tensions and conditions of equilibrium of the constituent molecules; that, in short, the chemistry of the future must deal with molecular mechanics by the methods and in the strict language of mathematics, as the astronomy of Newton and Laplace dealt with the stars in their courses. We know how great a step was made towards this distant goal as Kant defined it, when van't Hoff laid the firm foundations of a mathematical chemistry, and earned his proud epitaph—*Physicam chemiae adiunxit.*

We need not wait for the full realisation of Kant's desire, to apply to the natural sciences the principle which he laid down. Though chemistry fall short of its ultimate goal in mathematical mechanics, nevertheless physiology is vastly strengthened and enlarged by making use of the chemistry, and of the physics, of the age. Little by little it draws nearer to our conception of a true science with each branch of physical science which it brings into relation with itself: with every physical law and mathematical theorem which it learns to take into its employ. Between the physiology of Haller, fine as it was, and that of Liebig, Helmholtz, Ludwig, Claude Bernard, there was all the difference in the world.[1]

As soon as we adventure on the paths of the physicist, we learn

---

[1] It is well within my own memory how Thomson and Tait, and Klein and Sylvester had to lay stress on the mathematical aspect, and urge the mathematical study, of physical science itself!

to *weigh* and to *measure*, to deal with time and space and mass and their related concepts, and to find more and more our knowledge expressed and our needs satisfied through the concept of *number*, as in the dreams and visions of Plato and Pythagoras; for modern chemistry would have gladdened the hearts of those great philosophic dreamers. Dreams apart, numerical precision is the very soul of science, and its attainment affords the best, perhaps the only criterion of the truth of theories and the correctness of experiments.[1] So said Sir John Herschel, a hundred years ago; and Kant had said that it was Nature herself, and not the mathematician, who brings mathematics into natural philosophy.

But the zoologist or morphologist has been slow, where the physiologist has long been eager, to invoke the aid of the physical or mathematical sciences; and the reasons for this difference lie deep, and are partly rooted in old tradition and partly in the diverse minds and temperaments of men. To treat the living body as a mechanism was repugnant, and seemed even ludicrous, to Pascal; and Goethe, lover of nature as he was, ruled mathematics out of place in natural history. Even now the zoologist has scarce begun to dream of defining in mathematical language even the simplest organic forms. When he meets with a simple geometrical construction, for instance in the honeycomb, he would fain refer it to psychical instinct, or to skill and ingenuity, rather than to the operation of physical forces or mathematical laws; when he sees in snail, or nautilus, or tiny foraminiferal or radiolarian shell a close approach to sphere or spiral, he is prone of old habit to believe that after all it is something more than a spiral or a sphere, and that in this 'something more' there lies what neither mathematics nor physics can explain. In short, he is deeply reluctant to compare the living with the dead, or to explain by geometry or by mechanics the things which have their part in the mystery of life. Moreover he is little inclined to feel the need of such explanations, or of such extension of his field of thought. He is not without some justification if he feels that in admiration of nature's handiwork he has an horizon before his eyes as wide as any man requires. He has the help of many fascinating theories within the bounds of his own science, which, though a little lacking in precision, serve the purpose of ordering his thoughts and of suggesting new objects of enquiry. His art of classification becomes an endless search

---

[1] Dr Johnson says that 'to count is a modern practice, the ancient method was to guess'; but Seneca was alive to the difference—'magnum esse solem philosophus probabit, quantus sit mathematicus'.

after the blood-relationships of things living and the pedigrees of things dead and gone. The facts of embryology record for him (as Wolff, von Baer and Fritz Müller proclaimed) not only the life-history of the individual but the ancient annals of its race. The facts of geographical distribution or even of the migration of birds lead on and on to speculations regarding lost continents, sunken islands, or bridges across ancient seas. Every nesting bird, every ant-hill or spider's web, displays its psychological problems of instinct or intelligence. Above all, in things both great and small, the naturalist is rightfully impressed and finally engrossed by the peculiar beauty which is manifested in apparent fitness or 'adaptation'—the flower for the bee, the berry for the bird.

Some lofty concepts, like space and number, involve truths remote from the category of causation; and here we must be content, as Aristotle says, if the mere facts be known. But natural history deals with ephemeral and accidental, not eternal nor universal things; their causes and effects thrust themselves on our curiosity, and become the ultimate relations to which our contemplation extends.

Time out of mind it has been by way of the 'final cause', by the teleological concept of end, of purpose or of 'design', in one of its many forms (for its moods are many), that men have been chiefly wont to explain the phenomena of the living world; and it will be so while men have eyes to see and ears to hear withal. With Galen, as with Aristotle, it was the physician's way; with John Ray, as with Aristotle, it was the naturalist's way; with Kant, as with Aristotle, it was the philosopher's way. It was the old Hebrew way, and has its splendid setting in the story that God made 'every plant of the field before it was in the earth, and every herb of the field before it grew'. It is a common way, and a great way; for it brings with it a glimpse of a great vision, and it lies deep as the love of nature in the hearts of men.

The argument of the final cause is conspicuous in eighteenth-century physics, half overshadowing the 'efficient' or physical cause in the hands of such men as Euler, or Fermat or Maupertuis, to whom Leibniz had passed it on. Half overshadowed by the mechanical concept, it runs through Claude Bernard's *Leçons sur les phénomènes de la Vie*, and abides in much of modern physiology. Inherited from Hegel, it dominated Oken's *Naturphilosophie* and lingered among his later disciples, who were wont to liken the course of organic evolution not to the straggling branches of a tree, but to the building of a temple, divinely planned, and the crowning of it with its polished minarets.

3

It is retained, somewhat crudely, in modern embryology, by those who see in the early processes of growth a significance 'rather prospective than retrospective', such that the embryonic phenomena must 'be referred directly to their usefulness in building up the body of the future animal':[1]—which is no more, and no less, than to say, with Aristotle, that the organism is the τέλος, or final cause, of its own processes of generation and development. It is writ large in that Entelechy which Driesch rediscovered, and which he made known to many who had neither learned of it from Aristotle, nor studied it with Leibniz, nor laughed at it with Rabelais and Voltaire. And, though it is in a very curious way, we are told that teleology was 'refounded, reformed and rehabilitated' by Darwin's concept of the origin of species;[2] for, just as the older naturalists held (as Addison puts it)[3] that 'the make of every kind of animal is different from that of every other kind; and yet there is not the least turn in the muscles, or twist in the fibres of any one, which does not render them more proper for that particular animal's way of life than any other cut or texture of them would have been': so, by the theory of natural selection, 'every variety of form and colour was urgently and absolutely called upon to produce its title to existence either as an active useful agent, or as a survival' of such active usefulness in the past. But in this last, and very important case, we have reached a teleology without a τέλος, as men like Butler and Janet have been prompt to show, an 'adaptation' without 'design', a teleology in which the final cause becomes little more, if anything, than the mere expression or resultant of a sifting out of the good from the bad, or of the better from the worse, in short of a process of mechanism. The apparent manifestations of purpose or adaptation become part of a mechanical philosophy, 'une forme méthodologique de connaissance', according to which 'la Nature agit toujours par les moyens les plus simples',[4] and 'chaque chose finit toujours par s'accommoder à son milieu', as in the Epicurean creed or aphorism that Nature *finds a use* for everything. In short, by a road which resembles but is not the same as Maupertuis's road, we find our way to the very world in which we are living, and find that, if it be not, it is ever tending to become, 'the best of all possible worlds'.[5]

[1] Conklin, 'Embryology of Crepidula', *J. Morph.* **13** (1897), 203; cf. F. R. Lillie, 'Adaptation in Cleavage', *Woods Hole Biol. Lectures* (1899), pp. 43–67.
[2] Ray Lankester, art. Zoology, *Encycl. Brit.* (9th ed. 1888), p. 806.
[3] *Spectator*, no. 120.
[4] Janet, *Les Causes Finales* (1876), p. 350.
[5] The phrase is Leibniz's, in his *Théodicée*: and harks back to Aristotle—If one way be better than another, that you may be sure is Nature's way; *Nic. Eth.* 1099 *b*, 23 *et al.*

But the use of the teleological principle is but one way, not the whole or the only way, by which we may seek to learn how things came to be, and to take their places in the harmonious complexity of the world. To seek not for ends but for antecedents is the way of the physicist, who finds 'causes' in what he has learned to recognise as fundamental properties, or inseparable concomitants, or unchanging laws, of matter and of energy. In Aristotle's parable, the house is there that men may live in it; but it is also there because the builders have laid one stone upon another. It is as a *mechanism*, or a mechanical construction, that the physicist looks upon the world; and Democritus, first of physicists and one of the greatest of the Greeks, chose to refer all natural phenomena to mechanism and set the final cause aside.

Still, all the while, like warp and woof, mechanism and teleology are interwoven together, and we must not cleave to the one nor despise the other; for their union is rooted in the very nature of totality. We may grow shy or weary of looking to a final cause for an explanation of our phenomena; but after we have accounted for these on the plainest principles of mechanical causation it may be useful and appropriate to see how the final cause would tally with the other, and lead towards the same conclusion. In our own day the philosopher neither minimises nor unduly magnifies the mechanical aspect of the Cosmos; nor need the naturalist either exaggerate or belittle the mechanical phenomena which are profoundly associated with Life, and inseparable from our understanding of Growth and Form.

Nevertheless, when philosophy bids us hearken and obey the lessons both of mechanical and of teleological interpretation, the precept is hard to follow: so that oftentimes it has come to pass, just as in Bacon's day, that a leaning to the side of the final cause 'hath intercepted the severe and diligent enquiry of all real and physical causes', and has brought it about that 'the search of the physical cause hath been neglected and passed in silence'. So long and so far as 'fortuitous variation'[1] and the 'survival of the fittest' remain engrained as fundamental and satisfactory hypotheses in the philosophy of biology, so long will these 'satisfactory and specious causes' tend to stay 'severe and diligent enquiry...to the great arrest and prejudice of future discovery'. Long before the great

[1] The reader will understand that I speak, not of the 'severe and diligent enquiry' of variation or of fortuity, but merely of the easy assumption that these phenomena are a sufficient basis on which to rest, with the all-powerful help of natural selection, a theory of definite and progressive evolution.

Lord Keeper wrote these words, Roger Bacon had shown how easy it is, and how vain, to survey the operations of Nature and idly refer her wondrous works to chance or accident, or to the immediate interposition of God.

The difficulties which surround the concept of ultimate or 'real' causation, in Bacon's or Newton's sense of the word, the insuperable difficulty of giving any just and tenable account of the relation of cause and effect from the empirical point of view, need scarcely hinder us in our physical enquiry. As students of mathematical and experimental physics we are content to deal with those antecedents, or concomitants, of our phenomena without which the phenomenon does not occur—with causes, in short, which, *aliae ex aliis aptae et necessitate nexae*, are no more, and no less, than conditions *sine qua non*. Our purpose is still adequately fulfilled: inasmuch as we are still enabled to correlate, and to equate, our particular phenomena with more and more of the physical phenomena around, and so to weave a web of connection and interdependence which shall serve our turn, though the metaphysician withhold from that interdependence the title of causality. We come in touch with what the schoolmen called a *ratio cognoscendi*, though the true *ratio efficiendi* is still enwrapped in many mysteries. And so handled, the quest of physical causes merges with another great Aristotelian theme—the search for relations between things apparently disconnected, and for 'similitude in things to common view unlike'.[1] Newton did not show the cause of the apple falling, but he showed a similitude ('the more to increase our wonder, with an apple') between the apple and the stars. By doing so he turned old facts into new knowledge; and was well content if he could bring diverse phenomena under 'two or three Principles of Motion' even 'though the Causes of these Principles were not yet discovered'.

Moreover, the naturalist and the physicist will continue to speak of 'causes', just as of old, though it may be with some mental reservations: for, as a French philosopher said in a kindred difficulty: 'ce sont là des manières de s'exprimer, et si elles sont interdites il faut renoncer à parler de ces choses'.

---

[1] 'Plurimum amo analogias, fidelissimos meos magistros, omnium Naturae arcanorum conscios', said Kepler; and Perrin speaks with admiration, in *Les Atomes*, of men like Galileo and Carnot, who 'possessed the power of perceiving analogies to an extraordinary degree'. Hume declared, and Mill said much the same thing, that all reasoning whatsoever depends on resemblance or analogy, and the power to recognise it. Comparative anatomy (as Vicq d'Azyr first called it), or comparative physics (to use a phrase of Mach's), are particular instances of a sustained search for analogy or similitude.

6

The search for differences or fundamental contrasts between the phenomena of organic and inorganic, of animate and inanimate things, has occupied many men's minds, while the search for community of principles or essential similitudes has been pursued by few; and the contrasts are apt to loom too large, great though they may be. M. Dunan, discussing the *Problème de la Vie*,[1] in an essay which M. Bergson greatly commends, declares that 'les lois physico-chimiques sont aveugles et brutales; là où elles règnent seules, au lieu d'un ordre et d'un concert, il ne peut y avoir qu'incohérence et chaos'. But the physicist proclaims aloud that the physical phenomena which meet us by the way have their forms not less beautiful and scarce less varied than those which move us to admiration among living things. The waves of the sea, the little ripples on the shore, the sweeping curve of the sandy bay between the headlands, the outline of the hills, the shape of the clouds, all these are so many riddles of form, so many problems of morphology, and all of them the physicist can more or less easily read and adequately solve: solving them by reference to their antecedent phenomena, in the material system of mechanical forces to which they belong, and to which we interpret them as being due. They have also, doubtless, their *immanent* teleological significance; but it is on another plane of thought from the physicist's that we contemplate their intrinsic harmony[2] and perfection, and 'see that they are good'.

Nor is it otherwise with the material forms of living things. Cell and tissue, shell and bone, leaf and flower, are so many portions of matter, and it is in obedience to the laws of physics that their particles have been moved, moulded and conformed.[3] They are no exception to the rule that Θεὸς ἀεὶ γεωμετρεῖ. Their problems of form are in the first instance mathematical problems, their problems

---

[1] *Revue Philosophique*, 33 (1892).

[2] What I understand by 'holism' is what the Greeks called ἁρμονία. This is something exhibited not only by a lyre in tune, but by all the handiwork of craftsmen, and by all that is 'put together' by art or nature. It is the 'compositeness of any composite whole'; and, like the cognate terms κρᾶσις or σύνθεσις, implies a balance or attunement. Cf. John Tate, in *Class. Rev.* (February 1939).

[3] This general principle was clearly grasped by Mr George Rainey many years ago, and expressed in such words as the following: 'It is illogical to suppose that in the case of vital organisms a distinct force exists to produce results perfectly within the reach of physical agencies, especially as in many instances no end could be attained were that the case, but that of opposing one force by another capable of effecting exactly the same purpose.' (On artificial calculi, *Quart. J. Micr. Sci. (Trans. Micr. Soc.)*, 6 (1858), 49.) (Mr George Rainey, a man of learning and originality, was demonstrator of anatomy at St Thomas's; he followed that modest calling to a great age, and is remembered by a few old pupils with peculiar affection.)

of growth are essentially physical problems, and the morphologist is, *ipso facto*, a student of physical science.

Apart from the physico-chemical problems of physiology, the road of physico-mathematical or dynamical investigation in morphology has found few to follow it; but the pathway is old. The way of the old Ionian physicians, of Anaxagoras,[1] of Empedocles and his disciples in the days before Aristotle, lay just by that highway side. It was Galileo's and Borelli's way; and Harvey's way, when he discovered the circulation of the blood. It was little trodden for long afterwards, but once in a while Swammerdam and Réaumur passed thereby. And of later years Moseley and Meyer, Berthold, Errera and Roux have been among the little band of travellers. We need not wonder if the way be hard to follow, and if these wayfarers have yet gathered little. A harvest has been reaped by others, and the gleaning of the grapes is slow.

It behoves us always to remember that in physics it has taken great men to discover simple things. They are very great names indeed which we couple with the explanation of the path of a stone, the droop of a chain, the tints of a bubble, the shadows in a cup. It is but the slightest adumbration of a dynamical morphology that we can hope to have until the physicist and the mathematician shall have made these problems of ours their own, or till a new Boscovich shall have written for the naturalist the new *Theoria Philosophiae Naturalis*.

How far even then mathematics will suffice to describe, and physics to explain, the fabric of the body, no man can foresee. It may be that all the laws of energy, and all the properties of matter, and all the chemistry of all the colloids are as powerless to explain the body as they are impotent to comprehend the soul. For my part, I think it is not so. Of how it is that the soul informs the body, physical science teaches me nothing; and that living matter influences and is influenced by mind is a mystery without a clue. Consciousness is not explained to my comprehension by all the nerve-paths and neurones of the physiologist; nor do I ask of physics how goodness shines in one man's face, and evil betrays itself in another. But of the construction and growth and working of the body, as of all else that is of the earth earthy, physical science is, in my humble opinion, our only teacher and guide.

Often and often it happens that our physical knowledge is in-

---

[1] Whereby he incurred the reproach of Socrates, in the *Phaedo*. See Clerk Maxwell on 'Anaxagoras as a Physicist', in *Phil. Mag.* (4), **46** (1873), 453–60.

adequate to explain the mechanical working of the organism; the phenomena are superlatively complex, the procedure is involved and entangled, and the investigation has occupied but a few short lives of men. When physical science falls short of explaining the order which reigns throughout these manifold phenomena—an order more characteristic in its totality than any of its phenomena in themselves —men hasten to invoke a guiding principle, an entelechy, or call it what you will. But all the while no physical law, any more than gravity itself, not even among the puzzles of stereo-chemistry or of physiological surface-action and osmosis, is known to be transgressed by the bodily mechanism.

Some physicists declare, as Maxwell did, that atoms or molecules more complicated by far than the chemist's hypotheses demand, are requisite to explain the phenomena of life. If what is implied be an explanation of psychical phenomena, let the point be granted at once; we may go yet further and decline, with Maxwell, to believe that anything of the nature of physical complexity, however exalted, could ever suffice. Other physicists, like Auerbach,[1] or Larmor,[2] or Joly,[3] assure us that our laws of thermodynamics do not suffice, or are inappropriate, to explain the maintenance, or (in Joly's phrase) the accelerative absorption, of the bodily energies, the retardation of entropy, and the long battle against the cold and darkness which is death. With these weighty problems I am not for the moment concerned. My sole purpose is to correlate with mathematical statement and physical law certain of the simpler outward phenomena of organic growth and structure or form, while all the while regarding the fabric of the organism, *ex hypothesi*, as a material and mechanical configuration. This is my purpose here. But I would not for the world be thought to believe that this is the only story which Life and her Children have to tell. One does not come by studying living things for a lifetime to suppose that physics and chemistry can account for them all.[4]

Physical science and philosophy stand side by side, and one upholds the other. Without something of the strength of physics philosophy would be weak; and without something of philosophy's wealth

[1] *Ektropismus oder die physikalische Theorie des Lebens* (Leipzig, 1810).
[2] Wilde Lecture, *Nature*, 12 March 1908; *ibid.* 6 September 1900; *Aether and Matter*, p. 288. Cf. also Kelvin, *Fortnightly Rev.* (1892), p. 313.
[3] The abundance of life, *Proc. Roy. Soc. Dublin*, 7 (1890); *Scientific Essays* (1915), pp. 60 *seq.*
[4] That mechanism has its share in the scheme of nature no philosopher has denied. Aristotle (or whosoever wrote the *De Mundo*) goes so far as to assert that in the most mechanical operations of nature we behold some of the divinest attributes of God.

physical science would be poor. 'Rien ne retirera du tissu de la science les fils d'or que la main du philosophe y a introduits.'[1] But there are fields where each, for a while at least, must work alone; and where physical science reaches its limitations physical science itself must help us to discover. Meanwhile the appropriate and legitimate postulate of the physicist, in approaching the physical problems of the living body, is that with these physical phenomena no alien influence interferes. But the postulate, though it is certainly legitimate, and though it is the proper and necessary prelude to scientific enquiry, may some day be proven to be untrue; and its disproof will not be to the physicist's confusion, but will come as his reward. In dealing with forms which are so concomitant with life that they are seemingly controlled by life, it is in no spirit of arrogant assertiveness if the physicist begins his argument, after the fashion of a most illustrious exemplar, with the old formula of scholastic challenge: *An Vita sit? Dico quod non.*

The terms Growth and Form, which make up the title of this book, are to be understood, as I need hardly say, in their relation to the study of organisms. We want to see how, in some cases at least, the forms of living things, and of the parts of living things, can be explained by physical considerations, and to realise that in general no organic forms exist save such as are in conformity with physical and mathematical laws. And while growth is a somewhat vague word for a very complex matter, which may depend on various things, from simple imbibition of water to the complicated results of the chemistry of nutrition, it deserves to be studied in relation to form: whether it proceed by simple increase of size without obvious alteration of form, or whether it so proceed as to bring about a gradual change of form and the slow development of a more or less complicated structure.

In the Newtonian language of elementary physics, force is recognised by its action in producing or in changing motion, or in preventing change of motion or in maintaining rest. When we deal with matter in the concrete, force does not, strictly speaking, enter into the question, for force, unlike matter, has no independent objective existence. It is energy in its various forms, known or unknown, that acts upon matter. But when we abstract our thoughts from the material to its form, or from the thing moved to its motions, when

[1] J. H. Fr Papillon, *Histoire de la philosophie moderne dans ses rapports avec le développement des sciences de la nature,* I (1876), 300.

we deal with the subjective conceptions of form, or movement, or the movements that change of form implies, then Force is the appropriate term for our conception of the causes by which these forms and changes of form are brought about. When we use the term force, we use it, as the physicist always does, for the sake of brevity, using a symbol for the magnitude and direction of an action in reference to the symbol or diagram of a material thing. It is a term as subjective and symbolic as form itself, and so is used appropriately in connection therewith.

The form, then, of any portion of matter, whether it be living or dead, and the changes of form which are apparent in its movements and in its growth, may in all cases alike be described as due to the action of force. In short, the form of an object is a 'diagram of forces', in this sense, at least, that from it we can judge of or deduce the forces that are acting or have acted upon it: in this strict and particular sense, it is a diagram—in the case of a solid, of the forces which *have* been impressed upon it when its conformation was produced, together with those which enable it to retain its conformation; in the case of a liquid (or of a gas) of the forces which are for the moment acting on it to restrain or balance its own inherent mobility. In an organism, great or small, it is not merely the nature of the *motions* of the living substance which we must interpret in terms of force (according to kinetics), but also the *conformation* of the organism itself, whose permanence or equilibrium is explained by the interaction or balance of forces, as described in statics.

If we look at the living cell of an Amoeba or a Spirogyra, we see a something which exhibits certain active movements, and a certain fluctuating, or more or less lasting, form; and its form at a given moment, just like its motions, is to be investigated by the help of physical methods, and explained by the invocation of the mathematical conception of force.

Now the state, including the shape or form, of a portion of matter is the resultant of a number of forces, which represent or symbolise the manifestations of various kinds of energy; and it is obvious, accordingly, that a great part of physical science must be understood or taken for granted as the necessary preliminary to the discussion on which we are engaged. But we may at least try to indicate, very briefly, the nature of the principal forces and the principal properties of matter with which our subject obliges us to deal. Let us imagine, for instance, the case of a so-called 'simple' organism, such as Amoeba; and if our short list of its physical properties and conditions

11

be helpful to our further discussion, we need not consider how far it be complete or adequate from the wider physical point of view.

This portion of matter, then, is kept together by the inter-molecular force of cohesion; in the movements of its particles relatively to one another, and in its own movements relative to adjacent matter, it meets with the opposing force of friction—without the help of which its creeping movements could not be performed. It is acted on by gravity, and this force tends (though slightly, owing to the Amoeba's small mass, and to the small difference between its density and that of the surrounding fluid) to flatten it down upon the solid substance on which it may be creeping. Our Amoeba tends, in the next place, to be deformed by any pressure from outside, even though slight, which may be applied to it, and this circumstance shows it to consist of matter in a fluid, or at least semi-fluid, state: which state is further indicated when we observe streaming or current motions in its interior. Like other fluid bodies, its surface,[1] whatsoever other substance—gas, liquid or solid—it be in contact with, and in varying degree according to the nature of that adjacent substance, is the seat of molecular force exhibiting itself as a surface-tension, from the action of which many important consequences follow, greatly affecting the form of the fluid surface.

While the protoplasm of the Amoeba reacts to the slightest pressure, and tends to 'flow', and while we therefore speak of it as a fluid,[2] it is evidently far less mobile than such a fluid (for instance) as water, but is rather like treacle in its slow creeping movements as it changes its shape in response to force. Such fluids are said to have a high viscosity, and this viscosity obviously acts in the way of resisting change of form, or in other words of retarding the effects of any disturbing action of force. When the viscous fluid is capable of being drawn out into fine threads, a property in which we know that some Amoebae differ greatly from others, we say that the fluid is also *viscid*, or exhibits viscidity. Again, not by virtue of our Amoeba being liquid, but at the same time in vastly greater measure than if it were a solid (though far less rapidly than if it were a gas), a process of molecular diffusion is constantly going on within its substance, by which its particles interchange their places within the mass, while

---

[1] Whether an animal cell has a membrane, or only a pellicle or *zona limitans*, was once deemed of great importance, and played a big part in the early controversies between the cell-theory of Schwann and the protoplasma-theory of Max Schultze and others.

[2] One of the first statements which Dujardin made about protoplasm (or, as he called it, *sarcode*) was that it was *not* a fluid; and he relied greatly on this fact to show that it was a living, or an organised, structure.

surrounding fluids, gases and solids in solution diffuse into and out of it. In so far as the outer wall of the cell is different in character from the interior, whether it be a mere pellicle as in Amoeba or a firm cell-wall as in Protococcus, the diffusion which takes place *through* this wall is sometimes distinguished under the term *osmosis*.

Within the cell, chemical forces are at work, and so also in all probability (to judge by analogy) are electrical forces; and the organism reacts also to forces from without, that have their origin in chemical, electrical and thermal influences. The processes of diffusion and of chemical activity within the cell result, by the drawing in of water, salts, and food-material with or without chemical transformation into protoplasm, in *growth*, and this complex phenomenon we shall usually, without discussing its nature and origin, describe and picture as a *force*. Indeed we shall manifestly be inclined to use the term growth in two senses, just indeed as we do in the case of attraction or gravitation, on the one hand as a *process*, and on the other as a *force*.

In the phenomena of cell-division, in the attractions or repulsions of the parts of the dividing nucleus, and in the 'caryokinetic' figures which appear in connection with it, we seem to see in operation forces and the effects of forces which have, to say the least of it, a close analogy with known physical phenomena. But though they resemble known physical phenomena, their nature is still the subject of much dubiety and discussion, and neither the forms produced nor the forces at work can yet be satisfactorily and simply explained. We may readily admit then, that, besides phenomena which are obviously physical in their nature, there are actions visible as well as invisible taking place within living cells which our knowledge does not permit us to ascribe with certainty to any known physical force; and it may or may not be that these phenomena will yield in time to the methods of physical investigation. Whether they do or no, it is plain that we have no clear rule or guidance as to what is 'vital' and what is not; the whole assemblage of so-called vital phenomena, or properties of the organism, cannot be clearly classified into those that are physical in origin and those that are *sui generis* and peculiar to living things. All we can do meanwhile is to analyse, bit by bit, those parts of the whole to which the ordinary laws of the physical forces more or less obviously and clearly and indubitably apply.

But even the ordinary laws of the physical forces are by no means simple and plain. In the winding up of a clock (so Kelvin once said), and in the properties of matter which it involves, there is enough and

13

more than enough of mystery for our limited understanding: 'a watchspring is much farther beyond our understanding than a gaseous nebula.' We learn and learn, but never know all, about the smallest, humblest thing. So said St Bonaventure: 'Si per multos annos viveres, adhuc naturam unius festucae seu muscae seu minimae creaturae de mundo ad plenum cognoscere non valeres.'[1] There is a certain fascination in such ignorance; and we learn (like the Abbé Galiani) without discouragement that Science is 'plutôt destiné à étudier qu'à connaître, à chercher qu'à trouver la vérité'.

Morphology is not only a study of material things and of the forms of material things, but has its dynamical aspect, under which we deal with the interpretation, in terms of force, of the operations of Energy.[2]* And here it is well worth while to remark that, in dealing with the facts of embryology or the phenomena of inheritance, the common language of the books seems to deal too much with the *material* elements concerned, as the causes of development, of variation or of hereditary transmission. Matter as such produces nothing, changes nothing, does nothing; and however convenient it may afterwards be to abbreviate our nomenclature and our descriptions, we must most carefully realise in the outset that the spermatozoon, the nucleus, the chromosomes or the germ-plasma can never *act* as matter alone, but only as seats of energy and as centres of force. And this is but an adaptation (in the light, or rather in the conventional symbolism, of modern science) of the old saying of the philosopher: ἀρχὰ γὰρ ἡ φύσις μᾶλλον τῆς ὕλης.

[1] *Op.* v, 541; *cit.* E. Gilson.
[2] This is a great theme. Boltzmann, writing in 1886 on the second law of thermodynamics, declared that available energy was the main object at stake in the struggle for existence and the evolution of the world. Cf. Lotka, 'The Energetics of Evolution', *Proc. Nat. Acad. Sci.* (1922), p. 147.

* See also H. F. Blum, *Time's Arrow and Evolution*, 2nd ed. (Princeton University Press, 1955).

CHAPTER II

ON MAGNITUDE

The Principle of Similitude

To terms of magnitude, and of direction, must we refer all our conceptions of Form. For the form of an object is defined when we know its magnitude, actual or relative, in various directions; and Growth involves the same concepts of magnitude and direction, related to the further concept, or 'dimension', of Time. Before we proceed to the consideration of specific form, it will be well to consider certain general phenomena of spatial magnitude, or of the extension of a body in the several dimensions of space.

We are taught by elementary mathematics—and by Archimedes himself—that in similar figures the surface increases as the square, and the volume as the cube, of the linear dimensions. If we take the simple case of a sphere, with radius r, the area of its surface is equal to $4\pi r^2$, and its volume to $\frac{4}{3}\pi r^3$; from which it follows that the ratio of its volume to surface, of V/S, is $\frac{1}{3}r$. That is to say, V/S *varies* as r; or, in other words, the larger the sphere by so much the greater will be its volume (or its mass, if it be uniformly dense throughout) in comparison with its superficial area. And, taking L to represent any linear dimension, we may write the general equations in the form

$$S \propto L^2, \qquad V \propto L^3,$$

or
$$S = kL^2 \quad \text{and} \quad V = k'L^3,$$

where k, k' are 'factors of proportion',

and
$$\frac{V}{S} \propto L \quad \text{or} \quad \frac{V}{S} = \frac{k}{k'}L = KL.$$

So, in Lilliput, 'His Majesty's Ministers, finding that Gulliver's stature exceeded theirs in the proportion of twelve to one, concluded from the similarity of their bodies that his must contain at least 1728 [or 12^3] of theirs, and must needs be rationed accordingly'.[1]

---

[1] Likewise Gulliver had a whole Lilliputian hogshead for his half-pint of wine: in the due proportion of 1728 half-pints, or 108 gallons, equal to one pipe or double-hogshead. But Gilbert White of Selborne could not see what was plain to the Lilliputians; for finding that a certain little long-legged bird, the stilt, weighed $4\frac{1}{4}$ oz and had legs 8 in. long, he thought that a flamingo, weighing 4 lb, should have legs 10 ft long, to be in the same proportion as the stilt's. But it is obvious to us that, as the weights of the two

From these elementary principles a great many consequences follow, all more or less interesting, and some of them of great importance. In the first place, though growth in length (let us say) and growth in volume (which is usually tantamount to mass or weight) are parts of one and the same process or phenomenon, the one attracts our *attention* by its increase very much more than the other. For instance a fish, in doubling its length, multiplies its weight no less than eight times; and it all but doubles its weight in growing from 4 in. to 5 in. long.

In the second place, we see that an understanding of the correlation between length and weight in any particular species of animal, in other words a determination of k in the formula $W = k.L^3$, enables us at any time to translate the one magnitude into the other, and (so to speak) to weigh the animal with a measuring-rod; this, however, being always subject to the condition that the animal shall in no way have altered its form, nor its specific gravity. That its specific gravity or density should materially or rapidly alter is not very likely; but as long as growth lasts changes of form, even though inappreciable to the eye, are apt and likely to occur. Now weighing is a far easier and far more accurate operation than measuring; and the measurements which would reveal slight and otherwise imperceptible changes in the form of a fish—slight relative differences between length, breadth and depth, for instance—would need to be very delicate indeed. But if we can make fairly accurate determinations of the length, which is much the easiest linear dimension to measure, and correlate it with the weight, then the value of k, whether it varies or remains constant, will tell us at once whether there has or has not been a tendency to alteration in the general form, or, in other words, a difference in the rates of growth in different directions.

We are accustomed to think of magnitude as a purely relative matter. We call a thing *big* or *little* with reference to what it is wont to be, as when we speak of a small elephant or a large rat; and we are apt accordingly to suppose that size makes no other or more essential difference, and that Lilliput and Brobdingnag[1]* are all alike, according

birds are as 1:15, so the legs (or other linear dimensions) should be as the cube-roots of these numbers, or nearly as 1:2¼. And on this scale the flamingo's legs should be, as they actually are, about 20 in. long.

[1] Swift paid close attention to the arithmetic of magnitude, but none to its physical aspect. See De Morgan, on Lilliput, in *N. and Q.* (2), vi, pp. 123–125, 1858. On relative magnitude see also Berkeley, in his *Essay towards a New Theory of Vision*, 1709.

* See also F. Moog, *Sci. Amer.* **179**, no. 5 (1948), 52.

as we look at them through one end of the glass or the other. Gulliver himself declared, in Brobdingnag, that 'undoubtedly philosophers are in the right when they tell us that nothing is great and little otherwise than by comparison': and Oliver Heaviside used to say, in like manner, that there is no absolute scale of size in the Universe, for it is boundless towards the great and also boundless towards the small. It is of the very essence of the Newtonian philosophy that we should be able to extend our concepts and deductions from the one extreme of magnitude to the other; and Sir John Herschel said that 'the student must lay his account to finding the distinction of great and little altogether annihilated in nature.'

All this is true of *number*, and of *relative magnitude*. The Universe has its endless gamut of great and small, of near and far, of many and few. Nevertheless, in physical science the scale of absolute magnitude becomes a very real and important thing; and a new and deeper interest arises out of the changing ratio of dimensions when we come to consider the inevitable changes of physical relations with which it is bound up. The effect of *scale* depends not on a thing in itself, but in relation to its whole environment or milieu; it is in conformity with the thing's 'place in Nature', its field of action and reaction in the Universe. Everywhere Nature works true to scale, and everything has its proper size accordingly. Men and trees, birds and fishes, stars and star-systems, have their appropriate dimensions, and their more or less narrow range of absolute magnitudes. The scale of human observation and experience lies within the narrow bounds of inches, feet or miles, all measured in terms drawn from our own selves or our own doings. Scales which include light-years, parsecs, Ångström units, or atomic and sub-atomic magnitudes, belong to other orders of things and other principles of cognition.

A common effect of scale is due to the fact that, of the physical forces, some act either directly at the surface of a body, or otherwise in proportion to its surface or area; while others, and above all gravity, act on all particles, internal and external alike, and exert a force which is proportional to the mass, and so usually to the volume of the body.

A simple case is that of two similar weights hung by two similar wires. The forces exerted by the weights are proportional to their masses, and these to their volumes, and so to the cubes of the several linear dimensions, including the diameters of the wires. But the areas of cross-section of the wires are as the squares of the said linear dimensions; therefore the stresses in the wires *per unit area* are not

17

identical, but increase in the ratio of the linear dimensions, and the larger the structure the more severe the strain becomes:

$$\frac{\text{Force}}{\text{Area}} \propto \frac{l^3}{l^2} \propto l,$$

and the less the wires are capable of supporting it.

In short, it often happens that of the forces in action in a system some vary as one power and some as another, of the masses, distances or other magnitudes involved; the 'dimensions' remain the same in our equations of equilibrium, but the relative values alter with the scale. This is known as the 'Principle of Similitude', or of dynamical similarity, and it and its consequences are of great importance. In a handful of matter cohesion, capillarity, chemical affinity, electric charge are all potent; across the solar system gravitation rules supreme;[1] in the mysterious region of the nebulae, it may haply be that gravitation grows negligible again.

To come back to homelier things, the strength of an iron girder obviously varies with the cross-section of its members, and each cross-section varies as the square of a linear dimension; but the weight of the whole structure varies as the cube of its linear dimensions. It follows at once that, if we build two bridges geometrically similar, the larger is the weaker of the two,[2] and is so in the ratio of their linear dimensions. It was elementary engineering experience such as this that led Herbert Spencer to apply the principle of similitude to biology.[3]

But here, before we go further, let us take careful note that increased weakness is no necessary concomitant of increasing size. There are exceptions to the rule, in those exceptional cases where we have to deal only with forces which vary merely with the *area* on which they impinge. If in a big and a little ship two similar masts carry two similar sails, the two sails will be similarly strained, and equally

[1] In the early days of the theory of gravitation, it was deemed especially remarkable that the action of gravity 'is proportional to the quantity of solid matter in bodies, and not to their surfaces as is usual in mechanical causes; this power, therefore, seems to surpass mere mechanism' (Colin Maclaurin, on *Sir Isaac Newton's Philosophical Discoveries*, IV, 9).

[2] The subject is treated from the engineer's point of view by Prof. James Thomson, 'Comparison of Similar Structures as to Elasticity, Strength and Stability', *Coll. Papers* (1912), pp. 361–72, and *Trans. Inst. Eng. Scot.* (1876); also by Prof. A. Barr, *ibid.* (1899). See also Rayleigh, *Nature*, 22 April 1915; Sir G. Greenhill, 'On Mechanical Similitude', *Math. Gaz.* (March 1916), *Coll. Works*, VI, 300. For a mathematical account, see (e.g.) P. W. Bridgeman, *Dimensional Analysis* (2nd ed., 1931), or F. W. Lanchester, *The Theory of Dimensions* (1936).

[3] Herbert Spencer, 'The Form of the Earth, etc.', *Phil. Mag.* 30 (1847), 194–6; also *Principles of Biology*, pt. II (1864), pp. 123 *seq.*

stressed at homologous places, and alike suitable for resisting the force of the same wind. Two similar umbrellas, however differing in size, will serve alike in the same weather; and the expanse (though not the leverage) of a bird's wing may be enlarged with little alteration.

The principle of similitude had been admirably applied in a few clear instances by Lesage, a celebrated eighteenth-century physician, in an unfinished and unpublished work.[1] Lesage argued, for example, that the larger ratio of surface to mass in a small animal would lead to excessive transpiration, were the skin as 'porous' as our own; and that we may thus account for the hardened or thickened skins of insects and many other small terrestrial animals. Again, since the weight of a fruit increases as the cube of its linear dimensions, while the strength of the stalk increases as the square, it follows that the stalk must needs grow out of apparent due proportion to the fruit: or, alternatively, that tall trees should not bear large fruit on slender branches, and that melons and pumpkins must lie upon the ground. And yet again, that in quadrupeds a large head must be supported on a neck which is either excessively thick and strong like a bull's, or very short like an elephant's.[2]

But it was Galileo who, wellnigh three hundred years ago, had first laid down this general principle of similitude; and he did so with the utmost possible clearness, and with a great wealth of illustration drawn from structures living and dead.[3] He said that if we tried building ships, palaces or temples of enormous size, yards, beams and bolts would cease to hold together; nor can Nature grow a tree nor construct an animal beyond a certain size, while retaining the proportions and employing the materials which suffice in the case of a smaller structure.[4] The thing will fall to pieces of its own weight unless we either change its relative proportions, which will at length cause it to become clumsy, monstrous and inefficient, or else we must find new material, harder and stronger than was used before. Both processes are familiar to us in Nature and in

[1] See Pierre Prévost, *Notices de la vie et des écrits de Lesage* (1805). George Louis Lesage, born at Geneva in 1724, devoted sixty-three years of a life of eighty to a mechanical theory of gravitation; see W. Thomson (Lord Kelvin), 'On the Ultramundane Corpuscles of Lesage', *Proc. Roy. Soc. Edinb.* 7 (1872), 577–89; *Phil. Mag.* 45 (1873), 321–45; and Clerk Maxwell, art. 'Atom', *Encycl. Brit.* (9), p. 46.

[2] Cf. W. Walton, 'On the Debility of Large Animals and Trees', *Quart. J. Math.* 9 (1868), 179–84; also L. J. Henderson, 'On Volume in Biology', *Proc. Amer. Acad. Sci.* 2 (1916), 654–8; etc.

[3] *Discorsi e Dimostrazioni matematiche, intorno à due nuove scienze attenenti alla Mecanica ed ai Muovimenti Locali*: appresso gli Elzevirii (1638); *Opere*, ed. Favoro, VIII, 169 *seq.* Transl. by Henry Crew and A. de Salvio (1914), p. 130.

[4] So Werner remarked that Michelangelo and Bramanti could not have built of gypsum at Paris on the scale they built of travertin at Rome.

art, and practical applications, undreamed of by Galileo, meet us at every turn in this modern age of cement and steel.[1]

Again, as Galileo was also careful to explain, besides the questions of pure stress and strain, of the strength of muscles to lift an increasing weight or of bones to resist its crushing stress, we have the important question of *bending moments*. This enters, more or less, into our whole range of problems; it affects the whole form of the skeleton, and sets a limit to the height of a tall tree.[2]

We learn in elementary mechanics the simple case of two similar beams, supported at both ends and carrying no other weight than their own. Within the limits of their elasticity they tend to be deflected, or to sag downwards, in proportion to the squares of their linear dimensions; if a matchstick be 2 in. long and a similar beam 6 ft (or 36 times as long), the latter will sag under its own weight thirteen hundred times as much as the other. To counteract this tendency, as the size of an animal increases, the limbs tend to become thicker and shorter and the whole skeleton bulkier and heavier; bones make up some 8 per cent of the body of mouse or wren, 13 or 14 per cent of goose or dog, and 17 or 18 per cent of the body of a man. Elephant and hippopotamus have grown clumsy as well as big, and the elk is of necessity less graceful than the gazelle. It is of high interest, on the other hand, to observe how little the skeletal proportions differ in a little porpoise and a great whale, even in the limbs and limbbones; for the whole influence of gravity has become negligible, or nearly so, in both of these.

In the problem of the tall tree we have to determine the point at which the tree will begin to bend under its own weight if it be ever so little displaced from the perpendicular.[3] In such an investigation we have to make certain assumptions—for instance that the trunk tapers uniformly, and that the sectional area of the branches varies according to some definite law, or (as Ruskin assumed) tends to be constant in any horizontal plane; and the mathematical treat-

---

[1] The Chrysler and Empire State Buildings, the latter 1048 ft high to the foot of its 200 ft 'mooring mast', are the last word, at present, in this brobdingnagian architecture.

[2] It was Euler and Lagrange who first showed (about 1776–8) that a column of a certain height would merely be compressed, but one of a greater height would be bent by its own weight. See Euler, 'De altitudine columnarum etc.', *Acta Acad. Sci. Imp. Petropol.* (1778), pp. 163–93; G. Greenhill, 'Determination of the Greatest Height to which a Tree of given Proportions can Grow', *Proc. Camb. Phil. Soc.* 4 (1881), 65, and Chree, *ibid.* 7 (1892).

[3] In like manner the wheat-straw bends over under the weight of the loaded ear and the cat's tail bends over when held erect—not because they 'possess flexibility', but because they outstrip the dimensions within which stable equilibrium is possible in a vertical position. The kitten's tail, on the other hand, stands up spiky and straight.

ment is apt to be somewhat difficult. But Greenhill showed, on such assumptions as the above, that a certain British Columbian pine-tree, of which the Kew flag-staff, which is 221 ft high and 21 in. in diameter at the base, was made, could not possibly, by theory, have grown to more than about 300 ft. It is very curious that Galileo had suggested precisely the same height (*ducente braccie alta*) as the utmost limit of the altitude of a tree. In general, as Greenhill showed, the diameter of a tall homogeneous body must increase as the power 3/2 of its height, which accounts for the slender proportions of young trees compared with the squat or stunted appearance of old and large ones.[1] In short, as Goethe says in *Dichtung und Wahrheit*, 'Es ist dafür gesorgt, dass die Bäume nicht in den Himmel wachsen'.

But the tapering pine-tree is but a special case of a wider problem. The oak does not grow so tall as the pine-tree, but it carries a heavier load, and its boll, broad-based upon its spreading roots, shows a different contour. Smeaton took it for the pattern of his lighthouse, and Eiffel built his great tree of steel, a thousand feet high, to a similar but a stricter plan. Here the profile of tower or tree follows, or tends to follow, a logarithmic curve, giving equal strength throughout, according to a principle which we shall have occasion to discuss later on, when we come to treat of form and mechanical efficiency in the skeletons of animals. In the tree, moreover, anchoring roots form powerful wind-struts, and are most developed opposite to the direction of the prevailing winds; for the lifetime of a tree is affected by the frequency of storms, and its strength is related to the wind-pressure which it must needs withstand.[2]

Among animals we see, without the help of mathematics or of physics, how small birds and beasts are quick and agile, how slower and sedater movements come with larger size, and how exaggerated bulk brings with it a certain clumsiness, a certain inefficiency, an element of risk and hazard, a preponderance of disadvantage. The case was well put by Owen, in a passage which has an interest of its own as a premonition, somewhat like De Candolle's, of the 'struggle for existence'. Owen wrote as follows: 'In proportion to the bulk of a species is the difficulty of the contest which, as a living organised whole, the individual of each species has to maintain

---

[1] The stem of the giant bamboo may attain a height of 60 m while not more than about 40 cm in diameter near its base, which dimensions fall not far short of the theoretical limits; A. J. Ewart, *Phil. Trans.* **198** (1906), 71.

[2] Cf. (*int. al.*) T. Petch, 'On Buttress Tree-roots', *Ann. R. Bot. Gdns, Peradeniya*, **11** (1930), 277–85. Also an interesting paper by James Macdonald, 'On the Form of Coniferous Trees', *Forestry*, **6** (1931/2), 1 and 2.

against the surrounding agencies that are ever tending to dissolve the vital bond, and subjugate the living matter to the ordinary chemical and physical forces. Any changes, therefore, in such external conditions as a species may have been originally adapted to exist in, will militate against that existence in a degree proportionate, perhaps in a geometrical ratio, to the bulk of the species. If a dry season be greatly prolonged, the large mammal will suffer from the drought sooner than the small one; if any alteration of climate affect the quantity of vegetable food, the bulky Herbivore will be the first to feel the effects of stinted nourishment.'[1]

Speed and Size

But the principle of Galileo carries us further and along more certain lines. The strength of a muscle, like that of a rope or girder, varies with its cross-section; and the resistance of a bone to a crushing stress varies, again like our girder, with its cross-section. But in a terrestrial animal the weight which tends to crush its limbs, or which its muscles have to move, varies as the cube of its linear dimensions; and so, to the possible magnitude of an animal, living under the direct action of gravity, there is a definite limit set. The elephant, in the dimensions of its limb-bones, is already showing signs of a tendency to disproportionate thickness as compared with the smaller mammals; its movements are in many ways hampered and its agility diminished: it is already tending towards the maximal limit of size which the physical forces permit.[2] The spindleshanks of gnat or daddy-long-legs have their own factor of safety, conditional on the creature's exiguous bulk and weight; for after their own fashion even these small creatures tend towards an inevitable limitation of their natural size. But, as Galileo also saw, if the animal be wholly immersed in water like the whale, or if it be partly so, as was probably the case with the giant reptiles of the mesozoic age, then the weight is counterpoised to the extent of an equivalent volume of water, and is completely counterpoised if the density of the animal's body, with the included air, be identical (as a whale's very nearly is) with that of the water around. Under these circumstances there is no longer the same physical barrier to the indefinite growth of the animal. Indeed, in the case of the aquatic animal, there is, as Herbert Spencer pointed out, a distinct advantage, in that the larger it grows the greater is its

[1] *Trans. Zool. Soc.* 4 (1850), 27.
[2] Cf. A. Rauber, 'Galileo über Knochenformen', *Morph. Jb.* 7 (1882), 327.

speed. For its available energy depends on the mass of its muscles, while its motion through the water is opposed, not by gravity, but by 'skin-friction', which increases only as the square of the linear dimensions: whence, other things being equal, the bigger the ship or the bigger the fish the faster it tends to go, but only in the ratio of the square root of the increasing length. For the velocity (V) which the fish attains depends on the work (W) it can do and the resistance (R) it must overcome. Now we have seen that the dimensions of W are l^3, and of R are l^2; and by elementary mechanics

$$W \propto RV^2 \text{ or } V^2 \propto \frac{W}{R}.$$

Therefore
$$V^2 \propto \frac{l^3}{l^2} = l \text{ and } V \propto \sqrt{l}.$$

This is what is known as *Froude's Law*, of the correspondence of speeds—a simple and most elegant instance of 'dimensional theory'.[1]

But there is often another side to these questions, which makes them too complicated to answer in a word. For instance, the work (per stroke) of which two similar engines are capable should vary as the cubes of their linear dimensions, for it varies on the one hand with the *area* of the piston, and on the other with the *length* of the stroke; so is it likewise in the animal, where the corresponding ratio depends on the cross-section of the muscle, and on the distance through which it contracts. But in two similar engines, the available horse-power varies as the square of the linear dimensions, and not as the cube; and this for the reason that the actual *energy* developed depends on the heating-surface of the boiler.[2] So likewise must there be a similar tendency among animals for the rate of supply of kinetic energy to vary with the surface of the lung, that is to say (other things being equal) with the *square* of the linear dimensions of the animal; which means that, *caeteris paribus*, the small animal is stronger (having more power per unit weight) than a large one. We may of course (departing from the condition of similarity) increase the heating-surface of the boiler, by means of an internal system of tubes, without increasing its outward dimensions, and in this very way Nature increases the respiratory surface of a lung by a complex system of branching tubes and minute air-cells; but nevertheless in two similar and closely related animals, as also in two steam-engines

[1] Though, as Lanchester says, the great designer 'was not hampered by a knowledge of the theory of dimensions'.
[2] The analogy is not a very strict or complete one. We are not taking account, for instance, of the thickness of the boiler-plates.

of the same make, the law is bound to hold that the rate of working tends to vary with the square of the linear dimensions, according to Froude's *law of steamship comparison*. In the case of a very large ship, built for speed, the difficulty is got over by increasing the size and number of the boilers, till the ratio between boiler-room and engine-room is far beyond what is required in an ordinary small vessel;[1] but though we find lung-space increased among animals where greater rate of working is required, as in general among birds, I do not know that it can be shown to increase, as in the 'over-boilered' ship, with the size of the animal, and in a ratio which outstrips that of the other bodily dimensions. If it be the case then, that the working mechanism of the muscles should be able to exert a force proportionate to the cube of the linear bodily dimensions, while the respiratory mechanism can only supply a store of energy at a rate proportional to the square of the said dimensions, the singular result ought to follow that, in swimming for instance, the larger fish ought to be able to put on a spurt of speed far in excess of the smaller one; but the distance travelled by the year's end should be very much alike for both of them. And it should also follow that the curve of fatigue is a steeper one, and the staying power less, in the smaller than in the larger individual. This is the case in long-distance racing, where neither draws far ahead until the big winner puts on his big spurt at the end; on which is based an aphorism of the turf, that 'a good big 'un is better than a good little 'un'. For an analogous reason wise men know that in the 'Varsity boat-race it is prudent and judicious to bet on the heavier crew.

[1] Let L be the length, S the (wetted) surface, T the tonnage, D the displacement (or volume) of a ship; and let it cross the Atlantic at a speed V. Then, in comparing two ships, similarly constructed but of different magnitudes, we know that $L = V^2$, $S = L^2 = V^4$, $D = T = L^3 = V^6$; also R (resistance) $= S . V^2 = V^6$; H (horse-power) $= R . V = V^7$; and the coal (C) necessary for the voyage $= H/V = V^6$. That is to say, in ordinary engineering language, to increase the speed across the Atlantic by 1 per cent the ship's length must be increased 2 per cent, her tonnage or displacement 6 per cent, her coal-consumption also 6 per cent, her horse-power, and therefore her boiler-capacity, 7 per cent. Her bunkers, accordingly, keep pace with the enlargement of the ship, but her boilers tend to increase out of proportion to the space available. Suppose a steamer 400 ft long, of 2000 tons, 2000 h.p. and a speed of 14 knots. The corresponding vessel of 800 ft long should develop a speed of 20 knots ($1:2::14^2:20^2$), her tonnage would be 16,000, her h.p. 25,000 or thereby. Such a vessel would probably be driven by four propellers instead of one, each carrying 8000 h.p. See (*int. al.*) W. J. Millar, 'On the Most Economical Speed to Drive a Steamer', *Proc. Edinb. Math. Soc.* **7** (1889), 27–9; Sir James R. Napier, 'On the Most Profitable Speed for a Fully Laden Cargo Steamer for a Given Voyage', *Proc. R. Phil. Soc. Glasg.* **6** (1865), 33–8.

Size and Heat Gain and Loss

Let us consider more closely the actual energies of the body. A hundred years ago, in Strasbourg, a physiologist and a mathematician were studying the temperature of warm-blooded animals.[1] The heat lost must, they said, be proportional to the surface of the animal: and the gain must be equal to the loss, since the temperature of the body keeps constant. It would seem, therefore, that the heat lost by radiation and that gained by oxidation vary both alike, as the surface-area, or the square of the linear dimensions, of the animal. But this result is paradoxical; for whereas the heat lost may well vary as the surface-area, that produced by oxidation ought rather to vary as the bulk of the animal: one should vary as the square and the other as the cube of the linear dimensions. Therefore the ratio of loss to gain, like that of surface to volume, ought to increase as the size of the creature diminishes. Another physiologist, Carl Bergmann, took the case a step further.[2] It was he, by the way, who first said that the real distinction was not between warm-blooded and cold-blooded animals, but between those of constant and those of variable temperature: and who coined the terms *homœothermic* and *poecilothermic* which we use today. He was driven to the conclusion that the smaller animal does produce more heat (per unit of mass) than the large one, in order to keep pace with surface-loss; and that this extra heat-production means more energy spent, more food consumed, more work done.[3] That the smaller animal needs more food is certain and obvious. The amount of food and oxygen consumed by a small flying insect is enormous; and bees and flies and hawkmoths and humming-birds live on nectar, the richest and most concentrated of foods.[4] Man consumes a fiftieth part of his own weight of food daily, but a mouse will eat half its own weight in a day; its rate of

[1] MM. Rameaux et Sarrus, *Bull. Acad. Méd.* 3 (1838–9), 1094–1100.

[2] Carl Bergmann, 'Verhältnisse der Wärmeökonomie der Tiere zu ihrer Grösse', *Göttinger Studien*, I (1847), 594–708—a very original paper.

[3] The metabolic activity of sundry mammals, per 24 hours, has been estimated as follows:

	Weight (kg)	Calories per kg
Guinea-pig	0·7	223
Rabbit	2	58
Man	70	33
Horse	600	22
Elephant	4,000	13
Whale	150,000	*circa* 1·7

[4] Cf. R. A. Davies and G. Fraenkel, 'The Oxygen-consumption of Flies during Flight', *J. Exp. Biol.* 17 (1940), 402–7.

living is faster, it breeds faster, and old age comes to it much sooner than to man.* A warm-blooded animal much smaller than a mouse becomes an impossibility; it could neither obtain nor yet digest the food required to maintain its constant temperature, and hence no mammals and no birds are as small as the smallest frogs or fishes. The disadvantage of small size is all the greater when loss of heat is accelerated by conduction as in the Arctic, or by convection as in the sea. The far north is a home of large birds but not of small; bears but not mice live through an Arctic winter; the least of the dolphins live in warm waters, and there are no small mammals in the sea. This principle is sometimes spoken of as *Bergmann's Law*.

This is an inadequate statement of what remains today a doubtful or at least controversial principle. For a recent critical discussion see P. F. Scholander and his debate on the subject with E. Mayr.† The main point is that the law must be stated in terms of the size of races within a species, and then, as Rensch has shown for many birds, as one moves north, the varieties of a particular species often tend to increase in size. But difficulties arise in other species, for instance the otter, which becomes smaller the more northerly its range. And furthermore, as Scholander argues, various physiological adaptations, all of considerable interest, tend to counteract the small-size problem in the Arctic where we find lemmings, weasels and delicate snow buntings; while the tropics, beginning with the elephant, surely hold all the records for the larger land animals.

Size and Jumping

Leaving aside the question of the supply of energy, and keeping to that of the mechanical efficiency of the machine, we may find endless biological illustrations of the principle of similitude. All through the physiology of locomotion we meet with it in various ways: as, for instance, when we see a cockchafer carry a plate many times its own weight upon its back, or a flea jump many inches high. 'A dog,' says Galileo, 'could probably carry two or three such dogs upon his back; but I believe that a horse could not carry even one of his own size.'

Such problems were admirably treated by Galileo and Borelli, but many writers remained ignorant of their work. Linnaeus remarked that if an elephant were as strong in proportion as a stag-beetle, it would be able to pull up rocks and level mountains; and Kirby and Spence have a well-known passage directed to show that

---

* For a general reference to the points considered here see S. Brody, *Bioenergetics and Growth* (Reinhold, New York, 1945).
† *Evolution*, **9** (1955), 15; **10** (1956), 105 and 339.

such powers as have been conferred upon the insect have been with-
held from the higher animals, for the reason that had these latter
been endued therewith they would have 'caused the early desolation
of the world'.[1]

Such problems as that presented by the flea's jumping powers,[2]
though essentially physiological in their nature, have their interest
for us here: because a steady, progressive diminution of activity
with increasing size would tend to set limits to the possible growth
in magnitude of an animal just as surely as those factors which tend
to break and crush the living fabric under its own weight. In the
case of a leap, we have to do rather with a sudden impulse than with
a continued strain, and this impulse should be measured in terms of
the velocity imparted. The velocity is proportional to the impulse (x),
and inversely proportional to the mass (M) moved: $V = x/M$. But,
according to what we still speak of as 'Borelli's law', the impulse
(i.e. the work of the impulse) is proportional to the volume of the
muscle by which it is produced,[3] that is to say (in similarly constructed
animals) to the mass of the whole body; for the impulse is proportional on the one hand to the cross-section of the muscle, and on the
other to the distance through which it contracts. It follows from this
that the velocity is constant, whatever be the size of the animal.

Putting it still more simply, the work done in leaping is proportional to the mass and to the height to which it is raised, $W \propto mH$.
But the muscular power available for this work is proportional to
the mass of muscle, or (in similarly constructed animals) to the mass
of the animal, $W \propto m$. It follows that H is, or tends to be, a constant.
In other words, all animals, provided always that they are similarly
fashioned, with their various levers in like proportion, ought to jump
not to the same relative but to the same *actual* height.[4] The grasshopper seems to be as well planned for jumping as the flea, and the

---

[1] *Introduction to Entomology*, II (1826), 190. Kirby and Spence, like many less
learned authors, are fond of popular illustrations of the 'wonders of Nature', to the
neglect of dynamical principles. They suggest that if a white ant were as big as a man,
its tunnels would be 'magnificent cylinders of more than three hundred feet in diameter';
and that if a certain noisy Brazilian insect were as big as a man, its voice would be heard
all the world over, 'so that Stentor becomes a mute when compared with these insects!'
It is an easy consequence of anthropomorphism, and hence a common characteristic
of fairy-tales, to neglect the dynamical and dwell on the geometrical aspect of similarity.

[2] The flea is a very clever jumper; he jumps backwards, is stream-lined accordingly,
and alights on his two long hind-legs. Cf. G. I. Watson, in *Nature, Lond.* (21 May 1938).

[3] That is to say, the available energy of muscle, in ft-lb per lb of muscle, is the same
for all animals: a postulate which requires considerable qualification when we come
to compare very different kinds of muscle, such as the insect's and the mammal's.

[4] Borelli, Prop. CLXXVII. Animalia minora et minus ponderosa majores saltus
efficiunt respectu sui corporis, si caetera fuerint paria.

27

actual heights to which they jump are much of a muchness; but the flea's jump is about 200 times its own height, the grasshopper's at most 20–30 times; and neither flea nor grasshopper is a better but rather a worse jumper than a horse or a man.[1]

As a matter of fact, Borelli is careful to point out that in the act of leaping the impulse is not actually instantaneous, like the blow of a hammer, but takes some little time, during which the levers are being extended by which the animal is being propelled forwards; and this interval of time will be longer in the case of the longer levers of the larger animal. To some extent, then, this principle acts as a corrective to the more general one, and tends to leave a certain balance of advantage in regard to leaping power on the side of the larger animal.[2] But on the other hand, the question of strength of materials comes in once more, and the factors of stress and strain and bending moment make it more and more difficult for Nature to endow the larger animal with the length of lever with which she has provided the grasshopper or the flea. To Kirby and Spence it seemed that 'This wonderful strength of insects is doubtless the result of something peculiar in the structure and arrangement of their muscles, and principally their extraordinary power of contraction'. This hypothesis, which is so easily seen on physical grounds to be unnecessary, has been amply disproved in a series of excellent papers by Felix Plateau.[3]

Walking

An apparently simple problem, much less simple than it looks, lies in the act of walking, where there will evidently be great economy of work if the leg swing with the help of gravity, that is to say, at a *pendulum-rate*. The conical shape and jointing of the limb, the time spent with the foot upon the ground, these and other mechanical differences complicate the case, and make the rate hard to define or calculate. Nevertheless, we may convince ourselves by counting our steps, that the leg does actually tend to swing, as a pendulum does,

[1] The high-jump is nowadays a highly skilled performance. For the jumper contrives that his centre of gravity goes *under* the bar, while his body, bit by bit, goes *over* it.

[2] See also (*int. al.*), John Bernoulli, *De Motu Musculorum* (Basil., 1694); Chabry, 'Mécanisme du saut', *J. Anat.*, *Paris*, **19** (1883); 'Sur la longueur des membres des animaux sauteurs', *ibid.* **21** (1885), 356; Le Hello, 'De l'action des organes locomoteurs, etc.', *ibid.* **29** (1893), 65–93; etc.

[3] 'Recherches sur la force absolue des muscles des Invertébrés', *Bull. Acad. R. Belg.* (3), **6**, **7** (1883–4): see also *ibid.* (2), **20** (1865); **22** (1866); *Ann. Mag. N.H.* **17** (1866), 139; **19** (1867), 95. Cf. M. Radau, 'Sur la force musculaire des insectes', *Revue des deux Mondes*, **64** (1866), p. 770. The subject had been well treated by Strauss-Dürckheim, in his *Considérations générales sur l'anatomie comparée des animaux articulés* (1828).

at a certain definite rate.[1] So on the same principle, but to the slower beat of a longer pendulum, the scythe swings smoothly in the mower's hands.

To walk quicker, we 'step out'; we cause the leg-pendulum to describe a greater arc, but it does not swing or vibrate faster until we shorten the pendulum and begin to run. Now let two similar individuals, A and B, walk in a similar fashion, that is to say with a similar *angle* of swing (Fig. 1). The *arc* through which the leg swings, or the *amplitude* of each step, will then vary as the length of leg (say as a/b), and so as the height or other linear dimension (l) of the man.[2] But the time of swing varies inversely as the square root of the pendulum-length, or $\sqrt{a}/\sqrt{b}$. Therefore the velocity, which is measured by amplitude/time, or $a/b \times \sqrt{b}/\sqrt{a}$, will also vary as the square root of the linear dimensions; which is Froude's law over again.

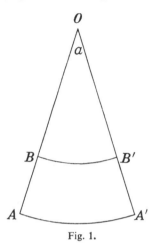

Fig. 1.

The smaller man, or smaller animal, goes slower than the larger, but only in the ratio of the square roots of their linear dimensions; whereas, if the limbs moved alike, irrespective of the size of the animal—if the limbs of the mouse swung no faster than those of the horse—then the mouse would be as slow in its gait or slower than the tortoise. M. Delisle saw a fly walk 3 in. in half-a-second;[3] this was good steady walking. When we walk 5 miles an hour we go about 88 in. in a second, or $88/6 = 14\cdot7$ times the pace of M. Delisle's fly. We should walk at just about the fly's pace if our stature were $1/(14\cdot7)^2$, or $1/216$ of our present height—say $72/216$ in., or one-third of an inch high. Let us note in passing that the number of legs does not matter, any more than the number of wheels to a coach; the centipede runs none the faster for all his hundred legs.

But the leg comprises a complicated system of levers, by whose

---

[1] The assertion that the limb tends to swing in pendulum-time was first made by the brothers Weber (*Mechanik der menschl. Gehwerkzeuge*, Göttingen, 1836). Some later writers have criticised the statement (e.g. Fischer, 'Die Kinematik des Beinschwingens etc.', *Abh. math. phys. Kl. k. Sächs. Ges.* **25–28**, 1899–1903), but for all that, with proper and large qualifications, it remains substantially true.

[2] So the stride of a Brobdingnagian was 10 yards long, or just twelve times the 2 ft 6 in. which make the average stride or half-pace of a man.

[3] Quoted in Mr John Bishop's interesting article in Todd's *Cyclopaedia*, III, 443.

various exercise we obtain very different results. For instance, by being careful to rise upon our instep we increase the length or amplitude of our stride, and improve our speed very materially; and it is curious to see how Nature lengthens this metatarsal joint, or instep-lever, in horse[1] and hare and greyhound, in ostrich and in kangaroo, and in every speedy animal. Furthermore, in running we bend and so shorten the leg, in order to accommodate it to a quicker rate of pendulum-swing.[2] In short the jointed structure of the leg permits us to use it as the shortest possible lever while it is swinging, and as the longest possible lever when it is exerting its propulsive force.

Flying

The bird's case is of peculiar interest.* In running, walking or swimming, we consider the speed which an animal *can attain*, and the increase of speed which increasing size permits of. But in flight there is a certain necessary speed—a speed (relative to the air) which the bird *must attain* in order to maintain itself aloft, and which *must* increase as its size increases. It is highly probable, as Lanchester remarks, that Lilienthal met his untimely death (in August 1896) not so much from any intrinsic fault in the design or construction of his machine, but simply because his speed fell somewhat short of that necessary for stability.

The principle of *necessary speed*, or the inevitable relation between the dimensions of a flying object and the minimum velocity at which its flight is stable, accounts for a considerable number of observed phenomena. It tells us why the larger birds have a marked difficulty in rising from the ground, that is to say, in acquiring to begin with the horizontal velocity necessary for their support; and why accordingly, as Mouillard[3] and others have observed, the heavier birds, even those weighing no more than a pound or two, can be effectually caged in small enclosures open to the sky. It explains why, as Mr

[1] The 'cannon-bones' are not only relatively longer, but may even be actually longer in a little racehorse than a great carthorse.

[2] There is probably another factor involved here: for in bending and thus shortening the leg, we bring its centre of gravity nearer to the pivot, that is to say to the joint, and so the muscle tends to move it the more quickly. After all, we know that the pendulum theory is not the whole story, but only an important first approximation to a complex phenomenon.

[3] Mouillard, *L'empire de l'air; essai d'ornithologie appliquée à l'aviation* (1881); transl. in *Annual Report of the Smithsonian Institution* (1892).

* For some recent discussions of bird and insect flight (with references) see H. I. Fisher's chapter in *Recent Studies in Avian Biology* (Univ. of Illinois Press, 1955), p. 82, and J. Maynard Smith, *New Biology*, 14 (1953), 64.

Abel Chapman says, 'all ponderous birds, wild swans and geese, great bustard and capercailzie, even blackcock, fly faster than they appear to do', while 'light-built types with a big wing-area, such as herons and harriers, possess no turn of speed at all'.[1] For the fact is that the heavy birds must fly quickly, or not at all. It tells us why very small birds, especially those as small as humming-birds, and *a fortiori* the still smaller insects, are capable of 'stationary flight', a very slight and scarcely perceptible velocity relatively to the air being sufficient for their support and stability. And again, since it is in all these cases velocity relative to the air which we are speaking of, we comprehend the reason why one may always tell which way the wind blows by watching the direction in which a bird starts to fly.

The whole matter of the relation of the speed of animals to their size has been greatly advanced by the contribution of A. V. Hill,* and we must not only add to what D'Arcy Thompson has to say, but alter it to some extent. To begin with, Hill stresses the important principle of the intrinsic speed of a muscle and shows that each muscle has a particular speed of contraction (at which it attains maximum power and efficiency) and that in general the smaller muscles are faster than the larger ones. This is especially well illustrated in bird flight where the wing-beat frequency is approximately proportionate to the linear dimensions of the bird. For instance, a humming-bird beats its wings about 75 times a second, a sparrow 15 times and a stork 2–3 times a second. Because of this proportion between the linear size of an animal and the speed of contraction of its muscle, he comes to the conclusion that animals of different size should all have about the same maximal speed, or, as D'Arcy Thompson has pointed out, jump about the same height.† Hill states the matter in its simplest terms by saying 'if one animal is 1000 times as heavy as another, its linear dimensions will be 10 times as great, it will take 10 times as long for one movement, but since that movement is 10 times as great its linear speed over the ground will be the same'. But there remains, as we have already heard from D'Arcy Thompson, an advantage in size, and Hill concludes that it will take 10 times as long for the larger animal to become exhausted in a maximum effort.

The conclusion from Froude's law was that speed varies as the square root of the linear dimensions, and now Hill informs us that for animals of similar construction but of different size the speed should be the same. Part of the reason for the discrepancy is that Froude's law is an over-

---

[1] On wing-area in relation to weight of bird see Lendenfeld in *Naturw. Wochenschr.* (November 1904), transl. in *Smithsonian Inst. Rep.* (1904); also E. H. Hankin, *Animal Flight* (1913); etc.

---

* *Sci. Progr.* **38** (1950), 209.

† Hill presents some interesting tables to substantiate these points.

simplification that fails, among other things, to take into account the intrinsic speed of muscle. This is no doubt related to the fact that in large organisms the rate of metabolism per unit weight of tissue decreases, which makes the analogy of boilers to any energy-transmitting structures in animals somewhat doubtful and further lessens the usefulness of Froude's Law.

But the case of swimming animals has other complications which are of special interest. J. Gray was able to show some years ago that if there is any turbulence at all about a dolphin (as one would surely expect if it were an inert model being pulled through the water) then the amount of work needed to swim 15 knots (which it can accomplish for extended periods of time) would be impossibly large.* In fact the only conclusion left is that the dolphin must have laminar flow over its surface, yet from work with models this seemed most unlikely. By a fortunate circumstance during the war, an observer on a corvette reported to Hill that he noticed, in water full of bioluminescent organisms, that there was no turbulence in the wake of a dolphin. Apparently, as Gray has suggested, the undulating rhythmic movements of the body in motion might somehow damp out the turbulence and perform the remarkable service of keeping complete laminar flow, that is, perfect streamlining.

Gravity and other Size-limiting Factors

From all the foregoing discussion we learn that, as Crookes once upon a time remarked,[1] the forms as well as the actions of our bodies are entirely conditioned (save for certain exceptions in the case of aquatic animals) by the strength of gravity upon this globe; or, as Sir Charles Bell had put it some sixty years before, the very animals which move upon the surface of the earth are proportioned to its magnitude. Were the force of gravity to be doubled our bipedal form would be a failure, and the majority of terrestrial animals would resemble short-legged saurians, or else serpents. Birds and insects would suffer likewise, though with some compensation in the increased density of the air. On the other hand, if gravity were halved, we should get a lighter, slenderer, more active type, needing less energy, less heat, less heart, less lungs, less blood. Gravity not only controls the actions but also influences the forms of all save the least of organisms. The tree under its burden of leaves or fruit has changed its every curve and outline since its boughs were bare, and a mantle of snow will alter its configuration again. Sagging wrinkles, hanging breasts and many another sign of age are part of gravitation's slow relentless handiwork.

[1] *Proc. Soc. Psych. Res., Lond.* **12** (1897), 338–55.

* *J. Exp. Biol.* **13** (1936), 170.

There are other physical factors besides gravity which help to limit the size to which an animal may grow and to define the conditions under which it may live. The small insects skating on a pool have their movements controlled and their freedom limited by the surface-tension between water and air, and the measure of that tension determines the magnitude which they may attain. A man coming wet from his bath carries a few ounces of water, and is perhaps 1 per cent heavier than before; but a wet fly weighs twice as much as a dry one, and becomes a helpless thing. A small insect finds itself imprisoned in a drop of water, and a fly with two feet in one drop finds it hard to extricate them.

The mechanical construction of insect or crustacean is highly efficient up to a certain size, but even crab and lobster never exceed certain moderate dimensions, perfect within these narrow bounds as their construction seems to be. Their body lies within a hollow shell, the stresses within which increase much faster than the mere scale of size; every hollow structure, every dome or cylinder, grows weaker as it grows larger, and a tin canister is easy to make but a great boiler is a complicated affair. The boiler has to be strengthened by 'stiffening rings' or ridges, and so has the lobster's shell; but there is a limit even to this method of counteracting the weakening effect of size. An ordinary girder-bridge may be made efficient up to a span of 200 ft or so; but it is physically incapable of spanning the Firth of Forth. The great Japanese spider-crab, *Macrocheira*, has a span of some 12 ft across; but Nature meets the difficulty and solves the problem by keeping the body small, and building up the long and slender legs out of short lengths of narrow tubes. A hollow shell is admirable for small animals, but Nature does not and cannot make use of it for the large.

In the case of insects, other causes help to keep them of small dimensions. In their peculiar respiratory system blood does not carry oxygen to the tissues, but innumerable fine tubules or tracheae lead air into the interstices of the body. If we imagine them growing even to the size of crab or lobster, a vast complication of tracheal tubules would be necessary, within which friction would increase and diffusion be retarded, and which would soon be an inefficient and inappropriate mechanism.

Hearing and Size

The vibration of vocal chords and auditory drums has this in common with the pendulum-like motion of a limb: that its rate also tends to vary inversely as the square root of the linear dimensions. We know by common experience of fiddle, drum or organ, that pitch rises, or the frequency of vibration increases, as the dimensions of pipe or membrane or string diminish; and in like manner we expect to hear a bass note from the great beasts and a piping treble from the small. The rate of vibration (N) of a stretched string depends on its tension and its density; these being equal, it varies inversely as its own length and as its diameter. For similar strings, $N \propto 1/l^2$, and for a circular membrane, of radius r and thickness e, $N \propto 1/(r^2 \sqrt{e})$.

But the delicate drums or tympana of various animals seem to vary much less in thickness than in diameter, and we may be content to write, once more, $N \propto 1/r^2$.

Suppose one animal to be fifty times less than another, vocal chords and all: the one's voice will be pitched 2500 times as many beats, or some ten or eleven octaves, above the other's; and the same comparison, or the same contrast, will apply to the tympanic membranes by which the vibrations are received. But our own perception of musical notes only reaches to 4000 vibrations per second, or thereby; a squeaking mouse or bat is heard by few, and to vibrations of 10,000 per second we are all of us stone-deaf. Structure apart, mere size is enough to give the lesser birds and beasts a music quite different to our own: the humming-bird, for aught we know, may be singing all day long. A minute insect may utter and receive vibrations of prodigious rapidity; even its little wings may beat hundreds of times a second.[1][*] Far more things happen to it in a second than to us; a thousandth part of a second is no longer negligible, and time itself seems to run a different course to ours.

Vision and Size

The eye and its retinal elements have ranges of magnitude and limitations of magnitude of their own. A big dog's eye is hardly bigger than a little dog's; a squirrel's is much larger, proportionately, than

---

[1] The wing-beats are said to be as follows: dragonfly 28 per sec., bee 190, housefly 330; cf. Erhard, *Deutsche Zool. ges. Verh.* 23 (1913), p. 206.

[*] For recent discussions of sound emission and hearing in bats and insects see D. R. Griffin, *Listening in the Dark* (Yale Univ. Press, 1958), and P. T. Haskel, *New Biology*, 23 (1957), 29.

an elephant's; and a robin's is but little less than a pigeon's or a crow's. For the rods and cones do not vary with the size of the animal, but have their dimensions optically limited by the interference-patterns of the waves of light, which set bounds to the production of clear retinal images. True, the larger animal may want a larger field of view; but this makes little difference, for but a small area of the retina is ever needed or used. The eye, in short, can never be very small and need never be very big; it has its own conditions and limitations apart from the size of the animal. But the insect's eye tells another story. If a fly had an eye like ours, the pupil would be so small that diffraction would render a clear image impossible. The only alternative is to unite a number of small and optically isolated simple eyes into a compound eye, and in the insect Nature adopts this alternative possibility.[1]

Surface keeps pace with Volume

Another phenomenon, and one which is visible throughout the whole field of morphology, is the tendency (referable doubtless in each case to some definite physical cause) for mere bodily *surface* to keep pace with *volume*, through some alteration of its form. The development of villi on the lining of the intestine (which increase its surface much as we enlarge the effective surface of a bath-towel), the various valvular folds of the intestinal lining, including the remarkable 'spiral valve' of the shark's gut, the lobulation of the kidney in large animals,[2] the vast increase of respiratory surface in the air-sacs and alveoli of the lung, the development of gills in the larger crustacea and worms though the general surface of the body suffices for respiration in the smaller species—all these and many more are cases in which a more or less constant ratio tends to be maintained between mass and surface, which ratio would have been more and more departed from with increasing size, had it not been for such alteration of surface-form.[3] A leafy wood, a grassy sward,[*] a piece of sponge, a reef of coral, are all instances of a like phenomenon. In fact, a deal

[1] Cf. C. J. van der Horst, 'The Optics of the Insect Eye', *Acta Zool.* (1933), p. 108.

[2] Cf. R. Anthony, *C.R.* **69** (1919), 1174, etc. Cf. also A. Pütter, 'Studien über physiologische Ähnlichkeit', *Pflüg. Arch. ges. Physiol.* **168** (1917), 209–46.

[3] For various calculations of the increase of surface due to histological and anatomical subdivision, see E. Babak, 'Ueber die Oberflächenentwickelung bei Organismen', *Biol. Zbl.* **30** (1910), 225–39, 257–67.

[*] It should be added here that F. O. Bower has made a large and important study of this phenomenon, especially on the arrangement of vascular tissue in ferns, in his *Size and Form in Plants* (Macmillan, 1930).

of evolution is involved in keeping due balance between surface and mass as growth goes on.

In the case of very small animals, and of individual cells, the principle becomes especially important, in consequence of the molecular forces whose resultant action is limited to the superficial layer. In the cases just mentioned, action is *facilitated* by increase of surface: diffusion, for instance, of nutrient liquids or respiratory gases is rendered more rapid by the greater area of surface; but there are other cases in which the ratio of surface to mass may change the whole condition of the system. Iron rusts when exposed to moist air, but it rusts ever so much faster, and is soon eaten away, if the iron be first reduced to a heap of small filings; this is a mere difference of degree. But the spherical surface of the rain-drop and the spherical surface of the ocean (though both happen to be alike in mathematical form) are two totally different phenomena, the one due to surface-energy, and the other to that form of mass-energy which we ascribe to gravity. The contrast is still more clearly seen in the case of waves: for the little ripple, whose form and manner of propagation are governed by surface-tension, is found to travel with a velocity which is inversely as the square root of its length; while the ordinary big waves, controlled by gravitation, have a velocity directly proportional to the square root of their wave-length. In like manner we shall find that the form of all very small organisms is independent of gravity, and largely if not mainly due to the force of surface-tension: either as the direct result of the continued action of surface-tension on the semi-fluid body, or else as the result of its action at a prior stage of development, in bringing about a form which subsequent chemical changes have rendered rigid and lasting. In either case, we shall find a great tendency in small organisms to assume either the spherical form or other simple forms related to ordinary inanimate surface-tension phenomena, which forms do not recur in the external morphology of large animals.

Now this is a very important matter, and is a notable illustration of that principle of similitude which we have already discussed in regard to several of its manifestations. We are coming to a conclusion which will affect the whole course of our argument throughout this book, namely that there is an essential difference in kind between the phenomena of form in the larger and the smaller organisms. I have called this book a study of *Growth and Form*, because in the most familiar illustrations of organic form, as in our own bodies for example, these two factors are inseparably associated, and because we are here justified in thinking of form as the direct resultant and consequence of growth: of growth, whose varying rate in one direction or another has produced, by its gradual and unequal increments, the successive stages of development and the final configuration of the whole material structure. But it is by no means true that form and growth are in this direct and simple fashion correlative or complementary in the case of minute portions of living matter. For in the smaller organisms, and in the individual cells of the larger, we have reached an order of magnitude in which the intermolecular forces strive under favourable conditions with, and at length altogether outweigh, the force of gravity, and also those other forces leading to movements of convection which are the prevailing factors in the larger material aggregate.

Cell-size

The living cell is a very complex field of energy, and of energy of many kinds, of which surface-energy is not the least. Now the whole surface-energy of the cell is by no means restricted to its *outer* surface; for the cell is a very heterogeneous structure, and all its protoplasmic alveoli and other visible (as well as invisible) heterogeneities make up a great system of internal surfaces, at every part of which one 'phase' comes in contact with another 'phase,' and surface-energy is manifested accordingly. But still, the external surface is a definite portion of the system, with a definite 'phase' of its own, and however little we may know of the distribution of the total energy of the system, it is at least plain that the conditions which favour equilibrium will be greatly altered by the changed ratio of external surface to mass which a mere change of magnitude produces in the cell. In short, the phenomenon of division of the growing cell, however it be brought about, will be precisely what is wanted to keep fairly constant the ratio between surface and mass, and to retain or restore the balance between surface-energy and the other forces of the system.[1] But when a germ-cell divides or 'segments' into two, it does not increase in mass; at least if there be some slight alleged tendency for the egg to increase in mass or volume during segmentation it is very slight indeed, generally imperceptible, and wholly denied by some.[2] The

[1] Certain cells of the cucumber were found to divide when they had grown to a volume half as large again as that of the 'resting cells'. Thus the volumes of resting, dividing and daughter cells were as $1:1\cdot5:0\cdot75$; and their surfaces, being as the power 2/3 of these figures, were, roughly, as $1:1\cdot3:0\cdot8$. The ratio of S/V was then as $1:0\cdot9:1\cdot1$, or much nearer equality. Cf. F. T. Lewis, *Anat. Rec.* **47** (1930), 59–99.

[2] Though the entire egg is not increasing in mass, that is not to say that its living protoplasm is not increasing all the while at the expense of the reserve material.

37

growth or development of the egg from a one-celled stage to stages of two or many cells is thus a somewhat peculiar kind of growth; it is growth limited to change of form and increase of surface, unaccompanied by growth in volume or in mass. In the case of a soap-bubble, by the way, if it divide into two bubbles the volume is actually diminished, while the surface-area is greatly increased;[1] the diminution being due to a cause which we shall have to study later, namely to the increased pressure due to the greater curvature of the smaller bubbles.

An immediate and remarkable result of the principles just described is a tendency on the part of all cells, according to their kind, to vary but little about a certain mean size, and to have in fact certain absolute limitations of magnitude. The diameter of a large parenchymatous cell is perhaps tenfold that of a little one; but the tallest phanerogams are ten thousand times the height of the least. In short, Nature has her materials of predeterminate dimensions, and keeps to the same bricks whether she build a great house or a small. Even ordinary drops tend towards a certain fixed size, which size is a function of the surface-tension, and may be used (as Quincke used it) as a measure thereof. In a shower of rain the principle is curiously illustrated, as Wilding Köller and V. Bjerknes tell us. The drops are of graded sizes, *each twice as big as another*, beginning with the minute and uniform droplets of an impalpable mist. They rotate as they fall, and if two rotate in contrary directions they draw together and presently coalesce; but this only happens when two drops are falling side by side, and since the rate of fall depends on the size it always is a pair of coequal drops which so meet, approach and join together. A supreme instance of constancy or quasi-constancy of size, remote from but yet analogous to the size-limitation of a rain-drop or a cell, is the fact that the stars of heaven (however else one differeth from another), and even the nebulae themselves, are all wellnigh co-equal in *mass*. Gravity draws matter together, condensing it into a world or into a star; but ethereal pressure is an opponent force leading to disruption, negligible on the small scale but potent on the large. High up in the scale of magnitude, from about 10^{33} to 10^{35} g of matter, these two great cosmic forces balance one another; and all the magnitudes of all the stars lie within or hard by these narrow limits.

In the living cell, Sachs pointed out (in 1895) that there is a tendency for each nucleus to gather around itself a certain definite

[1] Cf. P. G. Tait, *Proc. Roy. Soc. Edinb.* 5 (1866) and 6 (1868).

amount of protoplasm.[1] Driesch, a little later, found it possible, by artificial subdivision of the egg, to rear dwarf sea-urchin larvae, one-half, one-quarter or even one-eighth of their usual size; which dwarf larvae were composed of only a half, a quarter or an eighth of the normal number of cells.[2] These observations have been often repeated and amply confirmed: and Loeb found the sea-urchin eggs capable of reduction to a certain size, but no further.

In the development of *Crepidula* (an American 'slipper-limpet', now much at home on our oyster-beds), Conklin has succeeded in rearing dwarf and giant individuals, of which the latter may be five-and-twenty times as big as the former.[3] But the individual cells, of skin, gut, liver, muscle and other tissues, are just the same size in one as in the other, in dwarf and in giant.[4] In like manner the leaf-cells are found to be of the same size in an ordinary water-lily, in the great *Victoria regia*, and in the still huger leaf, nearly 3 m long, of *Euryale ferox* in Japan.[5] Driesch has laid particular stress upon this principle of a 'fixed cell-size', which has, however, its own limitations and exceptions. Among these exceptions, or apparent exceptions, are the giant frond-like cell of a Caulerpa or the great undivided plasmodium of a Myxomycete. The flattening of the one and the branching of the other serve (or help) to increase the ratio of surface to content, the nuclei tend to multiply, and streaming currents keep the interior and exterior of the mass in touch with one another.

[1] *Physiologische Notizen* (9) (1895), p. 425. Cf. Amelung, *Flora* (1893); Strasbürger, 'Ueber die Wirkungssphäre der Kerne und die Zellgrösse', *Histol. Beitr.* (5) (1893), pp. 95–129; R. Hertwig, 'Ueber Korrelation von Zell- und Kerngrösse (Kernplasma-relation)', *Biol. Zbl.* **18** (1903), 49–62, 108–19; G. Levi and T. Terni, 'Le variazioni dell' indice plasmatico-nucleare durante l' intercinesi', *Arch. Ital. Anat. Embriol.* **10** (1911), 545; also E. le Breton and G. Schaeffer, *Variations biochimiques du rapport nucléo-plasmatique* (Strasbourg, 1923). [2] *Arch. Entw. Mech.* **4** (1898), 75, 247. [3] E. G. Conklin, 'Cell-size and Nuclear Size', *J. Exp. Zool.* **12** (1912), 1–98; 'Body-size and Cell-size', *J. Morph.* **23** (1912), 159–88. Cf. M. Popoff, 'Ueber die Zellgrösse', *Arch. Zellforsch.* **3** (1909). [4] Thus the fibres of the crystalline lens are of the same size in large and small dogs, Rabl, *Z. wiss. Zool.* **67** (1899). Cf. (*int. al.*) Pearson, 'On the Size of the Blood-Corpuscles in Rana', *Biometrika*, **6** (1909), 403. Dr Thomas Young caught sight of the phenomenon early in last century: 'The solid particles of the blood do not by any means vary in magnitude in the same ratio with the bulk of the animal', *Natural Philosophy* (ed. 1845), p. 466; and Leeuwenhoek and Stephen Hales were aware of it nearly two hundred years before. Leeuwenhoek, indeed, had a very good idea of the size of a human blood-corpuscle, and was in the habit of using its diameter—about 1/3000 of an inch—as a standard of comparison. But though the blood-corpuscles show no relation of magnitude to the size of the animal, they are related without doubt to its activity; for the corpuscles in the sluggish Amphibia are much the largest known to us, while the smallest are found among the deer and other agile and speedy animals (cf. Gulliver, *P.Z.S.* (1875), p. 474, etc.). This correlation is explained by the surface condensation or adsorption of oxygen in the blood-corpuscles, a process greatly facilitated and intensified by the increase of surface due to their minuteness. [5] Okada and Yonosuke, in *Sci. Rep. Tôhoku Univ.* **3** (1928), 271–8.

We get a good and even a familiar illustration of the principle of size-limitation in comparing the brain-cells or ganglion-cells, whether of the lower or of the higher animals.[1] In Fig. 2 we show certain identical nerve-cells from various mammals, from mouse to elephant, all drawn to the same scale of magnification; and we see that they are all of much the same *order* of magnitude. The nerve-cell of the elephant is about twice that of the mouse in linear dimensions, and therefore about eight times greater in volume or in mass. But making due allowance for difference of shape, the linear dimensions of the elephant are to those of the mouse as not less than one to fifty; and the bulk of the larger animal is something like 125,000 times that

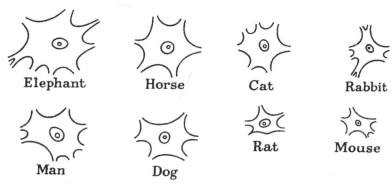

Fig. 2. Motor ganglion-cells, from the cervical spinal cord.
From Minot, after Irving Hardesty.

of the less. It follows, if the size of the nerve-cells are as eight to one, that, in corresponding parts of the nervous system, there are more than 15,000 times as many individual cells in one animal as in the other. In short we may (with Enriques) lay it down as a general law that among animals, large or small, the ganglion-cells vary in size within narrow limits; and that, amidst all the great variety of structure observed in the nervous system of different classes of animals, it is always found that the smaller species have simpler ganglia than the larger, that is to say ganglia containing a smaller number of cellular elements.[2]* The bearing of such facts as this upon

[1] Cf. P. Enriques, 'La forma come funzione della grandezza: Ricerche sui gangli nervosi degli invertebrati', *Arch. Entw. Mech.* 25 (1907–8), 655.
[2] While the difference in cell-volume is vastly less than that between the volumes, and very much less also than that between the surfaces, of the respective animals,

* See also a most interesting study by G. Fankhauser *et al.* (*Science,* 122, 1955, 692) on the relation between learning ability, neuron size and the number of neurons in diploid and triploid newt larvae.

the cell-theory in general is not to be disregarded; and the warning is especially clear against exaggerated attempts to correlate physiological processes with the visible mechanism of associated cells, rather than with the system of energies, or the field of force, which is associated with them. For the life of the body is more than the *sum* of the properties of the cells of which it is composed: as Goethe said, 'Das Lebendige ist zwar in Elemente zerlegt, aber man kann es aus diesen nicht wieder zusammenstellen und beleben'.

Among certain microscopic organisms such as the Rotifera (which have the least average size and the narrowest range of size of all the Metazoa), we are still more palpably struck by the small number of cells which go to constitute a usually complex organ, such as kidney, stomach or ovary; we can sometimes number them in a few units, in place of the many thousands which make up such an organ in larger, if not always higher, animals. We could mention the Fairy-flies, a few score of which would hardly weigh down one of the larger rotifers, and a hundred thousand would weigh less than one honey-bee. Their form is complex and their little bodies exquisitely beautiful; but I feel sure that their cells are few, and their organs of great histological simplicity. These considerations help, I think, to show that, however important and advantageous the subdivision of the tissues into cells may be from the constructional, or from the dynamic, point of view, the phenomenon has less fundamental importance than was once, and is often still, assigned to it.

Just as Sachs showed there was a limit to the amount of cytoplasm which could gather round a nucleus, so Boveri has demonstrated that the nucleus itself has its own limitations of size, and that, in cell-division after fertilisation, each new nucleus has the same size as its parent nucleus;[1] we may nowadays transfer the statement to the chromosomes. It may be that a bacterium lacks a nucleus for the simple reason that it is too small to hold one, and that the same is true of such small plants as the Cyanophyceae, or blue-green algae.

yet there *is* a certain difference; and this it has been attempted to correlate with the need for each cell in the many-celled ganglion of the larger animal to possess a more complex 'exchange-system' of branches, for intercommunication with its more numerous neighbours. Another explanation is based on the fact that, while such cells as continue to divide throughout life tend to uniformity of size in all mammals, those which do not do so, and in particular the ganglion cells, continue to grow, and their size becomes, therefore, a function of the duration of life. Cf. G. Levi, 'Studii sulla grandezza delle cellule', *Arch. Ital. Anat. Embriol.* 5 (1906), 291; cf. also A. Berezowski, 'Studien über die Zellgrösse', *Arch. Zellforsch.* 5 (1910), 375–84.

[1] Boveri, *Zellenstudien*, 'V: Ueber die Abhängigkeit, der Kerngrösse und Zellenzahl von der Chromosomenzahl der Ausgangszellen' (Jena, 1905). Cf. also (*int. al.*) H. Voss, 'Kerngrössenverhältnisse in der Leber etc.', *Z. Zellforsch.* 7 (1928), 187–200.

Even a chromatophore with its 'pyrenoids' seems to be impossible below a certain size.[1]*

Body-size

Size of body is no mere accident. Man, respiring as he does, cannot be as small as an insect, nor vice versa; only now and then, as in the Goliath beetle, do the sizes of mouse and beetle meet and overlap. The descending scale of mammals stops short at a weight of about 5 g, that of beetles at a length of about half a millimetre, and every group of animals has its upper and its lower limitations of size. So, not far from the lower limit of our vision, does the long series of bacteria come to an end. There remain still smaller particles which the ultra-microscope in part reveals; and here or hereabouts are said to come the so-called viruses or 'filter-passers', brought within our ken by the maladies, such as hydrophobia, or foot-and-mouth disease, or the mosaic diseases of tobacco and potato, to which they give rise. These minute particles, of the order of one-tenth the diameter of our smallest bacteria, have no diffusible contents, no included water— whereby they differ from every living thing. They appear to be inert colloidal (or even crystalloid) aggregates of a nucleo-protein, of perhaps ten times the diameter of an ordinary protein-molecule, and not much larger than the giant molecules of haemoglobin or haemo-cyanin.

Beijerinck called such a virus a *contagium vivum*; 'infective nucleo-protein' is a newer name. We have stepped down, by a single step, from living to non-living things, from bacterial dimensions to the molecular magnitudes of protein chemistry.

But, after all, a simple tabulation is all we need to show how nearly the least of organisms approach to molecular magnitudes. In fact each main group of animals has its mean and characteristic size, and a range on either side, sometimes greater and sometimes less. A certain range, and a narrow one, contains mouse and elephant, and all whose business it is to walk and run; this is our own world, with

[1] The size of the nucleus may be affected, even determined, by the number of chromosomes it contains. There are giant races of *Oenothera*, *Primula* and *Solanum* whose cell-nuclei contain twice the normal number of chromosomes, and a dwarf race of a little freshwater crustacean, *Cyclops*, has half the usual number. The cytoplasm in turn varies with the amount of nuclear matter, the whole cell is unusually large or unusually small; and in these exceptional cases we see a direct relation between the size of the organism and the size of the cell. Cf. (*int. al.*) R. P. Gregory, *Proc. Camb. Phil. Soc.* **15** (1909), 239–46; F. Keeble, *J. Genet.* **2** (1912), 163–88.

* For a review of the recent work on animals see G. Fankhauser's chapter in *Analysis of Development*, ed. by Willier, Weiss and Hamburger (Saunders, 1955), p. 126.

whose dimensions our lives, our limbs, our senses are in tune. The great whales grow out of this range by throwing the burden of their bulk upon the waters; the dinosaurs wallowed in the swamp, and the hippopotamus, the sea-elephant and Steller's great sea-cow pass or passed their lives in the rivers or the sea. The things which fly are smaller than the things which walk and run; the flying birds are never as large as the larger mammals, the lesser birds and mammals are much of a muchness, but insects come down a step in the scale and more. The lessening influence of gravity facilitates flight, but makes it less easy to walk and run; first claws, then hooks and suckers and glandular hairs help to secure a foothold, until to creep upon wall or ceiling becomes as easy as to walk upon the ground. Fishes, by evading gravity, increase their range of magnitude both above and below that of terrestrial animals. Smaller than all these, passing out of our range of vision and going down to the least dimensions of living things, are Protozoa, rotifers, spores, pollen-grains and bacteria. All save the largest of these float rather than swim; they are buoyed up by air or water, and fall (as Stokes's law explains) with exceeding slowness.

Physical Limitations in small Organisms

While considerations based on the chemical composition of the organism have taught us that there must be a definite lower limit to its magnitude, other considerations of a purely physical kind lead us to the same conclusion. For our discussion of the principle of similitude has already taught us that long before we reach these all but infinitesimal magnitudes the dwindling organism will have experienced great changes in all its physical relations, and must at length arrive at conditions surely incompatible with life, or what we understand as life, in its ordinary development and manifestation.

Surface-tension

We are told, for instance, that the powerful force of surface-tension, or capillarity, begins to act within a range of about 1/500,000 of an inch, or say 0·05 μ. A soap film, or a film of oil on water, may be attenuated to far less magnitudes than this; the black spots on a soap bubble are known, by various concordant methods of measurement, to be only about 6×10^{-7} cm, or about 6 mμ thick, and Lord Rayleigh and M. Devaux have obtained films of oil of 2 mμ, or even

1 mμ in thickness. But while it is possible for a fluid film to exist of these molecular dimensions, it is certain that long before we reach these magnitudes there arise conditions of which we have little knowledge, and which it is not easy to imagine. A bacillus lives in a world, or on the borders of a world, far other than our own, and preconceptions drawn from our experience are not valid there.

Viscosity

There are many other ways in which, when we make a long excursion into space, we find our ordinary rules of physical behaviour upset. A very familiar case, analysed by Stokes, is that the viscosity of the surrounding medium has a relatively powerful effect upon bodies below a certain size. A droplet of water, a thousandth of an inch (25 μ) in diameter, cannot fall in still air quicker than about an inch and a half per second; as its size decreases, its resistance varies as the radius, not (as with larger bodies) as the surface; and its 'critical' or terminal velocity varies as the square of the radius, or as the surface of the drop. A minute drop in a misty cloud may be one-tenth that size, and will fall a hundred times slower, say an inch a minute; and one again a tenth of this diameter (say 0·25 μ, or about twice as big as a small micrococcus) will scarcely fall an inch in two hours.[1] Not only do dust-particles, spores[2] and bacteria fall, by reason of this principle, very slowly through the air, but all minute bodies meet with great proportionate resistance to their movements through a fluid. In salt water they have the added influence of a larger coefficient of friction than in fresh;[3] and even such comparatively large organisms as the diatoms and the foraminifera, laden though they are with a heavy shell of flint or lime, seem to be poised in the waters of the ocean, and fall with exceeding slowness.

Brownian Movement

The Brownian movement has also to be reckoned with—that remarkable phenomenon studied more than a century ago by Robert

[1] The resistance depends on the radius of the particle, the viscosity, and the rate of fall (V); the effective weight by which this resistance is to be overcome depends on gravity, on the density of the particle compared with that of the medium, and on the mass, which varies as r^3. Resistance $= krV$, and effective weight $= k'r^3$; when these two equal one another we have the critical or terminal velocity, and $V \propto r^2$.

[2] A. R. H. Buller found the spores of a fungus (*Collybia*), measuring $5 \times 3\mu$, to fall at the rate of half a millimetre per second, or rather more than an inch a minute; *Studies on Fungi* (1909).

[3] Cf. W. Krause, *Biol. Zbl.* 1 (1881), 578; Flügel, *Meteorol. Ztschr.* (1881), p. 321.

Brown,[1] Humboldt's *facile princeps botanicorum*, and discoverer of the nucleus of the cell.[2] It is the chief of those fundamental phenomena which the biologists have contributed, or helped to contribute, to the science of physics.

The quivering motion, accompanied by rotation and even by translation, manifested by the fine granular particle issuing from a crushed pollen-grain, and which Brown proved to have no vital significance but to be manifested by all minute particles whatsoever, was for many years unexplained. Thirty years and more after Brown wrote, it was said to be 'due, either directly to some calorical changes continually taking place in the fluid, or to some obscure chemical action between the solid particles and the fluid which is indirectly promoted by heat'.[3] Soon after these words were written it was ascribed by Christian Wiener to molecular movements within the fluid, and was hailed as visible proof of the atomistic (or molecular) constitution of the same.[4] We now know that it is indeed due to the impact or bombardment of molecules upon a body so small that these impacts do not average out, for the moment, to approximate equality on all sides.[5] The movement becomes manifest with particles of somewhere about 20 μ, and is better displayed by those of about 10 μ, and especially well by certain colloid suspensions or emulsions whose particles are just below 1 μ in diameter. The bombardment causes our particles to behave just like molecules of unusual size, and this behaviour is manifested in several ways.[6]

[1] *A Brief Description of Microscopical Observations...on the Particles contained in the Pollen of Plants; and on the General Existence of Active Molecules in Organic and Inorganic Bodies* (London, 1828). See also *Edinb. New Philosoph. J.* 5 (1828), 358; *Edinb. J. Sci.* 1 (1829), 314; *Ann. Sci. Nat.* 14 (1828), 341–62; etc. The Brownian movement was hailed by some as supporting Leibniz's theory of Monads, a theory once so deeply rooted and so widely believed that even under Schwann's cell-theory Johannes Müller and Henle spoke of the cells as 'organische Monaden'; cf. Emil du Bois Reymond, 'Leibnizische Gedanken in der neueren Naturwissenschaft', *Mber. Akad. Wiss., Berl.* (1870).

[2] The 'nucleus' was first seen in the epidermis of orchids; but 'this areola, or nucleus of the cell as perhaps it might be termed, is not confined to the epidermis', etc. See his paper on 'Fecundation in Orchideae and Asclepiadae' *Trans. Linn. Soc.* 16 (1829–33).

[3] Carpenter, *The Microscope* (ed. 1862), p. 185.

[4] In *Poggendorff's Annalen*, 118 (1863), 79–94. For an account of this remarkable man, see *Naturwissenschaften*, 15 (1927); cf. also Sigmund Exner, 'Ueber Brown's Molecularbewegung', *S.B. Akad. Wien*, 56 (1867), 116.

[5] Perrin, 'Les preuves de la réalité moléculaire', *Ann. Phys.* 17 (1905), 549; 19 (1906), 571. The actual molecular collisions are unimaginably frequent; we see only the residual fluctuations.

[6] For a full, but still elementary, account, see J. Perrin, *Les Atomes*; cf. also Th. Svedberg, *Die Existenz der Moleküle* (1912); R. A. Millikan, *The Electron* (1917), etc. The modern literature of the Brownian movement (by Einstein, Perrin, de Broglie, Smoluchowski and Millikan) is very large, chiefly owing to the value which the

It is only in regard to particles of the simplest form that these phenomena have been theoretically investigated, and we may take it as certain that more complex particles, such as the twisted body of a Spirillum, would show other and still more complicated manifestations.[1] It is at least clear that, just as the early microscopists in the days before Robert Brown never doubted but that these phenomena were purely vital, so we also may still be apt to confuse, in certain cases, the one phenomenon with the other. We cannot, indeed, without the most careful scrutiny, decide whether the movements of our minutest organisms are intrinsically 'vital' (in the sense of being beyond a physical mechanism, or working model) or not. For example, Schaudinn has suggested that the undulating movements of *Spirochaete pallida* must be due to the presence of a minute, unseen, 'undulating membrane'; and Doflein says of the same species that 'sie verharrt oft mit eigenthümlich zitternden Bewegungen an einem Orte'. Both movements, the trembling or quivering movement described by Doflein, and the undulating or rotating movement described by Schaudinn, are just such as may be easily and naturally interpreted as part and parcel of the Brownian phenomenon.

While the Brownian movement may thus simulate in a deceptive way the active movements of an organism, the reverse statement also to a certain extent holds good. One sometimes lies awake of a summer's morning watching the flies as they dance under the ceiling. It is a very remarkable dance. The dancers do not whirl or gyrate, either in company or alone; but they advance and retire; they seem to jostle and rebound; between the rebounds they dart hither or thither in short straight snatches of hurried flight, and turn again sharply in a new rebound at the end of each little rush.[2] Their motions are erratic, independent of one another, and devoid of common purpose.[3] This is nothing else than a vastly magnified picture, or simulacrum, of the Brownian movement; the parallel between the

phenomenon is shown to have in determining the size of the atom or the charge on an electron, and of giving, as Ostwald said, experimental proof of the atomic theory.

[1] Cf. R. Gans, 'Wie fallen Stäbe und Scheiben in einer reibenden Flüssigkeit?' *Münchener Bericht* (1911), p. 191; K. Przibram, 'Ueber die Brown'sche Bewegung nicht kugelförmiger Teilchen', *Wiener Bericht* (1912), p. 2339; (1913), pp. 1895–1912.

[2] As Clerk Maxwell put it to the British Association at Bradford in 1873, 'We cannot do better than observe a swarm of bees, where every individual bee is flying furiously, first in one direction and then in another, while the swarm as a whole is either at rest or sails slowly through the air.'

[3] Nevertheless, there may be a certain amount of bias or direction in these seemingly random divagations: cf. J. Brownlee, *Proc. Roy. Soc. Edinb.* **31** (1910–11), 262; F. H. Edgeworth, *Metron*, 1 (1920), 75; Lotka, *Elem. of Physical Biology* (1925), p. 344.

46

two cases lies in their complete irregularity, but this in itself implies a close resemblance. One might see the same thing in a crowded market-place, always provided that the bustling crowd had no *business* whatsoever. In like manner Lucretius, and Epicurus before him, watched the dust-motes quivering in the beam, and saw in them a mimic representation, *rei simulacrum et imago*, of the eternal motions of the atoms. Again the same phenomenon may be witnessed under the microscope, in a drop of water swarming with Paramoecia or such-like Infusoria; and here the analogy has been put to a numerical test. Following with a pencil the track of each little swimmer, and dotting its place every few seconds (to the beat of a metronome), Karl Przibram found that the mean successive distances from a common base-line obeyed with great exactitude the 'Einstein formula', that is to say, the particular form of the 'law of chance' which is applicable to the case of the Brownian movement.[1] The phenomenon is (of course) merely analogous, and by no means identical with the Brownian movement; for the range of motion of the little active organisms, whether they be gnats or infusoria, is vastly greater than that of the minute particles which are passive under bombardment; nevertheless Przibram is inclined to think that even his comparatively large infusoria are small enough for the molecular bombardment to be a stimulus, even though not the actual cause, of their irregular and interrupted movements.[2]

George Johnstone Stoney, the remarkable man to whom we owe the name and concept of the *electron*, went further than this; for he supposed that molecular bombardment might be the source of the life-energy of the bacteria. He conceived the swifter moving molecules to dive deep into the minute body of the organism, and this in turn to be able to make use of these importations of energy.[3]

We draw near the end of this discussion. We found, to begin with, that 'scale' had a marked effect on physical phenomena, and that increase or diminution of magnitude might mean a complete change of statical or dynamical equilibrium. In the end we begin to see that there are discontinuities in the scale, defining phases in which different

---

[1] That is to say, the mean square of the displacements of a particle, in any direction, is proportional to the interval of time. Cf. K. Przibram, 'Ueber die ungeordnete Bewegung niederer Tiere', *Pflüg. Arch. ges. Physiol.* **153** (1913), 401–5; *Arch. Entw. Mech.* **43** (1917), 20–27.

[2] All that is actually proven is that 'pure chance' has governed the movements of the little organism. Przibram has made the analogous observation that infusoria, when not too crowded together, spread or diffuse through an aperture from one vessel to another at a rate very closely comparable to the ordinary laws of molecular diffusion.

[3] *Phil. Mag.* (April 1890).

forces predominate and different conditions prevail. Life has a range of magnitude narrow indeed compared to that with which physical science deals; but it is wide enough to include three such discrepant conditions as those in which a man, an insect and a bacillus have their being and play their several roles. Man is ruled by gravitation, and rests on mother earth. A water-beetle finds the surface of a pool a matter of life and death, a perilous entanglement or an indispensable support. In a third world, where the bacillus lives, gravitation is forgotten, and the viscosity of the liquid, the resistance defined by Stokes's law, the molecular shocks of the Brownian movement, doubtless also the electric charges of the ionised medium, make up the physical environment and have their potent and immediate influence on the organism. The predominant factors are no longer those of our scale; we have come to the edge of a world of which we have no experience, and where all our preconceptions must be recast.

CHAPTER III

THE FORMS OF CELLS

This chapter (and the beginning of the next) possesses a curious blind spot: it is primarily concerned with the effect of surface-tension on the form of cells, yet it completely ignores all the beautiful experimental work, begun in the early 1930's by E. N. Harvey, K. C. Cole and others on the actual measurement of the surface-tension of cells. This is all excellently reviewed in a recent paper by Harvey;* let me just state two of the important conclusions here. In the first place there are a number of clear demonstrations that the cell boundary primarily exerts membrane tension and not true surface-tension. In fact it is a composite of the two and the sum of all these tensions is referred to by Harvey as 'tension at the surface'. The second point is that these tensions are extremely low, far too low to account by themselves for cell shape. Instead, then, we must look to the micro-structure of the cell periphery for an understanding of cell form. Despite the fact that no reference is made to this considerable body of experimental work, it must be admitted that D'Arcy Thompson seems, in certain passages, to be aware of the difficulties.†

But this omission need not mar the value of the chapter if we think of D'Arcy Thompson's presentation as a model rather than a reality. The formula of Laplace remains a useful description of the sites of forces playing upon a cell, even though those forces are not surface-tension alone. In fact all the subjects—the surfaces of revolution of Plateau, the limits of equilibrium, the analogy to splashes and drops—are either useful descriptions or stimulating models. The work of Harvey and the other experimenters has been in the main current of scientific papers and this is a perfect example of how the work of D'Arcy Thompson is completely individual and untouched. Provided we recognise it for what it is, analogy and description, the objections may be laid to one side.

Surface-tension

Surface-tension is due to molecular force: to force, that is to say, arising from the action of one molecule upon another; and since we can only ascribe a small 'sphere of action' to each several molecule, this force is manifested only within a narrow range. Within the interior of the liquid mass we imagine that such molecular interactions negative one another; but at and near the free surface, within a layer

* *Protoplasmatologia*, **2** (1954), 1. † See p. 87

or film approximately equal to the range of the molecular force— or to the radius of the aforesaid 'sphere of action'—there is a lack of equilibrium and a consequent manifestation of force.

The action of the molecular forces has been variously explained. But one simple explanation (or mode of statement) is that the molecules of the surface-layer are being constantly attracted into the interior by such as are just a little more deeply situated; the surface shrinks as molecules keep quitting it for the interior, and this *surface-shrinkage* exhibits itself as a *surface-tension*. The process continues till it can go no farther, that is to say until the surface itself becomes a 'minimal area'. This is a sufficient description of the phenomenon in cases where a portion of liquid is subject to no other than its own molecular forces, and (since the sphere has, of all solids, the least surface for a given volume) it accounts for the spherical form of the raindrop,[1] of the grain of shot, or of the living cell in innumerable simple organisms. It accounts also, as we shall presently see, for many much more complicated forms, manifested under less simple conditions.

Surface-tension Forms

Let us now make some enquiry into the forms which a fluid surface can assume under the mere influence of surface-tension. In doing so we are limited to conditions under which other forces are relatively unimportant, that is to say where the surface energy is a considerable fraction of the whole energy of the system; and in general this will be the case when we are dealing with portions of liquid so small that their dimensions come within or near to what we have called the molecular range, or, more generally, in which the 'specific surface' is large. In other words it is the small or minute organisms, or small cellular elements of larger organisms, whose forms will be governed by surface-tension; while the forms of the larger organisms are due to other and non-molecular forces. A large surface of water sets itself level because here gravity is predominant; but the surface of water in a narrow tube is curved, for the reason that we are here dealing with particles which lie within the range of each

---

[1] Raindrops must be spherical, or they would not produce a rainbow; and the fact that the upper part of the bow is the brightest and sharpest shows that the higher raindrops are more truly spherical, as well as smaller than the lower ones. So also the smallest dewdrops are found to be more iridescent than the large, showing that they also are the more truly spherical; cf. T. W. Backhouse, in *Mon. Meteorol. Mag.* (March 1879). Mercury has a high surface-tension, and its globules are very nearly round.

other's molecular forces. The like is the case with the cell-surfaces and cell-partitions which we are about to study, and the effect of gravity will be especially counteracted and concealed when the object is immersed in a liquid of nearly its own density.

We have already learned, as a fundamental law of 'capillarity', that a liquid film *in equilibrium* assumes a form which gives it a minimal area under the conditions to which it is subject. These conditions include (1) the form of the boundary, if such exist, and (2) the pressure, if any, to which the film is subject: which pressure is closely related to the volume of air, or of liquid, that the film (if it be a closed one) may have to contain. In the simplest of cases, as when we take up a soap-film on a plane wire ring, the film is exposed to equal atmospheric pressure on both sides, and it obviously has its minimal area in the form of a plane. So long as our wire ring lies in one plane (however irregular in outline), the film stretched across it will still be in a plane; but if we bend the ring so that it lies no longer in a plane, then our film will become curved into a surface which may be extremely complicated, but is still the smallest possible surface which can be drawn continuously across the uneven boundary.

The question of pressure involves not only external pressures acting on the film, but also that which the film is capable of exerting. For we have seen that the film is always contracting to the utmost; and when the film is curved, this leads to a pressure directed inwards—perpendicular, that is to say, to the surface of the film. In the case of the soap-bubble, the uniform contraction of whose surface has led to its spherical form, this pressure is balanced by the pressure of the air within; and if an outlet be given for this air, then the bubble contracts with perceptible force until it stretches across the mouth of the tube, for instance across the mouth of the pipe through which we have blown the bubble. A precisely similar pressure, directed inwards, is exercised by the surface layer of a drop of water or a globule of mercury, or by the surface pellicle on a portion or 'drop' of protoplasm. Only we must always remember that in the soap-bubble, or the bubble which a glass-blower blows, there is a twofold pressure as compared with that which the surface-film exercises on the drop of liquid of which it is a part; for the bubble consists (unless it be so thin as to consist of a mere layer of molecules) of a liquid layer,[1] with a free surface within and another without, and

---

[1] Or, more strictly speaking, unless its thickness be less than twice the range of the molecular forces.

each of these two surfaces exercises its own independent and co-equal tension and its corresponding pressure.[1]

If we stretch a tape upon a flat table, whatever be the tension of the tape it obviously exercises no pressure upon the table below. But if we stretch it over a *curved* surface, a cylinder for instance, it does exercise a downward pressure; and the more curved the surface the greater is this pressure, that is to say the greater is this share of the entire force of tension which is resolved in the downward direction. In mathematical language, the pressure (p) varies directly as the tension (T), and inversely as the radius of curvature (R): that is to say, $p = T/R$, per unit of surface.

If instead of a cylinder, whose curvature lies only in one direction, we take a case of curvature in two dimensions (as for instance a sphere), then the effects of these two curvatures must be added together to give the resulting pressure p: which becomes equal to $T/R + T/R'$, or

$$p = \frac{1}{R} + \frac{1}{R'}.$$

And if in addition to the pressure p, which is due to surface-tension, we have to take into account other pressures, p', p'', etc., due to gravity or other forces, then we may say that the total pressure

$$P = p' + p'' + T\left(\frac{1}{R} + \frac{1}{R'}\right).$$

This simple but immensely important formula is due to Laplace.[2]

We may have to take account of the extraneous pressures in some cases, as when we come to speak of the shape of a bird's egg; but in this first part of our subject we are able for the most part to neglect them.

Our equation is an equation of equilibrium. The resistance to compression—the pressure outwards—of our fluid mass is a constant quantity (P); the pressure inwards, $T(1/R + 1/R')$, is also constant; and if the surface (unlike that of the mobile amoeba) be homogeneous, so that T is everywhere equal, it follows that $1/R + 1/R' = C$ (a constant), throughout the whole surface in question.

Now equilibrium is reached after the surface-contraction has done its utmost, that is to say when it has reduced the surface to the least possible area. So we arrive at the conclusion, from the physical side,

---

[1] It follows that the tension of a bubble, depending only on the surface-conditions, is independent of the thickness of the film.

[2] Laplace, *Mécanique Céleste*, Bk x, suppl. *Théorie de l'action capillaire*, 1806.

that a surface such that $1/R + 1/R' = C$, in other words a surface which has the same *mean curvature* at all points, is equivalent to a surface of minimal area for the volume enclosed;[1] and to the same conclusion we may also come by ways purely mathematical. The plane and the sphere are two obvious examples of such surfaces, for in both the radius of curvature is everywhere constant.

Plateau's Surfaces of Revolution

From the fact that we may extend a soap-film across any ring of wire, however fantastically the wire be bent, we see that there is no end to the number of surfaces of minimal area which may be constructed or imagined.[2] While some of these are very complicated indeed, others, such as a spiral or helicoid screw, are relatively simple. But if we limit ourselves to *surfaces of revolution* (that is to say, to surfaces symmetrical about an axis), we find, as Plateau was the first to show, that those which meet the case are few in number. They are six in all, namely the plane, the sphere, the cylinder, the catenoid, the unduloid, and a curious surface which Plateau called the nodoid.

These several surfaces are all closely related, and the passage from one to another is generally easy. Their mathematical interrelation is expressed by the fact (first shown by Delaunay,[3] in 1841) that the plane curves by whose revolution they are generated are themselves generated as 'roulettes' of the conic sections.

Let us imagine a straight line, or axis, on which a circle, ellipse or other conic section rolls; the focus of the conic section will describe a line in some relation to the fixed axis, and this line (or roulette), when we rotate it around the axis, will describe in space one or another of the six surfaces of revolution of which we are speaking.

If we imagine an ellipse so to roll on a base-line, either of its foci

---

[1] A surface may be 'minimal' in respect of the area occupied, or of the volume enclosed: the former being such as the surface which a soap-film forms when it fills up a ring, whether plane or no. The geometers are apt to restrict the term 'minimal surface' to such as these, or, more generally, to all cases where the mean curvature is *nil*; the others, being only minimal with respect to the volume contained, they call 'surfaces of constant mean curvature'.

[2] To fit a minimal surface to the boundary of any given closed curve in space is a problem formulated by Lagrange, and commonly known as the 'problem of Plateau', who solved it with his soap-films.

[3] 'Sur la surface de révolution dont la courbure moyenne est constante', *J. de Math.* Ed. by J. Liouville, **6** (1841), 309. Cf. (*int. al.*) J. Clerk Maxwell, 'On the Theory of Rolling Curves', *Trans. Roy. Soc. Edinb.* **16** (1849), 519–40; J. K. Wittemore, 'Minimal Surfaces of Rotation', *Ann. Math.* (2), **19** (1917); *Amer. J. Math.* **40** (1918), 69; Gino Loria, *Courbes planes spéciales, théorie et histoire* (Milan, 1930, 574 pp.)

will describe a sinuous or wavy line (Fig. 3, *b*) at a distance alternately maximal and minimal from the axis; this wavy line, by rotation about the axis, becomes the meridional line of the surface which we call the *unduloid*, and the more unequal the two axes are of our ellipse, the more pronounced will be the undulating sinuosity of the roulette. If the two axes be equal, then our ellipse becomes a circle; the path described by its rolling centre is a straight line parallel to the axis (*a*), and the solid of revolution generated therefrom will be a *cylinder*: in other words, the cylinder is a 'limiting case' of the unduloid. If one axis of our ellipse vanish, while the other remains of finite length, then the ellipse is reduced to a straight line with its foci at the two ends, and its roulette will appear as a succession of

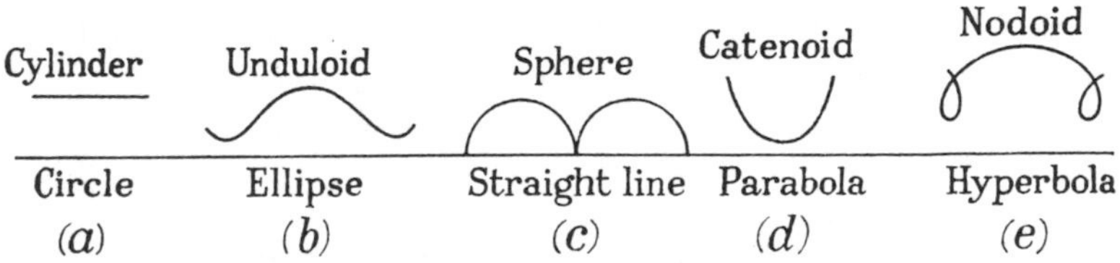

Cylinder	Unduloid	Sphere	Catenoid	Nodoid
Circle	Ellipse	Straight line	Parabola	Hyperbola
(*a*)	(*b*)	(*c*)	(*d*)	(*e*)

Fig. 3. Roulettes of the conic sections.

semicircles touching one another upon the axis (*c*); the solid of revolution will be a series of equal *spheres*. If as before one axis of the ellipse vanish, but the other be infinitely long, then the roulette described by the focus of this ellipse will be a circular arc at an infinite distance; i.e. it will be a straight line normal to the axis, and the surface of revolution traced by this straight line turning about the axis will be a *plane*. If we imagine one focus of our ellipse to remain at a given distance from the axis, but the other to become infinitely remote, that is tantamount to saying that the ellipse becomes transformed into a parabola; and by the rolling of this curve along the axis there is described a catenary (*d*), whose solid of revolution is the *catenoid*.

Lastly, but this is more difficult to imagine, we have the case of the hyperbola. We cannot well imagine the hyperbola rolling upon a fixed straight line so that its focus shall describe a continuous curve. But let us suppose that the fixed line is, to begin with, asymptotic to one branch of the hyperbola, and that the rolling proceeds until the line is now asymptotic to the other branch, that is to say touching it at an infinite distance; there will then be mathematical continuity if we recommence rolling with this second branch, and so in turn with the other, when each has run its course. We shall see,

54

on reflection, that the line traced by one and the same focus will be an 'elastic curve', describing a succession of kinks or knots (e), and the solid of revolution described by this meridional line about the axis is the so-called *nodoid*.

The physical transition of one of these surfaces into another can be experimentally illustrated by means of soap-bubbles, or better still, after another method of Plateau's, by means of a large globule of oil, supported when necessary by wire rings, and lying in a fluid of specific gravity equal to its own.

To prepare a mixture of alcohol and water of a density precisely equal to that of the oil-globule is a troublesome matter, and a method devised by Mr C. R. Darling is a great improvement on Plateau's.[1] Mr Darling used the oily liquid orthotoluidene, which does not mix with water, has a beautiful and conspicuous red colour, and has precisely the same density as water when both are kept at a temperature of 24° C. We have, therefore, only to run the liquid into water at this temperature in order to produce beautifully spherical drops of any required size: and by adding a little salt to the lower layers of water, the drop may be made to rest or float upon the denser liquid.

We have seen that the soap-bubble, spherical to begin with, is transformed into a plane when we release its internal pressure and let the film shrink back upon the orifice of the pipe. If we blow a bubble and then catch it up on a second pipe, so that it stretches between, we may draw the two pipes apart, with the result that the spheroidal surface will be gradually flattened in a longitudinal direction, and the bubble will be transformed into a cylinder. But if we draw the pipes yet farther apart, the cylinder narrows in the middle into a sort of hour-glass form, the increasing curvature of its transverse section being balanced by a gradually increasing *negative* curvature in the longitudinal section; the cylinder has, in turn, been converted into an unduloid. When we hold a soft glass tube in the flame and 'draw it out', we are in the same identical fashion converting a cylinder into an unduloid (Fig 4, *a*); when on the other hand we stop the end and blow, we again convert the cylinder into an unduloid (*b*), but into one which is now positively, while the former was negatively, curved. The two figures are essentially the same, save that the two halves of the one change places in the other.

That spheres, cylinders and unduloids are of the commonest

[1] See *Liquid Drops and Globules* (1914), p. 11. Robert Boyle used turpentine in much the same way; for other methods see Plateau, *op. cit.* p. 154.

occurrence among the forms of small unicellular organisms or of individual cells in the simpler aggregates, and that in the processes of growth, reproduction and development transitions are frequent from one of these forms to another, is obvious to the naturalist,[1] and we shall deal presently with a few of these phenomena. But before we go further in this enquiry we must consider, to some small extent at least, the *curvatures* of the six different surfaces, so far as to determine what modification is required, in each case, of the general equation which applies to them all. We shall find that with this question is closely connected the question of the *pressures* exercised by or impinging on the film, and also the very important question of the limiting conditions which, from the nature of the case, set bounds to the extensions of certain of the figures. The whole subject is mathematical, and we shall only deal with it in the most elementary way.

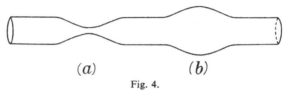

(a) (b)

Fig. 4.

We have seen that, in our general formula, the expression $1/R + 1/R' = C$, a constant; and that this is, in all cases, the condition of our surface being one of minimal area. That is to say, it is always true for one and all of the six surfaces which we have to consider; but the constant C may have any value, positive, negative or nil.

In the case of the plane, where R and R' are both infinite, $1/R + 1/R' = 0$. The expression therefore vanishes, and our dynamical equation of equilibrium becomes $P = p$. In short, we can only have a plane film, or we shall only find a plane surface in our cell, when on either side thereof we have equal pressures or no pressure at all; a simple case is the plane partition between two equal and similar cells, as in a filament of *Spirogyra*.

In the sphere the radii are all equal, $R = R'$; they are also positive, and $T(1/R + 1/R')$, or $2T/R$, is a positive quantity, involving a constant positive pressure P, on the other side of the equation.

In the cylinder one radius of curvature has the finite and positive value R; but the other is infinite. Our formula becomes T/R, to

---

[1] They tend to reappear, no less obviously, in those precipitated structures which simulate organic form in the experiments of Leduc, Herrera and Lillie.

which corresponds a positive pressure P, supplied by the surface-tension as in the case of the sphere, but evidently of just half the magnitude.

In plane, sphere and cylinder the two principal curvatures are constant, separately and together; but in the unduloid the curvatures change from one point to another. At the middle of one of the swollen 'beads' or bubbles, the curvatures are both positive; the expression $(1/R + 1/R')$ is therefore positive, and it is also finite. The film exercises (like the cylinder) a positive pressure inwards, to be compensated by an equivalent outward pressure from within. Between two adjacent beads, at the middle of one of the narrow necks, there is obviously a much stronger curvature in the transverse direction; but the total pressure is unchanged, and we now see that a negative curvature *along* the unduloid balances the increased curvature in the transverse direction. The sum of the two must remain positive as well as constant; therefore the convex or positive curvature must always be greater than the concave or negative curvature at the same point, and this is plainly the case in our figure of the unduloid.

The catenoid, in this respect a limiting case of the unduloid, has its curvature in one direction equal and opposite to its curvature in the other, this property holding good for all points of the surface; $R = -R'$; and the expression becomes

$$(1/R + 1/R') = (1/R - 1/R) = 0.$$

That is to say, the mean curvature is zero, and the catenoid, like the plane itself, has *no curvature*, and exerts no pressure. None of the other surfaces save these two share this remarkable property; and it follows that we may have at times the plane and the catenoid co-existing as parts of one and the same boundary system, just as the cylinder or the unduloid may be capped by portions of spheres. It follows also that if we stretch a soap-film between two rings, and so form an annular surface open at both ends, that surface is catenoid: the simplest case being when the rings are parallel and normal to the axis of the figure.[1]

[1] A topsail bellied out by the wind is not a catenoid surface, but in vertical section it is everywhere a catenary curve; and Dürer shows beautiful catenary curves in the wrinkles under an old man's eyes. A simple experiment is to invert a small funnel in a large one, wet them with soap-solution, and draw them apart; the film which develops between them is a catenoid surface, set perpendicularly to the two funnels. On this and other geometrical illustrations of the fact that a soap-film sets itself at right angles to a solid boundary, see an elegant paper by Mary E. Sinclair, in *Annals of Mathematics*, **8** (1907).

The nodoid is, like the unduloid, a continuous curve which keeps altering its curvature as it alters its distance from the axis; but in this case the resultant pressure inwards is negative instead of positive. But this curve is a complicated one, and its full mathematical treatment is too hard for us.

In one of Plateau's experiments, a bubble of oil (protected from gravity by a fluid of equal density to its own) is balanced between annuli; and by adjusting the distance apart of these, it may be brought to assume the form of Fig. 5, that is to say, of a cylinder with spherical ends; there is then everywhere a pressure inwards on the fluid contents of the bubble, a pressure due to the convexity of the

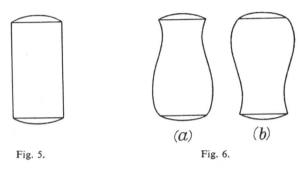

(a)   (b)

Fig. 5.   Fig. 6.

surface film. This cylinder may be converted into an unduloid, either by drawing the rings farther apart or by abstracting some of the oil, until at length rupture ensues, and the cylinder breaks up into two spherical drops. Or again, if the surrounding liquid be made ever so little heavier or lighter than that which constitutes the drop, then gravity comes into play, the conditions of equilibrium are modified accordingly, and the cylinder becomes part of an unduloid, with its dilated portion above or below as the case may be (Fig. 6).

In all cases the unduloid, like the original cylinder, is capped by spherical ends, the sign and the consequence of a positive pressure produced by the curved walls of the unduloid. But if our initial cylinder, instead of being tall, be a flat or dumpy one (with certain definite relations of height to breadth), then new phenomena may occur. For now, if oil be cautiously withdrawn from the mass by help of a small syringe, the cylinder may be made to flatten down so that its upper and lower surfaces become plane: which is of itself a sufficient indication that the pressure inwards is now *nil*. But at the very moment when the upper and lower surfaces become plane, it will be found that the sides curve inwards, in the fashion shown in

58

Fig 7, *b*. This figure is a catenoid, which, as we have seen, is, like the plane itself, a surface exercising no pressure, and which therefore may coexist with the plane as part of one and the same system.

We may continue to withdraw more oil from our bubble, drop by drop, and now the upper and lower surfaces dimple down into concave portions of spheres, as the result of the *negative* internal

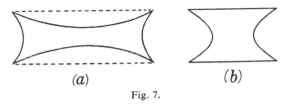

(a)   *(b)*

Fig. 7.

pressure; and thereupon the peripheral catenoid surface alters its form (perhaps, on this small scale, imperceptibly), and becomes a portion of a nodoid. It represents, in fact, that portion of the nodoid which in Fig 8 lies between such points as *O*, *P*. While it is easy to draw the outline, or meridional section, of the nodoid, it is obvious that the solid of revolution to be derived from it can never be realised in its entirety: for one part of the solid figure would cut, or entangle with, another. All that we can ever do, accordingly, is to realise isolated portions of the nodoid.[1]

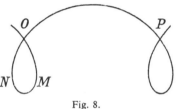

Fig. 8.

In all these cases the ring or annulus is not merely a means of mechanical restraint, controlling the form of the drop or bubble; it also marks the boundary, or 'locus of discontinuity', between one surface and another.

If, in a sequel to the preceding experiment of Plateau's, we use solid discs instead of annuli, we may exert pressure on our oil-globule as we exerted traction before. We begin again by adjusting the pressure of these discs so that the oil assumes the form of a cylinder: our discs, that is to say, are adjusted to exercise a mechanical pressure just equal to what in the former case was supplied by the surface-tension of the spherical caps or ends of the bubble. If we now

[1] This curve resembles the looped Elastic Curve (see Thomson and Tait, II, p. 148, fig. 7), but has its axis on the other side of the curve. The nodoid was represented upside-down in the first edition of this book, a mistake into which others have fallen, including no less a person than Clerk Maxwell, in his article 'Capillarity' in the *Encycl. Brit.* (9th ed.).

increase the pressure slightly, the peripheral walls become convexly curved, exercising a precisely corresponding pressure; the form assumed by the sides of our figure is now that of a portion of an unduloid. If we increase the pressure, the peripheral surface of oil will bulge out more and more, and will presently constitute a portion of a sphere. But we may continue the process yet further, and find within certain limits the system remaining perfectly stable. What is this new curved surface which has arisen out of the sphere, as the latter was produced from the unduloid? It is no other than a portion of a nodoid, that part which in Fig. 8 lies between M and N. But this surface, which is concave in both directions towards the surface of the oil within, is exerting a pressure upon the latter, just as did the sphere out of which a moment ago it was transformed; and we had just stated, in considering the previous experiment, that the pressure inwards exerted by the nodoid was a negative one. The explanation of this seeming discrepancy lies in the simple fact that, if we follow the outline of our nodoid curve in Fig. 8, from OP, the surface concerned in the former case, to MN, that concerned in the present, we shall see that in the two experiments the surface of the liquid is not the same, but lies on the positive side of the curve in the one case, and on the negative side in the other.

These capillary surfaces of Plateau's form a beautiful example of the 'materialisation' of mathematical law. Theory leads to certain equations which determine the position of points in a system, and these points we may then plot as curves on a co-ordinate diagram; but a drop or a bubble may realise in an instant the whole result of our calculations, and materialise our whole apparatus of curves. Such a case is what Bacon calls a 'collective instance', bearing witness to the fact that one common law is obeyed by every point or particle of the system. Where the underlying equations are unknown to us, as happens in so many natural configurations, we may still rest assured that kindred mathematical laws are being automatically followed, and rigorously obeyed, and sometimes half-revealed.

Of all the surfaces which we have been describing, the sphere is the only one which can enclose space of itself; the others can only help to do so, in combination with one another or with the sphere. Moreover, the sphere is also, of all possible figures, that which encloses the greatest volume with the least area of surface;[1] it is strictly

---

[1] On the circle and sphere as giving the smallest boundary for a given content, see, for example, Jacob Steiner, 'Einfache Beweise der isopermetrischen Hauptsätze', *Berlin. Abhandlungen* (1836), pp. 123–32.

and absolutely the surface of minimal area, and it is, *ipso facto*, the form which will be assumed by a unicellular organism (just as by a raindrop), if it be practically homogeneous and if, like *Orbulina* floating in the ocean, its surroundings be likewise homogeneous and its field of force symmetrical.[1] It is only relatively speaking that the rest of these configurations are surfaces *minimae areae*; for they are so under conditions which involve various pressures or restraints. Such restraints are imposed by the pipe or annulus which supports and confines our oil-globule or soap-bubble; and in the case of the organic cell, similar restraints are supplied by solidifications partial or complete, or other modifications local or general, of the cell-surface or cell-wall.

One thing we must not fail to bear in mind. In the case of the soap-bubble we look for stability or instability, equilibrium or non-equilibrium, in its several configurations. But the living cell is seldom in equilibrium. It is continually using or expending energy; and this ceaseless flow of energy gives rise to a 'steady state', taking the place of and simulating equilibrium. In like manner the hardly changing outline of a jet or waterfall is but in pseudo-equilibrium; it is in a steady state, dynamically speaking. Many puzzling and apparent paradoxes of physiology, such (to take a single instance) as the maintenance of a constant osmotic pressure on either side of a cell-membrane, are accounted for by the fact that energy is being spent and work done, and a *steady state* or pseudo-equilibrium maintained thereby.

Limits of Equilibrium

Among the many theoretical discoveries which we owe to Plateau, one is of peculiar importance: namely that, with the exception of the sphere and the plane, the surfaces with which we have been dealing are only in complete equilibrium within certain dimensional limits, or in other words, have a certain definite limit of stability; only the plane and the sphere, or any portion of a sphere, are perfectly stable, because they are perfectly symmetrical figures.

Perhaps it were better to say that their symmetry is such that any small disturbance will probably readjust itself, and leave the plane or spherical surface as it was before, while in the other configurations

---

[1] The essential conditions of homogeneity and symmetry are none too common, and a spherical organism is only to be looked for among simple things. The floating (or pelagic) eggs of fishes, the spores of red seaweeds, the oospheres of *Fucus* or *Oedogonium*, the plasma-masses escaping from the cells of *Vaucheria*, are among the instances which come to mind.

the chances are that a disturbance once set up will travel in one direction or another, increasing as it goes. For equilibrium and probability (as Boltzmann told us) are nearly allied: so nearly that that state of a system which is most likely to occur, or most likely to endure, is precisely that which we call the state of equilibrium.

For experimental demonstration, the case of the cylinder is the simplest. If we construct a liquid cylinder, either by drawing out a bubble or by supporting a globule of oil between two rings, the experiment proceeds easily until the length of the cylinder becomes just about three times as great as its diameter. But soon afterwards

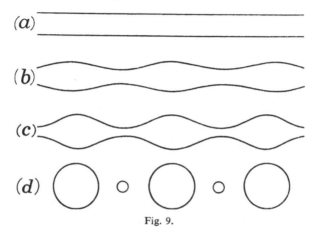

Fig. 9.

instability begins, and the cylinder alters its form; it narrows at the waist, so passing into an unduloid, and the deformation progresses quickly until our cylinder breaks in two, and its two halves become portions of spheres. This physical change of one surface into another corresponds to what the mathematicians call a 'discontinuous solution' of a problem of minima. The theoretical limit of stability, according to Plateau, is when the length of the cylinder is equal to its circumference, that is to say, when $L = 2\pi r$, or when the ratio of length to diameter is represented by π.

The fact is that any small disturbance takes the form of a wave, and travels along the cylinder. Short waves do not affect the stability of the system; but waves whose length exceeds that of the circumference tend to grow in amplitude: until, contracting here, expanding there, the cylinder turns into a pronounced unduloid, and soon breaks into two parts or more. Thus the cylinder is a stable figure until it becomes longer than its own circumference, and then the risk of rupture may be said to begin. But Rayleigh showed that still longer waves, leading to still greater instability, are needed to break down material resistance.[1] For, as Plateau knew well, his was a theoretical result, to be departed from under material conditions; it is affected largely by viscosity, and, as in the case of a flowing cylinder or jet, by inertia. When inertia plays a leading part, viscosity being small, the node of maximum instability corresponds to nearly half as much again as in the simple or theoretical case: and this result is very near to what Plateau himself had deduced from Savart's experiments on jets of water.[2] When the fluid is external (as when the cylinder is of air) the wavelength of maximal instability is longer still. Lastly, when viscosity is very large, and becomes paramount, then the wave-length between regions of maximal instability may become very long indeed: so that (as Rayleigh put it) 'long (viscid) threads do not tend to divide themselves into drops at mutual distances comparable with the diameter of the cylinder, but rather to give way by attenuation at few and distant places'. It is this that renders possible the making of long glass tubes, or the spinning of threads of 'viscose' and like materials; but while these latter preserve their continuity, the principle of Plateau tends to give them something of a wavy, unduloid surface, to the great enhancement of their beauty. We are prepared, then, to find that such cylinders and unduloids as occur in organic nature seldom approach in regularity to those which theory prescribes or a soap-film may be made to show; but rather exhibit all manner of gradations, from something exquisitely neat and regular to a coarse and distant approximation to the ideal thing.[3]

In a spider's web we find exemplified several of the principles of surface-tension which we have now explained. The thread is spun out of a glandular secretion which issues from the spider's body as a semi-fluid cylinder, the force of expulsion giving it its length and that of surface-tension giving it its circular section. It is too viscid, and too soon hardened on exposure to the air, to break up into drops or spherules; but it is otherwise with another sticky secretion which, coming from another gland, is simultaneously poured over the slacker cross-threads as they issue to form the spiral portion of the

---

[1] Rayleigh, 'On the Instability of Fluid Surfaces', *Sci. Papers*, 3, 594.
[2] Cf. E. Tylor, *Phil. Mag.* 16 (1933), 504–18.
[3] Cf. F. Savart, 'Sur la constitution des veines liquides lancées par des orifices, etc.', *Ann. Chimie*, 53 (1833), 337–86. Rayleigh, 'On the Instability of a Cylinder of Viscous Liquid', etc., *Phil. Mag.* (5), 34 (1892), or *Sci. Papers*, 1, 361. See also Larmor, 'On the Nature of Viscid Fluid Threads', *Nature, Lond.* (11 July 1936), p. 74.

web. This latter secretion is more fluid than the first, and only dries up after several hours.[1] By capillarity it 'wets' the thread, spreading over it in an even film or liquid cylinder. As such it has its limits of stability, and tends to disrupt at points more distant than the theoretical wavelength, owing to the imperfect fluidity of the viscous film and still more to the frictional drag of the inner thread with which it is in contact. Save for this qualification the cylinder disrupts in the usual manner, passing first into the wavy outline of an unduloid, whose swollen internodes swell more and more till the necks between them break asunder, and leave a row of spherical drops or beads strung like dewdrops at regular intervals along the thread. If we try to varnish a thin taut wire we produce automatically the same identical result;[2] unless our varnish be such as to dry almost instantaneously it gathers into beads, and do what we will we fail to spread it smooth. It follows that, according to the drying qualities of our varnish, the process may stop at any point short of the formation of perfect spherules; and as our final stage we may only obtain half-formed beads or the wavy outlines of an unduloid. The beads may be helped to form by jerking the stretched thread, and so disturbing the unstable equilibrium of the viscid cylinder. This the spider has been said to do, but Dr G. T. Bennett assures me that she does nothing of the kind. She only draws her thread out a little, and leaves it a trifle slack; if the gum should break into droplets, well and good, but it matters little. The web with its sticky threads is not improved thereby. Another curious phenomenon here presents itself.

In Plateau's experimental separation of a cylinder of oil into two spherical halves, it was noticed that, when contact was nearly broken, that is to say when the narrow neck of the unduloid had become very thin, the two spherical bullae, instead of absorbing the fluid out of the narrow neck into themselves as they had done with the preceding portion, drew out this small remaining part of the liquid into a thin thread as they completed their spherical form and receded from one another: the reason being that, after the thread or 'neck' has reached a certain tenuity, internal friction prevents or retards a rapid exit of the fluid from the thread to the adjacent spherule. It is for the same reason that we are able to draw a glass rod or tube, which we

---

[1] When we see a web bespangled with dew of a morning, the dewdrops are not drops of pure water, but of water mixed with the sticky, gummy fluid of the cross-threads; the radii seldom if ever show dewdrops. See F. Strehlke, 'Beobachtungen an Spinnengewebe', *Ann. Phys., Lpz.* **40** (1937), 146.

[2] Felix Plateau recommends the use of a weighted thread or plumb-line, to be drawn up slowly out of a jar of water or oil; *Phil. Mag.* **34** (1867), 246.

have heated in the middle, into a long and uniform cylinder or thread by quickly separating the two ends. But in the case of the glass rod the long thin thread quickly cools and solidifies, while in the ordinary separation of a liquid cylinder the corresponding intermediate cylinder remains liquid; and therefore, like any other liquid cylinder, it is liable to break up, provided that its dimensions exceed the limit of stability. And its length is generally such that it breaks at two points, thus leaving two terminal portions continuous and confluent with the spheres, and one median portion which resolves itself into a tiny spherical drop, midway between the original and larger two. Occasionally, the same process of formation of a connecting thread repeats itself a second time, between the small intermediate spherule and the large spheres; and in this case we obtain two additional spherules, still smaller in size, and lying one on either side of our first little one. This whole phenomenon, of equal and regularly interspaced beads, often with little beads regularly interspaced between the larger ones, and now and then with a third order of still smaller beads regularly intercalated, may be easily observed in a spider's web, such as that of *Epeira*, very often with beautiful regularity—sometimes interrupted and disturbed by a slight want of homogeneity in the secreted fluid; and the same phenomenon is repeated on a grosser scale when the web is bespangled with dew, and its threads bestrung with pearls innumerable. To the older naturalists, these regularly arranged and beautifully formed globules on the spider's web were a frequent source of wonderment. Blackwall, counting some twenty globules in a tenth of an inch, calculated that a large garden-spider's web should comprise about 120,000 globules; the net was spun and finished in about forty minutes, and Blackwall was filled with admiration of the skill and quickness with which the spider manufactured these little beads. And no wonder, for according to the above estimate they had to be made at the rate of about 50 per second.[1]

Here we see exemplified what Plateau told us of the law of minimal areas transforming the cylinder into the unduloid and disrupting it into spheres. The little delicate beads which stud the long thin pseudopodia of a foraminifer, such as *Gromia*, or which appear in like manner on the film of protoplasm coating the long radiating spicules of *Globigerina*, represent an identical phenomenon. Indeed we may

---

[1] J. Blackwall, *Spiders of Great Britain* (Ray Society, 1859), p. 10; *Trans. Linn. Soc.* **16** (1833), 477. On the strength and elasticity of the spider's web, see J. R. Benton, *Amer. J. Sci.* **24** (1907), 75–8.

study in a protoplasmic filament the whole process of formation of such beads: if we squeeze out on a slide the viscid contents of a mistletoe-berry, the long sticky threads into which the substance runs show the whole phenomenon particularly well. True, many long cylindrical cells, such as are common in plants, show no sign of beading or disruption; but here the cell-walls are never fluid but harden as they grow, and the protoplasm within is kept in place and shape by its contact with the cell-wall. It was noticed many years ago by Hofmeister,[1] and afterwards explained by Berthold, that if we dip the long root-hairs of certain water-plants, such as *Hydrocharis* or

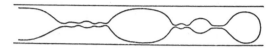

Fig. 10. Root-hair of *Trianea*, in glycerine. After Berthold.

Trianea, in a denser fluid (a little sugar-solution or dilute glycerine), the cell-sap tends to diffuse outwards, the protoplasm parts company with its surrounding and supporting wall, and then lies free as a protoplasmic cylinder in the interior of the cell. Thereupon it soon shows signs of instability, and commences to disrupt; it tends to gather into spheres, which however, as in our illustration, may be prevented by their narrow quarters from assuming the complete spherical form; and in between these spheres, we have more or less regularly alternate ones, of smaller size.[2]

In one of Robert Chambers's delicate experiments, a filament of protoplasm is drawn off, by a micro-needle, from the fluid surface of a starfish-egg. If drawn too far it breaks, and part returns within the protoplasm while the other rounds itself off on the needle's point. If drawn out less far, it looks like a row of beads or chain of droplets; if yet more relaxed, the droplets begin to fuse until the whole filament is withdrawn; if drawn out anew the process repeats itself. The whole story is a perfect description of the behaviour of a fluid jet or cylinder, of varying length and thickness.[3]

---

[1] *Lehrbuch von der Pflanzenzelle*, p. 71; cf. Nägeli, *Pflanzenphysiologische Untersuchungen (Spirogyra)*, III, 10.

[2] The intermediate spherules appear with great regularity and beauty whenever a liquid jet breaks up into drops. So a bursting soap-bubble scatters a shower of droplets all around, sometimes all alike, but often with a beautiful alternation of great and small. How the breaking-up of thread or jet into drops may be helped, regularised, and sometimes complicated, by external vibrations is another and by no means unimportant story.

[3] Cf. R. Chambers in *Colloid Chemistry, Theoretical and Applied* (1928), II, cap. 24; also *Ann. Physiol.* 6 (1930), 234; etc.

PLATE I

An instantaneous photograph of a 'splash' of milk. From Harold E. Edgerton, Massachusetts Institute of Technology

PLATE II

The latter phase of a splash: the crater has subsided, a columnar jet has risen up, and the jet is dividing into droplets. From Harold E. Edgerton, Massachusetts Institute of Technology

Splashes

We may take note here of a remarkable series of phenomena, which, though they seem at first sight to be of a very different order, are closely related to those which attend and which bring about the breaking-up of a liquid cylinder or thread.

In Mr Worthington's beautiful experiments on splashes, it was found that the fall of a round pebble into water from a height first formed a dip or hollow in the surface, and then caused a filmy 'cup' of water to rise up all round, opening out trumpet-fashion or closing in like a bubble, according to the height from which the pebble fell.[1] The cup or 'crater' tends to be fluted in alternate ridges and grooves, its edges get scolloped into corresponding lobes and notches, and the projecting lobes or prominences tend to break off or break up into drops or beads (see plates).[2] A similar appearance is seen on a great scale in the edge of a breaking wave: for the smooth edge becomes notched or sinuous, and the surface near by becomes ribbed or fluted, owing to the internal flow being helped here and hindered there by a viscous shear; and then all of a sudden the uneven edge shoots out an array of tiny jets, which break up into the countless droplets which constitute 'spray'. The naturalist may be reminded also of the beautifully symmetrical notching of the calycles of many hydroid zoophytes, which little cups had begun their existence as liquid or semi-liquid films before they became stiff and rigid. The next phase of the splash (with which we are less directly concerned) is that the crater subsides, and where it stood a tall column rises up, which also tends, if it be tall enough, to break up into drops. Lastly the column sinks down in its turn, and a ripple runs out from where it stood.

The edge of our little cup forms a liquid ring or annulus, comparable on the one hand to the edge of an advancing wave, and on the other to a liquid thread or cylinder if only we conceive the thread to be bent round into a ring; and accordingly, just as the thread segments first into an unduloid and then into separate spherical drops, so likewise will the edge of cup or annulus tend to do. This phase of notching, or beading, of the edge of the splash is beautifully seen in

[1] *A Study of Splashes* (1908), p. 38, etc.; also various papers in *Proc. Roy. Soc.* (1876–82), and *Phil. Trans.* A (1897 and 1900).

[2] We owe this picture to the kindness of Mr Harold E. Edgerton, of the Massachusetts Institute of Technology. It shows the splash caused by a drop falling into a thin layer of milk; a second drop of milk is seen above, following the first. The exposure-time was 1/50,000 of a second.

many of Worthington's experiments,[1] and still more beautifully in the recent work of Edgerton (see plates). In the second place the fact that the crater rises up means that liquid is flowing in from below; the segmentation of the rim means that channels of easier flow are being created, along which the liquid is led or driven into the protuberances; and these last are thereby exaggerated into the jets or streams which become conspicuous at the edge of the crater. In short any film or film-like fluid or semi-fluid cup will be unstable; its instability will tend to show itself in a fluting of the surface and a notching of the edge; and just such a fluting and notching are conspicuous features of many minute organic cup-like structures. In the hydroids (Fig. 11),

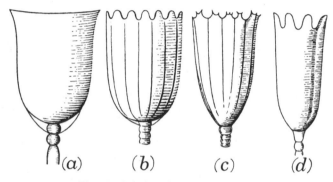

Fig. 11. Calycles of Campanularia spp.

we see that these common features of the cup and the annulation of the stem are phenomena of the same order. A cord-like thickening of the edge of the cup is a variant of the same order of phenomena; it is due to the checking at the rim of the flow of liquid from below, and a similar thickening is to be seen, not only in some hydroid calycles but also in many Vorticellae (cf. Fig. 19) and other cup-shaped organisms. And these are by no means the only manifestations of surface-tension in a splash which show resemblances and analogies to organic form.[2]

[1] Cf. *A Study of Splashes*, pp. 17, 77. The same phenomenon is often well seen in the splash of an oar. It is beautifully and continuously evident when a strong jet of water from a tap impinges on a curved surface and then shoots off again.

[2] The same phenomena are modified in various ways, and the drops are given off much more freely, when the splash takes place in an electric field—all owing to the general instability of an electrified liquid surface; and a study of this aspect of the subject might suggest yet more analogies with organic form. Cf. J. Zeleny, *Phys. Rev.* **10** (1917); J. P. Gott, *Proc. Camb. Phil. Soc.* **31** (1935); etc.

The phenomena of an ordinary liquid splash are so swiftly transitory that their study is only rendered possible by photography: but this excessive rapidity is not an essential part of the phenomenon. For instance, we can repeat and demonstrate many of the simpler phenomena, in a permanent or quasi-permanent form, by splashing water on to a surface of dry sand,[1] or by firing a bullet into a soft metal target. There is nothing, then, to prevent a slow and lasting manifestation, in a viscous medium such as a protoplasmic organism, of phenomena which appear and disappear with evanescent rapidity in a more mobile liquid. Nor is there anything peculiar in the splash itself; it is simply a convenient method of setting up certain motions or currents, and producing certain surface-forms, in a liquid medium —or even in such an imperfect fluid as a bed of sand. Accordingly, we have a large range of possible conditions under which the organism might conceivably display configurations analogous to, or identical with, those which Mr Worthington has shown us how to exhibit by one particular experimental method.

To one who has watched the potter at his wheel, it is plain that the potter's thumb, like the glass-blower's blast of air, depends for its efficacy upon the physical properties of the clay or 'slip' it works on, which for the time being is essentially a fluid. The cup and the saucer, like the tube and the bulb, display (in their simple and primitive forms) beautiful surfaces of equilibrium as manifested under certain limiting conditions. They are neither more nor less than glorified 'splashes', formed slowly, under conditions of restraint which enhance or reveal their mathematical symmetry. We have seen, and we shall see again before we are done, that the art of the glass-blower is full of lessons for the naturalist as also for the physicist: illustrating as it does the development of a host of mathematical configurations and organic conformations which depend essentially on the establishment of a constant and uniform pressure within a *closed* elastic shell or fluid envelope or bubble. In like manner the potter's art illustrates the somewhat obscurer and more complex problems (scarcely less frequent in biology) of a figure of equilibrium which is an *open* surface of revolution. The two series of problems are closely akin; for the glass-blower can make most things which the potter makes, by cutting off *portions* of his hollow ware; besides,

---

[1] We find now and then in certain brick-clays of glacial origin, hard, quoit-shaped rings, each with an equally indurated, round or flattened ball resting on it. These may be precisely imitated by splashing large drops of water on a smooth surface of fine dry sand. The ring corresponds, apparently, to the crater of the splash, and the ball (or its water content) to the pillar rising in the middle.

when this fails and the glass-blower, ceasing to blow, begins to use his rod to trim the sides or turn the edges of wineglass or of beaker, he is merely borrowing a trick from the still older craft of the potter.

It would seem venturesome to extend our comparison with these liquid surface-tension phenomena from the cup or calycle of the hydrozoon to the little hydroid polyp within: and yet there is something to be learned by such a comparison. The cylindrical body of the tiny polyp, the jet-like row of tentacles, the beaded annulations which these tentacles exhibit, the web-like film which sometimes (when they stand a little way apart) conjoins their bases, the thin annular film of tissue which surrounds the little organism's mouth, and the manner in which this annular 'peristome' contracts,[1] like a shrinking soap-bubble, to close the aperture, are every one of them features to which we may find a singular and striking parallel in the surface-tension phenomena of the splash.[2]

Falling Drops

Some seventy years ago much interest was aroused by Helmholtz's work (and also Kirchhoff's) on 'discontinuous motions of a fluid';[3] that is to say, on the movements of one body of fluid within another, and the resulting phenomena due to friction at the surfaces between. What Kelvin[4] called Helmholtz's 'admirable discovery of the law of vortex-motion in a perfect fluid' was the chief result of this investigation; and was followed by much experimental work, in order to illustrate and to extend the mathematical conclusions.

The drop, the bubble and the splash are parts of a long story; and a 'falling drop', or a drop moving through surrounding fluid, is a case deserving to be considered. A drop of water, tinged with fuchsin, is gently released (under a pressure of a couple of millimetres) at the bottom of a glass of water.[5] Its momentum enables it to rise through a few centimetres of the surrounding water, and in doing so it communicates motion to the water around. In front the rising drop *thrusts* its way through, almost like a solid body; behind it tends to *drag* the surrounding water after it, by fluid

---

[1] See *A Study of Splashes*, p. 54.

[2] There is little or no difference between a *splash* and a *burst bubble*. The craters of the moon have been compared with, and explained by, both of these.

[3] Helmholtz, in *Berl. Mber.* (1868), pp. 215–28; Kirchhoff, in *Crelle's J.* **70**, pp. 289–98; **71** (1869–70), 237–73.

[4] W. Thomson, in *Proc. Roy. Soc. Edinb.* **6** (1867), 94.

[5] See A. Overbeck, 'Ueber discontinuierliche Flüssigkeitsbewegungen', *Wiedemann's Annalen*, II, 1877; W. Bezold, 'Ueber Strömungsfiguren in Flüssigkeiten', *ibid.* XXIV (1885), 569–93; P. Czermak, *ibid.* **50** (1893), 329; etc.

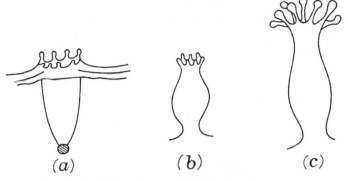

Fig. 12. (*a*, *b*) More phases of a splash, after Worthington;
(*c*) a hydroid polyp, after Allman.

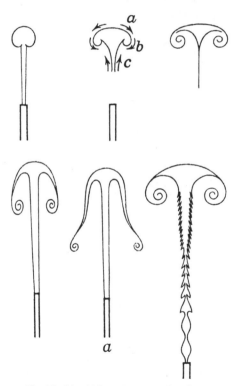

Fig. 13. Liquid jets. From A. Overbeck.

friction;[1] and these two motions together give rise to beautiful vorticoid configurations, the *Strömungspilze* or *Tintenpilze* of their first discoverers (Fig. 13). Under a higher and more continuous pressure the drop becomes a jet; the form of the vortex is modified thereby, and may be further modified by slight differences of temperature (i.e. of density), or by interrupting the rate of flow. To let a drop of

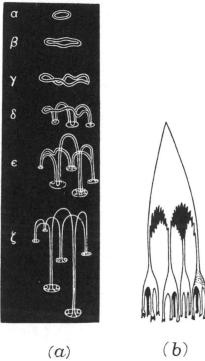

(a) (b)

Fig. 14. Falling drops. (*a*) ink in water, after J. J. Thomson and H. F. Newall; (*b*) fusel oil in paraffin, after Tomlinson.

ink fall into water is a simple and most beautiful experiment.[2] The effect is more violent than in the former case. The descending drop turns into a complete vortex-ring; it expands and attenuates; it waves about, and the descending loops again turn into incipient vortices (Fig. 14).

[1] The frictional drag on the hinder part of the drop is felt alike in the ship, the bird and the aeroplane, and tends to produce retarding vortices in them all. It is always minimised in one way or another, and it is automatically minimised in the present instance, as the drop thins off and tapers down.

[2] J. J. Thomson and H. F. Newall, 'On the Formation of Vortex-rings by Drops', *Proc. Roy. Soc.* **39** (1885), 417–36; Emil Hatschek, 'On Forms Assumed by a Gelatinising Liquid in Various Coagulating Solutions', *ibid.* A, **94** (1918), 303–16.

Lastly, instead of letting our drop rise or fall freely, we may use a hanging drop, which, while it sinks, remains suspended to the surface. Thus it cannot form a complete annulus, but only a partial vortex suspended by a thread or column—just as in Overbeck's jet-experiments; and the figure so produced, in either case, is closely analogous to that of a medusa or jellyfish, with its bell or 'umbrella', and its clapper or 'manubrium' as well. Some years ago Emil Hatschek made such vortex-drops as these of liquid gelatine dropped into a hardening fluid. These 'artificial medusae' sometimes show a symmetrical pattern of radial 'ribs', due to shrinkage, and this to

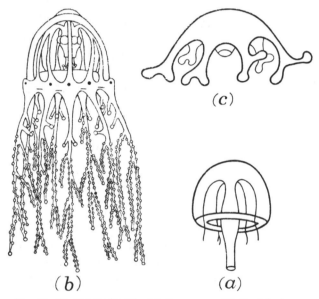

Fig. 15. Various medusoids: (a) *Syncoryme*; (b) *Cordylophora*; (c) *Cladonema* (after Allman).

dehydration by the coagulating fluid. An extremely curious result of Hatschek's experiments is to show how sensitive these vorticoid drops are to physical conditions. For using the same gelatine all the while, and merely varying the density of the fluid in the third decimal place, we obtain a whole range of configurations, from the ordinary hanging drop to the same with a ribbed pattern, and then to medusoid vortices of various graded forms.

The living medusa has a geometrical symmetry so marked and regular as to suggest a physical or mechanical element in the little creature's growth and construction. It has, to begin with, its vortex-like bell or umbrella, with its cylindrical handle or manubrium. The bell is traversed by radial canals, four or in multiples of four; its edge is beset with tentacles, smooth or often beaded, at regular intervals and of graded sizes; and certain sensory structures, including solid concretions or 'otoliths', are also symmetrically interspaced. No

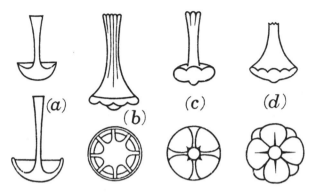

Fig. 16. 'Medusoid drops', of gelatine. After Hatschek.

sooner made, than it begins to pulsate; the little bell begins to 'ring'. Buds, miniature replicas of the parent-organism, are very apt to appear on the tentacles, or on the manubrium or sometimes on the edge of the bell; we seem to see one vortex producing others before our eyes. The development of a medusoid deserves to be studied without prejudice, from this point of view. Certain it is that the tiny medusoids of *Obelia*, for instance, are budded off with a rapidity and a complete perfection which suggests an automatic and all but instantaneous act of conformation, rather than a gradual process of growth.

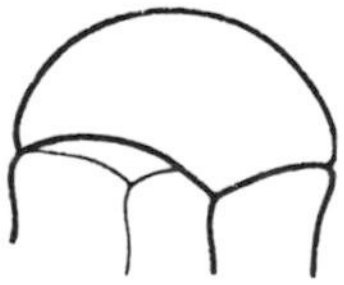

Moreover, not only do we recognise in a vorticoid drop a 'schema' or analogue of medusoid form, but we seem able to discover various actual phases of the splash or drop in the all but innumerable living types of jellyfish; in *Cladonema* we seem to see an early stage of a

Fig. 17. *Medusachloris*, a ciliate infusorian.

breaking drop, and in *Cordylophera* a beautiful picture of incipient vortices. It is hard indeed to say how much or little all these analogies imply. But they indicate, at the very least, how certain simple organic forms might be naturally assumed by one fluid mass within another, when gravity, surface tension and fluid friction play their part, under balanced conditions of temperature, density and chemical composition.

A little green infusorian from the Baltic Sea is, as near as may be, a medusa in miniature.[1] It is curious indeed to find the same medusoid, or as we may now call it vorticoid, configuration occurring in a form so much lower in the scale, and so much less in order of magnitude, than the ordinary medusae.

Viscous Threads

According to Plateau, the viscidity of the liquid, while it retards the breaking up of the cylinder and increases the length of the segments beyond that which theory demands, has nevertheless, less influence in this direction than we might have expected. On the other hand any external support or adhesion, or mere contact with a solid body, will be equivalent to a reduction of surface-tension and so will very greatly increase the stability of our cylinder. It is for this reason that the mercury in our thermometers seldom separates into drops: though it sometimes does so, much to our inconvenience. And again it is for this reason that the protoplasm in a long tubular or cylindrical cell need not divide into separate cells and internodes until the length of these far exceeds the theoretical limits.

An interesting case is that of a viscous drop immersed in another viscous fluid, and drawn out into a thread by a shearing motion of the latter. The thread seems stable at first, but when left to rest it breaks up into drops of a very definite and uniform size, the size of the drops, or wavelength of the unduloid of which they are made, depending on the relative viscosities of the two threads.[2]

Plateau's results, though discovered by way of experiment and though (as we have said) they illustrate the 'materialisation' of mathematical law, are nevertheless essentially theoretical results approached rather than realised in material systems. That a liquid cylinder begins to be unstable when its length exceeds $2\pi r$ is all but mathematically true of an all but immaterial soap-bubble; but very far from true, as Plateau himself was well aware, in a flowing jet, retarded by viscosity and by inertia. The principle is true and universal; but our living cylinders do not follow the abstract laws of mathematics, any more than do the drops and jets of ordinary fluids or the quickly drawn and quickly cooling tubes in the glass-worker's hands.

[1] *Medusachloris phiale*, of A. Pascher, *Biol. Zbl.* 37 (1917), 421–9.
[2] See especially Rayleigh, *Phil. Mag.* 34 (1892), 145, by whom the subject is carried much further than where Plateau left it. See also (*int. al.*) G. I. Taylor, *Proc. Roy. Soc.* A, 146 (1934), 501; S. Tomotika, *ibid.* 150 (1935), 322; etc.

Plateau says that in most liquids the influence of viscosity is such as to cause the cylinder to segment when its length is about four times, or even six times, its diameter, instead of a fraction over three times, as theory would demand of a perfect fluid. If we take it at four times, the resulting spherules would have a diameter of about 1·8 times, and their distance apart would be about 2·2 times, the original diameter of the cylinder; and the calculation is not difficult which would show how these dimensions are altered in the case of a cylinder formed around a solid core, as in the case of a spider's web. Plateau also observed that the *time* taken in the division of the cylinder is directly proportional to its diameter, while varying with the nature of the liquid. This question, of the time taken in the division of a cell or filament in relation to its dimensions, has not so far as I know been enquired into by biologists.

Asymmetry and Anisotropy

From the simple fact that the sphere is of all configurations that whose surface-area for a given volume is an absolute minimum, we have seen it to be the one figure of equilibrium assumed by a drop or vesicle when no disturbing factor is at hand; but such freedom from counter-influences is likely to be rare, and neither does the rain-drop nor the round world itself retain its primal sphericity. For one thing, gravity will always be at hand to drag and distort our drop or bubble, unless its dimensions be so minute that gravity becomes insignificant compared with capillarity. Even the soap-bubble will be flattened or elongated by gravity, according as we support it from below or from above; and the bubble which is thinned out almost to blackness will, from its small mass, be the one which remains most nearly spherical.[1]

Innumerable new conditions will be introduced, in the shape of complicated tensions and pressures, when one drop or bubble becomes associated with another, and when a system of intermediate films or partition-walls is developed between them. This subject we shall discuss later, in connection with cell-aggregates or tissues, and we shall find that further theoretical considerations are needed as a preliminary to any such enquiry. Meanwhile let us consider a few cases of the forms of cells, either solitary, or in such simple aggregates that their individual form is little disturbed thereby. Let us clearly

[1] Cf. Dewar, 'On Soap-bubbles of Long Duration', *Proc. Roy. Inst.* (19 January 1929).

understand that the cases we are about to consider are those where the perfect symmetry of the sphere is replaced by another symmetry, less complete, such as that of an ellipsoidal or cylindrical cell. The cases of asymmetrical deformation or displacement, such as are illustrated in the production of a bud or the development of a lateral branch, are much simpler; for here we need only assume a slight and localised variation of surface-tension, such as may be brought about in various ways through the heterogeneous chemistry of the cell. But such diffused and graded asymmetry as brings about, for instance, the ellipsoidal shape of a yeast-cell is another matter.

If the sphere be the one surface of complete symmetry and therefore of independent equilibrium, it follows that in every cell which is otherwise conformed there must be some definite cause of its departure from sphericity; and if this cause be the obvious one of resistance offered by a solidified envelope, such as an eggshell or firm cell-wall, we must still seek for the deforming force which was in action to bring about the given shape prior to the assumption of rigidity. Such a cause may be either external to, or may lie within, the cell itself. On the one hand it may be due to external pressure or some form of mechanical restraint, as when we submit our bubble to the partial restraint of discs or rings or more complicated cages of wire; on the other hand it may be due to intrinsic causes, which must come under the head either of differences of internal pressure, or of lack of homogeneity or isotropy in the surface of its envelope.[1]

Our formula of equilibrium, or equation to an elastic surface, is $P = p_e + (T/R + T'/R')$, where P is the internal pressure, p_e any extraneous pressure normal to the surface, R, R' the radii of curvature at a point, and T, T' the corresponding tensions, normal to one another, of the envelope.

Now in any given form which we seek to account for, R, R' are known quantities; but all the other factors of the equation are subject to enquiry. And somehow or other, by this formula, we

[1] A case which we have not specially considered, but which may be found to deserve consideration in biology, is that of a cell or drop suspended in a liquid of *varying* density, for instance, in the upper layers of a fluid (e.g. sea-water) at whose surface condensation is going on, so as to produce a steady density-gradient. In this case the normally spherical drop will be flattened into an oval form, with its maximum surface-curvature lying at the level where the densities of the drop and the surrounding liquid are just equal. The sectional outline of the drop has been shown to be not a true oval or ellipse, but a somewhat complicated quartic curve. (Rice, *Phil. Mag.* January 1915.) A more general case, which also may well deserve consideration by the biologist, is that of a charged bubble in (for instance) a uniform field of force: which will expand or elongate in the direction of the lines of force, and become a spheroidal surface in continuous transformation with the original sphere.

must account for the form of any solitary cell whatsoever (provided always that it be not formed by successive stages of solidification), the cylindrical cell of *Spirogyra*, the ellipsoidal yeast-cell, or, an even more obvious example, the egg of any bird. In using this formula hitherto we have taken it in a simplified form, that is to say we have made several limiting assumptions. We have assumed that P was the uniform hydrostatic pressure, equal in all directions, of a body of liquid; we have assumed likewise that the tension T was due to surface-tension in a homogeneous liquid film, and was therefore equal in all directions, so that $T = T'$; and we have only dealt with surfaces, or parts of a surface, where extraneous pressure, p_n, was non-existent. Now in the case of a bird's egg the external pressure p_n, that is to say the pressure exercised by the walls of the oviduct, will be found to be a very important factor; but in the case of the yeast-cell or the *Spirogyra*, wholly immersed in water, no such external pressure comes into play. We are accordingly left in such cases as these last with two hypotheses, namely that the departure from a spherical form is due to inequalities in the internal pressure P, or else to inequalities in the tension T, that is to say to a difference between T and T'. In other words, it is theoretically possible that the oval form of a yeast-cell is due to a greater internal pressure, a greater 'tendency to grow' in the direction of the longer axis of the ellipse, or alternatively, that with equal and symmetrical tendencies to growth there is associated a difference of external resistance in respect of the tension, and implicitly the molecular structure, of the cell-wall. Now the former hypothesis is not impossible. Protoplasm is far from being a perfect fluid; it is the seat of various internal forces, sometimes manifestly polar, and it is quite possible that the forces, osmotic and other, which lead to an increase of the content of the cell and are manifested in pressure outwardly directed upon its wall may be unsymmetrical, and such as to deform what would otherwise be a simple sphere. But while this hypothesis is not impossible, it is not very easy of acceptance. The protoplasm, though not a perfect fluid, has yet on the whole the properties of a fluid; within the small compass of the cell there is little room for the development of unsymmetrical pressures; and in such a case as *Spirogyra*, where most part of the cavity is filled by watery sap, the conditions are still more obviously, or more nearly, those under which a uniform hydrostatic pressure should be displayed. But in variations of T, that is to say of the specific surface-tension per unit area, we have an ample field for all the various deformations with which we shall have to

78

deal. Our condition now is, that $(T/R + T'/R') =$ a constant; but it no longer follows, though it may still often be the case, that this will represent a surface of absolute minimal area. As soon as T and T' become unequal, we are no longer dealing with a perfectly liquid surface film; but its departure from perfect fluidity may be of all degrees, from that of a slight non-isotropic viscosity to the state of a firm elastic membrane;[1] and it matters little whether this viscosity or semi-rigidity be manifested in the self-same layer which is still a part of the protoplasm of the cell, or in a layer which is completely differentiated into a distinct and separate membrane. As soon as, by secretion or adsorption, the molecular constitution of the surface-layer is altered, it is clearly conceivable that the alteration, or the secondary chemical changes which follow it, may be such as to produce an anisotropy, and to render the molecular forces less capable in one direction than another of exerting that contractile force by which they are striving to reduce to a minimum the surface area of the cell. A slight inequality in two opposite directions will produce the ellipsoid cell, and a great inequality will give rise to the cylindrical cell.

I take it therefore, that the cylindrical cell of *Spirogyra*, or any other cylindrical cell which grows in freedom from any manifest external restraint, has assumed that particular form simply by reason of the molecular constitution of its developing wall or membrane; and that this molecular constitution was anisotropous, in such a way as to render extension easier in one direction than another. Such a lack of homogeneity or of isotropy in the cell-wall is often rendered visible, especially in plant-cells, in the form of concentric lamellae, annular and spiral striations, and the like. But there exists yet another heterogeneity to help us account for the long threads, hairs, fibres, cylinders, which are so often formed. Carl Nägeli said many years ago that organised bodies, starch-grains, cellulose and protoplasm itself, consisted of invisible particles, each an aggregate of many molecules—he called them *micellae*; and these were isolated, or 'dispersed' as we should say, in a watery medium. This theory was, to begin with, an attempt to account for the colloid state; but at the same time, the particles were supposed to be so ordered and arranged as to render the substance anisotropic, to confer on it vectorial properties as we say nowadays, and so to account for the polarisation of

---

[1] Indeed any non-isotropic *stiffness*, even though T remained uniform, would simulate, and be indistinguishable from, a condition of non-stiffness and non-isotropic T.

light by a starch-grain or a hair. It was so criticised by Bütschli and von Ebner that it fell into disrepute, if not oblivion; but a great part of it was true. And the micellar structure of wool, cotton, silk and similar substances is now rendered clearly visible by the same X-ray methods as revealed the molecular orientation, or lattice-structure, of a crystal to von Laue.

Biological Examples of Plateau's Surfaces of Revolution: Spheres

Many unicellular forms, and a few other simple organisms, are spherical, and serve to illustrate in the simplest way the point at issue. Unicellular algae, such as *Protococcus* or *Halisphaera*, the

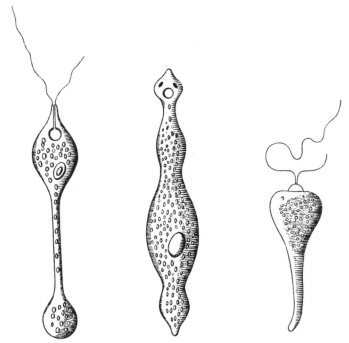

Fig. 18. A flagellate 'monad', *Distigma proteus* Ehr. After Saville Kent.

innumerable floating eggs of fishes, the floating unilocular foraminifer *Orbulina*, the lovely green multicellular *Volvox* of our ponds, all these in their several grades of simplicity or complication are so many round drops, spherical because no alien forces have deformed or mis-shapen them. But observe that, with the exception of *Volvox*,

80

whose spherical body is covered wholly and uniformly with minute cilia, all the above are passive or inactive forms; and in a 'resting' or encysted phase the spherical form is common and general in a great range of unicellular organisms.

Conversely, we see that those unicellular forms which depart markedly from sphericity—excluding for the moment the amoeboid forms and those provided with skeletons—are all ciliate or flagellate. Cilia and flagella are *sui generis*; we know nothing of them from the physical side, we cannot reproduce or imitate them in any non-living drop or fluid surface. But we can easily see that they have an influence on *form*, besides serving for locomotion. When our little *Monad* or *Euglena* develops a flagellum, that is in itself an indication of asymmetry or 'polarity', of non-homogeneity of the little cell; and in the various flagellate types the flagellum or its analogues always stand on prominent points, or ends, or edges of the cell—on parts, that is to say, where curvature is high and surface-tension may be expected to be low—for the product of surface-tension by mean curvature tends to be constant.

Cilia, like flagella, tend to occupy positions, or cover surfaces, which would otherwise be unstable; and often indeed (as in a trochosphere larva or even in a rotifer) a ring of cilia seems to play the very part of one of Plateau's wire rings, supporting and steadying the semi-fluid mass in its otherwise unstable configuration. Let us note here (in passing) what seems to be an analogous phenomenon. Chitinous hairs, spines or bristles are common and characteristic structures among the smaller Crustacea, and more or less generally among the Arthropods. We find them at every exposed point or corner; they fringe the sharp edge or border of a limb; as we draw the creature, we seem to know where to put them in! In short, they tend to occur, as the flagella do, just where the surface-tension would be lowest, if or when the surface was in a fluid condition.

Cylinders

Of the other surfaces of Plateau, we find cylinders enough and to spare in *Spirogyra* and a host of other filamentous algae and fungi. But it is to the vegetable kingdom that we go to find them, where a cellulose envelope enables the cylinder to develop beyond its ordinary limitations.

Unduloids

The unduloid makes its appearance whenever sphere or cylinder begin to give way. We see the transitory figure of an unduloid in the normal fission of a simple cell, or of the nucleus itself; and we have already seen it to perfection in the incipient beadings of a spider's web, or of a pseudopodial thread of protoplasm. A large number of infusoria have unduloid contours, in part at least; and this figure appears and reappears in a great variety of forms. The cups of various Vorticellae (Fig. 19), below the ciliated ring, look like a beautiful series of unduloids, in every gradation of form, from what is all but cylindrical to all but a perfect sphere; moreover successive phases in their life-history appear as mere graded changes of unduloid form. It has been shown lately, in one or two instances at least, that species of *Vorticella* may 'metamorphose' into one another: in other words, that contours supposed to characterise species are not 'specific'. These Vorticellid unduloids are not fully symmetrical; rather are they such

Fig. 19. Various species of *Vorticella*.

unduloids as develop when we suspend an oil-globule between two unequal rings, or blow a bubble between two unequal pipes. For our Vorticellid bell hangs by two terminal supports, the narrow stalk to which it is attached below, and the thickened ring from which spring its circumoral cilia; and it is most interesting to see how, when the bell leaves its stalk (as sometimes happens) and swims away, a new ring of cilia comes into being, to encircle and support its narrow end. Similar unduloids may be traced in even greater variety among other families or genera of the Infusoria.

Among the Foraminifera there is the solitary flask-shaped cell of *Lagena* (Fig. 22) as well as constricted cylinders, or successive unduloids, such as are represented in Fig. 23. In some of these, as in the arenaceous genus *Rheophax*, we have to do with the ordinary phenomenon of a partially segmenting cylinder. But in others, the structure is not developed out of a continuous protoplasmic cylinder, but, as we can see by examining the interior of the shell, it has been formed in successive stages, beginning with a simple unduloid

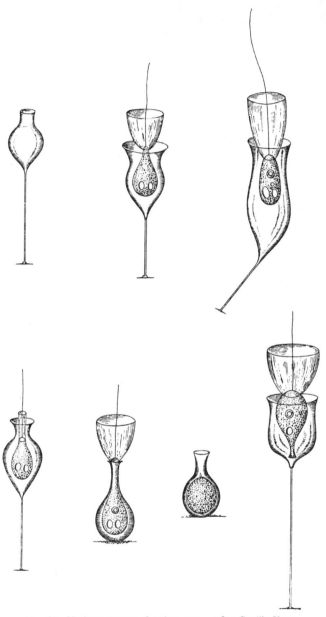

Fig. 20. Various species of *Salpingoeca*. After Saville Kent.

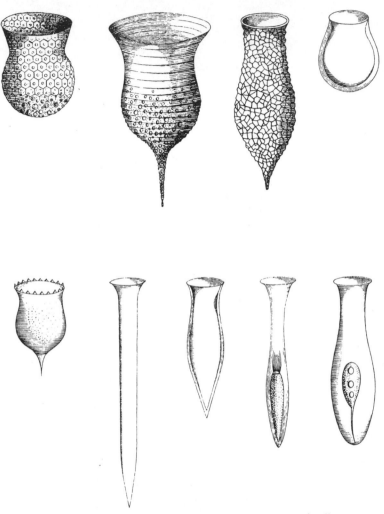

Fig. 21. Various species of *Tintinnus*, *Dinobryon* and *Codonella*.
After Saville Kent and others.

'*Lagena*', about which, after it solidified, another drop of protoplasm accumulated, and in turn assumed the unduloid or lagenoid
form. The chains of interconnected bubbles which Morey and Draper
made many years ago of melted resin are a similar if not identical
phenomenon.[1]

[1] See *Silliman's J*. **2** (1820), 179; and cf. Plateau, *op. cit.* II, 134, 461.

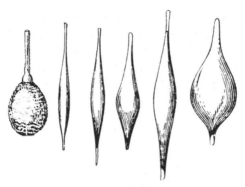

Fig. 22. Various species of *Lagena*. After Brady.

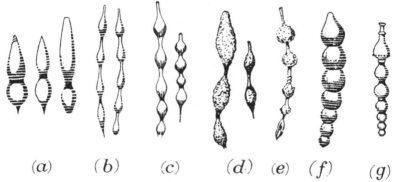

(a) (b) (c) (d) (e) (f) (g)

Fig. 23. Various species of *Nodosaria*, *Rheophax*, *Sagrina*. After Brady.

Other Examples of Surfaces of Revolution

We have found it easy to illustrate the sphere, the cylinder and the unduloid, three of the six 'surfaces of Plateau', all with an endless wealth of illustration among the simplest of organisms. The plane we need hardly look for among the finite outlines of a fluid body; and the catenoid, also a surface of zero mean curvature, can likewise only be a rare and transitory configuration.* One last surface still remains, namely the nodoid; and there also remains one very common but most remarkable protozoan configuration, that of the ciliate Infusoria, to the most characteristic feature of which we have not so far found a physical analogue. Here the curved contour seems to enter, re-enter, and disappear within the substance of the body,

* D'Arcy Thompson also suggested that the ciliate *Trichodina pediculus* approaches a catenoid, 1942 ed., p. 382.

85

so bounding a deep and twisted space or passage, which merges with the fluid contents and vanishes within the cell, and is called by naturalists the 'gullet'. This very peculiar and complicated structure is only kept in equilibrium, and in existence, by the constant activity of cilia over the general surface of the body and very especially in the said gullet or re-entrant portion of the surface. Now we have seen the

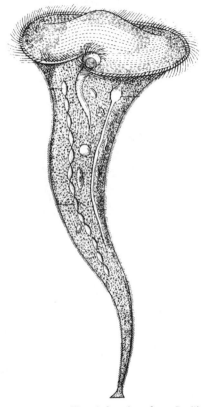

Fig. 24. *Stentor*, a ciliate infusorian: from Saville Kent.

nodoid to be a curved surface, re-entering on itself and endless; no method of support, by wire-rings or otherwise, enables us to construct or realise more than a small portion of it. But the typical ciliate, such as *Paramoecium*, looks just like what we might expect a nodoid surface to be, if we could only realise it (or a single segment of it) in a drop of fluid, and imagine it to be kept in quasi-equilibrium by continual ciliary activity. I suspect, indeed, that here is nothing more, and nothing less, than a partial realisation of the nodoid itself;

86

that the so-called gullet is but the characteristic inversion or 'kink' in that curve; and that the cilia, which normally clothe the surface and always line the gullet, are needed to realise and to maintain the unstable equilibrium of the figure. If this be so—it is a suggestion and no more—we shall have found among our simple organisms the complete realisation, in varying abundance, of each and all of the six surfaces of Plateau. On each and all of them we have a host of beautiful 'patterns' of various sorts; all of them so beautiful and so symmetrical that they *ought* to be capable of geometric representation —and all waiting for their interpreter!

Membrane-tension

The cases which we have just dealt with lead us to another considera-tion. In a semi-permeable membrane, through which water passes freely in and out, the conditions of a liquid surface are greatly modi-fied; in the ideal or ultimate case, there is neither surface nor surface-tension at all. And this would lead us somewhat to reconsider our position, and to enquire whether the true surface-tension of a liquid film is actually responsible for *all* that we have ascribed to it, or whether certain of the phenomena which we have assigned to that cause may not in part be due to the contractility of definite and elastic membranes. But to investigate this question, in particular cases, is rather for the physiologist: and the morphologist may go his way, paying little heed to what is no great difficulty. For in surface-tension we have the production of a film with the properties of an elastic membrane, and with the special peculiarity that contraction con-tinues with the same energy however far the process may have already gone; while the ordinary elastic membrane contracts to a certain extent, and contracts no more. But within wide limits the essential phenomena are the same in both cases. Our fundamental equations apply to both cases alike. And accordingly, so long as our purpose is *morphological*, so long as what we seek to explain is regularity and definiteness of form, it matters little if we should happen, here or there, to confuse surface-tension with elasticity, the contractile forces manifested at a liquid surface with those which come into play at the complex internal surfaces of an elastic solid.

CHAPTER IV

THE FORMS OF TISSUES, OR CELL-AGGREGATES

Surface-tension

We pass from the solitary cell to cells in contact with one another —to what we may call in the first instance 'cell-aggregates', through which we shall be led ultimately to the study of complex tissues. In this part of our subject, as in the preceding chapters, we shall have to consider the effect of various forces; but, as in the case of the solitary cell, we shall probably find, and we may at least begin by assuming, that the agency of surface-tension is especially manifest and important. The effect of this surface-tension will manifest itself in surfaces *minimae areae*: where, as Plateau was always careful to point out, we must understand by this expression not an absolute but a relative minimum, an area, that is to say, which approximates to an absolute minimum as nearly as the circumstances and material exigencies of the case permit.

Let us restate as follows, in terms of *Energy*, the general principle which underlies the theory of surface-tension or capillarity.[1]

When a fluid is in contact with another fluid, or with a solid or with a gas, a portion of the total energy of the system (that, namely, which we call *surface-energy*) is proportional to the area of the surface of contact; it is also proportional to a coefficient which is specific for each particular pair of substances and is constant for these, save only in so far as it may be modified by changes of temperature or of electrical charge. Equilibrium, which is the condition of *minimum potential energy* in the system, will accordingly be obtained, *caeteris paribus*, by the utmost possible reduction of the surfaces in contact.

When we have three bodies in contact with one another the same is true, but the case becomes a little more complex. Suppose a drop of some fluid, *A*, to float on another fluid, *B*, while both are exposed to air, *C*. Here are three surfaces of contact, that of the drop with the fluid on which it floats, and those of air with the one and other

---

[1] See Clerk Maxwell's famous article on 'Capillarity' in the ninth edition of the *Encyclopaedia Britannica*, revised by Lord Rayleigh in the tenth edition.

of these two; and the whole surface-energy, E, of the system consists of three parts resident in these three surfaces, or of three specific energies, E_{AB}, E_{AC}, E_{BC}. The condition of equilibrium, or minimal potential energy, will be reached by contracting those surfaces whose specific energy happens to be large and extending those where it is small—contraction leading to the production of a 'drop', and extension to a spreading 'film'. Floating on water, turpentine gathers into a drop, olive-oil spreads out in a film; and these, according to the several specific energies, are the ways by which the total energy of the system is diminished and equilibrium attained.

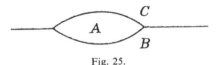

Fig. 25.

A drop will continue to exist provided its own two surface-energies exceed, per unit area, the specific energy of the water-air surface around: that is to say, provided (Fig. 25)

$$E_{AB} + E_{AC} > E_{BC}.$$

But if the one fluid happens to be oil and the other water, then the combined energy per unit-area of the oil-water and the oil-air surfaces together is less than that of the water-air surface:

$$E_{wa} > E_{oa} + E_{ow}.$$

Hence the oil-air and oil-water surfaces increase, the air-water surface contracts and disappears, the oil spreads over the water, and the 'drop' gives place to a 'film'. In both cases the total surface-area is a minimum under the circumstances of the case, and always provided that no external force, such as gravity, complicates the situation.

The surface-energy of which we are speaking here is manifested in that contractile force, or tension, of which we have had so much to say. In any part of the free water-surface, for instance, one surface-particle attracts another surface-particle, and the multitudinous attractions result in equilibrium. But a water-particle in the immediate neighbourhood of the drop may be pulled outwards, so to speak, by another water-particle, but find none on the other side to furnish the counter-pull; the pull required for equilibrium must therefore be provided by tensions existing in the *other two* surfaces of contact. In short, if we imagine a single particle placed at the very point of contact, it will be drawn upon by three different forces, whose directions lie in the three surface-planes and whose magnitudes are proportional to the specific tensions characteristic of the three 'interfacial' surfaces. Now for three forces acting at a point to be in equilibrium they must be capable of representation, in magnitude and direction, by the three sides of a triangle taken in order, in accordance with the theorem of the Triangle of Forces. So, if we know the form of our drop as it floats on the surface (Fig. 26),

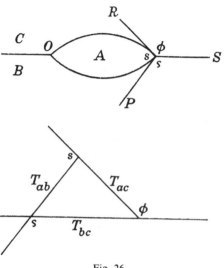

Fig. 26.

then by drawing tangents P, R, from O (the point of mutual contact), we determine the three angles of our triangle, and know therefore the relative magnitudes of the three surface-tensions proportional to its sides. Conversely, if we know the three tensions acting in the directions P, R, S (viz. T_{ab}, T_{ac}, T_{bc}) we know the three sides of the triangle, and know from its three angles the form of the section of the drop. All points round the edge of the drop being under similar conditions, the drop must be circular and its figure that of a solid of revolution.[1]

[1] Bubbles have many beautiful properties besides the more obvious ones. For instance, a floating bubble is always part of a sphere, but never more than a hemisphere; in fact it is always rather less, and a very small bubble is considerably less, than a hemisphere. Again, as we blow up a bubble, its thickness varies inversely as the square of its diameter; the bubble becomes a hundred and fifty times thinner as it grows from an inch in diameter to a foot. In an actual calculation we must always take account of the tensions *on both surfaces* of each film or membrane.

The principle of the triangle of forces is expanded, as follows, in an old seventeenth-century theorem, called Lamy's Theorem:

If three forces acting at a point be in equilibrium, each force is proportional to the sine of the angle contained between the directions of the other two. That is to say (in Fig. 26)

$$P:R:S = \sin \phi : \sin s : \sin s,$$

or
$$\frac{P}{\sin \phi} = \frac{R}{\sin s} = \frac{S}{\sin s}.$$

And from this, in turn, we derive the equivalent formulae by which each force is expressed in terms of the other two and of the angle between them: viz.

$$P^2 = R^2 + S^2 + 2RS \cos \phi, \text{ etc.}$$

From this and the foregoing, we learn the following important and useful deductions:

(1) The three forces can only be in equilibrium when each is less than the sum of the other two; otherwise the triangle is impossible. In the case of a drop of olive-oil on a clean water-surface, the relative magnitudes of the three tensions (at 15° C) are nearly as follows:

Water-air surface	59
Oil-air surface	25
Oil-water surface	16

No triangle having sides of these relative magnitudes is possible, and no such drop can remain in existence.[1]

(2) The three surfaces may be all alike: as when two soap-bubbles are joined together on either side of a partition-film. The three tensions then are all co-equal, and the three angles are co-equal; that is to say, when three similar liquid surfaces, or films, meet together, they always do so at identical angles of 120°. Whether our two conjoined soap-bubbles be equal or unequal, this is still the invariable rule; because the specific tension of a particular surface is independent of form or magnitude.

(3) If all three surfaces be different, as when a fluid drop lies between water and air, the three surface-tensions will (in all likelihood) be different, and the two surfaces of the drop will differ in their amount of curvature.

(4) If two only of the surfaces be alike, then two of the angles will be alike and the other will be unlike; and this last will be the

---

[1] Nevertheless, if the water-surface be contaminated by ever so thin a film of oil, the oil-drop may be made to float upon it. See Rayleigh on Foam, *Collected Works*, III, 351.

difference between 360° and the sum of the other two. A particular case is when a film is stretched between solid and parallel walls, like a soap-film within a cylindrical tube. Here, so long as no external pressure is applied to either side, so long as both ends of the tube are open or closed, the angles on either side of the film will be equal, that is to say the film will set itself at right angles to the sides. Many years ago Sachs laid it down as a principle, which has become celebrated in botany under the name of Sachs's Rule, that one cell-wall always tends to set itself at right angles to another cell-wall. But this rule only applies to the case we have just illustrated; and such validity as it possesses is due to the fact that among plant-tissues it commonly happens that one cell-wall has become solid and rigid before another partition-wall impinges upon it.

Fig. 27.

(5) Another important principle arises, not out of our equations but out of the general considerations which led to them. We saw in the soap-bubble that at and near the point of contact between our several surfaces, there is a continued balance of forces, carried (so to speak) across the interval; in other words, there is *physical continuity* between one surface and another and it follows that the surfaces merge one into another by a continuous curve. Whatever be the form of our surfaces and whatever the angle between them, a small intervening curved surface is always there to bridge over the line of contact; and this little fillet, or 'bourrelet', as Plateau called it, is big enough to be a common and conspicuous feature in the microscopy of tissues (Fig. 28). A similar 'bourrelet' is clearly seen at the boundary between a floating bubble and the liquid on which it floats: in which case it constitutes a 'masse annulaire', whose mathematical properties and relation to the form of the *nearly* hemispherical bubble have been investigated by van der Mens-brugghe.[1] The superficial vacuoles in *Actinophrys* or *Actinosphaerium* present an identical phenomenon.

(6) It is a curious effect, or consequence, of the bourrelet that a 'horizontal' soap-film is never either horizontal or plane. For the

[1] Cf. Plateau, *op. cit.* p. 366.

bourrelet at its edge is deformed by gravity, and the film is correspondingly inclined upwards where it meets (Fig. 29, *b*).

(7) The bourrelet, a fluid mass connected with a fluid film, is no
mere passive phenomenon but has its active influence or dynamical
effect. This was pointed out by Willard Gibbs,[1] and Plateau's bourrelet is more often called, nowadays, 'Gibbs's Ring'. The ring is
continuous in phase with the interior of the film, and fluid is sucked
into it from the latter, which thins rapidly; and this, becoming a more

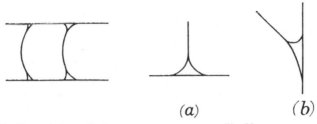

(a) (b)

Fig. 28. Plateau's *bourrelet*, in an
algal filament. After Berthold.

Fig. 29.

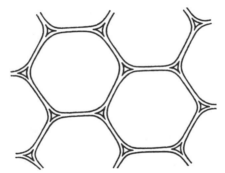

Fig. 30. Parenchyma of maize; showing intercellular spaces.

potent factor of unrest than gravity itself, leads presently to the
rupture of the film. Plateau's explanation of his bourrelet as a 'surface of continuity' is thus but a part, and a small part, of the story.

(8) In the succulent, or parenchymatous, tissue of a vegetable,
the cells have their internal corners rounded off (Fig. 30) in a way
which might suggest the bourrelet, but comes of another cause.
Where the angles are rounded off the cell-walls tend to split apart
from one another, and each cell seems tending to withdraw, as far
as it can, into a sphere; and this happens, not when the tissue is

[1] *Collected Works*, I, 309.

young and the cell-walls tender and quasi-fluid, but later on, when cellulose is forming freely at the surface of the cell. The cell-walls no longer meet as fluid films, but are stiffening into pellicles; the cells, which began as an association of bubbles, are now so many balls, in solid contact or partial detachment; and flexibility and elasticity have taken the place of the capillary forces of an earlier and more liquid phase.[1]

(9) Statically though not dynamically, that is to say as a line or surface of continuity in Plateau's sense, our bourrelet is analogous to the accumulation of sand seen where two nodal lines cross in a Chladni figure: 'Vers les endroits où des lignes nodales se coupent, elles s'élargissent toujours, de sorte que la forme des parties vibrantes près de ces endroits n'est pas angulaire mais plus ou moins arrondie, souvent en forme d'hyperbole.'[2] And in somewhat remoter analogy, we may look on the three *corpora Arantii* as so many *bourrelets*, helping to fill the angles where three semilunar valves meet at the base of the great arteries.

Cell Partitions

We may now illustrate some of the foregoing principles, constantly bearing in mind the principles set forth in our chapter on the Forms of Cells, and especially those relating to the pressure exercised by a curved film.

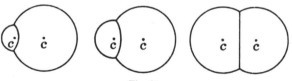

Fig. 31.

Let us look for a moment at the case presented by the partition-wall in a double soap-bubble. As we have just seen, the three films in contact (viz. the outer walls of the two bubbles and the partition-wall between) being all composed of the same substance and being all alike in contact with air, the three tensions must be equal, and the three films must, in all cases, meet at co-equal angles of 120°. But unless the two bubbles be of precisely equal size, and therefore of equal curvature, the tangents to the spheres will not meet the plane

[1] J. H. Priestley, 'Cell-growth...in the Flowering Plant', *New Phytol.* **28** (1929), 54–81.
[2] E. F. F. Chladni, *Traité d'acoustique* (1809), p. 127.

of their circle of contact at equal angles, and the partition-wall will of necessity be a curved, and indeed a spherical, surface; it is only plane when it divides two equal and symmetrical cells. It is obvious, from the symmetry of the figure, that the centres of the two bubbles and of the partition between are all on one and the same straight line.

The two bubbles exert a pressure inwards which is inversely proportional to their radii: that is to say, $p:p'::1/r:1/r'$; and the partition-wall must, for equilibrium, exert a pressure (P) which is equal to the difference between these two pressures, that is to say,

$$P = 1/R = 1/r' - 1/r = (r-r')/rr'.$$

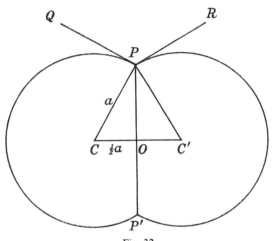

Fig. 32.

It follows that the curvature of the partition must be just such as is capable of exerting this pressure, that is to say, $R = rr'/(r-r')$. The partition, then, is a portion of a spherical surface, whose radius is equal to the product, divided by the difference, of the radii of the two bubbles; if the two bubbles be equal, the radius of curvature of the partition is infinitely great, that is to say the partition is (as we have already seen) a plane surface.

In the typical case of an evenly divided cell, such as a double and co-equal soap-bubble (Fig. 32), where partition-wall and outer walls are identical with one another and the same air is in contact with them all, we can easily determine the form of the system. For, at any point of the boundary of the partition, P, the tensions being equal, the angles QPP', RPP', QPR are all equal, and each is, therefore, an angle of 120°. But PQ, PR being tangents, the centres

95

of the two spheres (or circular arcs in the figure) lie on lines perpendicular to them; therefore the radii CP, $C'P$ meet at an angle of $60°$, and CPC' is an equilateral triangle. That is to say, the centre of each circle lies on the circumference of the other; the partition lies midway between the two centres; and the diameter of the partition-wall, PP', is

$$\frac{OP}{CP} = \sin 60° = \frac{\sqrt{3}}{2} = 0·866$$

times the diameter of each of the two cells. This gives us, then, the form of a combination of two co-equal spherical cells under uniform conditions.

In the case of unequal bubbles, the curvature of their partition-wall is easily determined, and is shown in Fig. 33. The three films

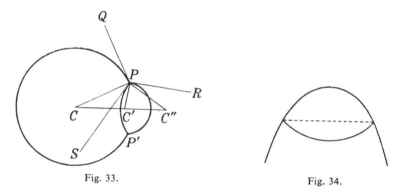

Fig. 33.   Fig. 34.

meeting in P being (as before) identical films, the three tangents, PQ, PR, PS, meet at co-equal angles of $120°$, and PS produced bisects the angle QPR. PQ, PR are tangents perpendicular to the radii CP, $C'P$; and $C''P$, the radius of the spherical partition PP', is found by drawing a perpendicular to PS in P. The centre C'' is, by the symmetry of the figure, in a straight line with C, C'.

Whether the partition be or be not a plane surface, it is obvious that its *line of junction* with the rest of the system lies in a plane, and is at right angles to the axis of symmetry. The actual curvature of the partition-wall is easily seen in optical section; but in surface view the line of junction is *projected* as a plane (Fig. 34), perpendicular to the axis, and this appearance has helped to lend support and authority to 'Sachs's Rule'.

As soon as the tensions of the cell-walls become unequal, whether from changes in their own substance or in the substances with which they are in contact, then the form alters. If the tension along the partition P diminishes, the partition itself enlarges and the angle QPR increases: until, when the tension p is very small compared with q or r, the whole figure becomes a sphere, and the partition-wall, dividing it into two hemispheres, stands at right angles to the outer wall. This is the case when the outer wall of the cell is practically solid. On the other hand, if p begins to increase relatively to q and r, then the partition-wall contracts, and the two adjacent cells become larger and larger segments of a sphere, until at length the system becomes divided into two separate cells.

To put the matter still more simply, let the annexed diagrams (Fig. 35) represent a system of three films, one being a partition-wall

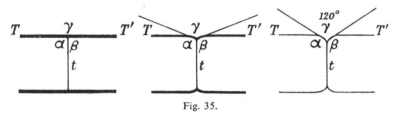

Fig. 35.

running between the other two; and where the partition t meets the outer wall TT', let the several tensions, or the tractions exerted on a point at their meeting-place, be proportional to T, T' and t. Let α, β, γ be, as in the figure, the opposite angles. Then:

(1) If T be equal to T', and t be relatively insignificant, the angles α, β will be of 90°.

(2) If $T = T'$, but be a little greater than t, then t will exert an appreciable traction, and α, β will be more than 90°, say for instance, 100°.

(3) If $T = T' = t$, then α, β, γ will all equal 120°.

The outer walls of the two cells on either side of the partition will be straight, as well as continuous, in the first case, and more or less curved in the other two. We have a vivid illustration (if a somewhat crude one) of the first case in a section of honey: where the waxen walls, which meet one another at 120°, meet the wooden sides of the box at 90°.

The wing of a dragon-fly shows a seemingly complicated system of veins which the foregoing considerations help much to simplify. The wing is traversed by a few strong 'veins', or ribs, more or less parallel to one another, between which finer veins make a meshwork of 'cells', these lesser veins being all much of a muchness, and exerting tensions insignificant compared with those of the greater veins. Where (*a*) two ribs run so near together that only one row of cells lies between, these cells are quadrangular in form, their thin partitions meeting the ribs at right angles on either side. Where (*b*) two rows of cells are intercalated between a pair of ribs, one row fits into the other by angles of 120°, the result of co-equal tensions; but both meet the ribs at right angles, as in the former case. Where (*c*) the cell-rows are numerous, all their angles in common tend to be co-equal angles of 120°, and the cells resolve, consequently, into a hexagonal mesh-work.*

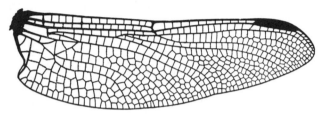

Fig. 36. A dragonfly's wing. From Needham and Westfall, *A Manual of the Dragonflies of North America* (University of California Press).

Many spherical cells, such as *Protococcus*, divide into two equal halves, separated by a plane partition. Among other lower Algae akin to *Protococcus*, such as the Nostocs and Oscillatoriae, in which the cells are embedded in a gelatinous matrix, we find a series of forms such as are represented in Fig. 37, which various conditions depend, according to what we have already learned, upon the relative magnitudes of the tensions at the surface of the cells and the boundary between them. In some cases (Fig. 37, *b*) the cells remain spherical, because they are merely embedded in the matrix, with no other physical continuity between them; even two soap-bubbles do not tend to unite, unless their surfaces be moist or we put a drop of soap-solution between them. In

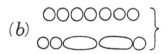

Fig. 37. Filaments, or chains of cells, in various lower Algae. (*a*) *Nostoc*; (*b*) *Anabaena*; (*c*) *Rivularia*; (*d*) *Oscillatoria*.

* When these veins are formed in development they are avenues of circulation and therefore their form might more appropriately be interpreted in terms of the principles affecting the branching of blood vessels, discussed at the end of this chapter.

certain other cases, the system consists of a relatively thick-walled tube, subdivided by more delicate partitions, which latter then tend (as in (d)) to become plane septa, set at right angles to the walls. Or again, side-walls and septa may be all alike, or nearly so; and then the configuration (as in (c), on Fig. 37) is that of a linear cluster of soap-bubbles.[1]

When we have three bubbles in contact, instead of two as in the case already considered, the phenomenon is strictly analogous to the former case. The three bubbles are separated by three partition surfaces, whose curvature will depend upon the relative size of the spheres, and which will be plane if the latter are all of equal size; but whether plane or curved, the three partitions will meet one another at angles of 120°, in an axial line.

When we have four bubbles meeting in a plane (Fig. 38), they would seem capable of arrangement in two symmetrical ways: either (a) with four partition-walls intersecting at right angles, or (b) with five partitions meeting, three and three, at angles of 120°. The latter arrangement is strictly analogous to the arrangement of *three* bubbles. Now, though both of these figures might seem, from their apparent symmetry, to be figures of equilibrium, yet in point of fact the latter turns out to be of stable and the former of unstable equilibrium. If we try to bring four bubbles into the form (a), that arrangement endures only for an instant; the partitions glide upon one another, an intermediate wall springs into existence, and the system assumes the form (b), with its two triple, instead of one quadruple, conjunction. In like manner, when four billiard-balls are packed close upon a table, two tend to come together and separate the other two.

Let us epitomise the Law of Minimal Areas and its chief clauses or corollaries in the particular case of an assemblage of fluid films, as was first done by Lamarle.[2] First, and in general: In every liquid system of thin films in stable equilibrium, the sum of the areas of the films is a minimum. From observation and experience, rather than by demonstration, it follows that (2) the area of *each* is a minimum under its own limiting conditions; and further that (3) the mean curvature of any film is constant throughout its whole area, null when the pressures are equal on either side and in other cases proportional to their difference. Less obvious, very important, and likewise subject (but none too easily) to rigorous mathematical proof,

---

[1] Cf. Dewar, 'Studies on Liquid Films', *Proc. Roy. Inst.* (1918), p. 359.

[2] Ernest Lamarle, 'Sur la stabilité des systèmes liquides en lames minces', *Mém. Acad. R. Belg.* **35, 36** (1864–7).

are the next two propositions, both of which had been laid down empirically by Plateau: (4) the films meeting in any one edge are three in number; (5) the crests or edges meeting in any one corner are four in number, neither more nor less. Lastly, and following easily from these: (6) the three films meeting in a crest or edge do so at co-equal angles, and the same is true of the four edges meeting in a corner.

Wherever we have a true cellular complex, an arrangement of cells in actual physical contact by means of their intervening boundary walls, we find these general principles in force; we must only bear in mind that, for their easy and perfect recognition, we must be able to view the object in a plane at right angles to the boundary walls. For instance, in any ordinary plane section of a vegetable

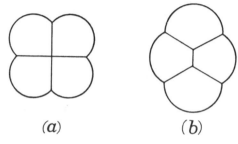

(a)   *(b)*

Fig. 38. (*a*) An unstable arrangement of four cells or bubbles. (*b*) The normal and stable configuration, showing the polar furrow.

parenchyma, we recognise the appearance of a 'froth', precisely resembling that which we can construct by imprisoning a mass of soap-bubbles in a narrow vessel with flat sides of glass; in both cases we see the cell-walls everywhere meeting, by threes, at angles of 120°, irrespective of the size of the individual cells; whose relative size, on the other hand, determines the *curvature* of the partition-walls. On the surface of a honeycomb we have precisely the same conjunction, between cell and cell, of three boundary walls, meeting at 120°. In embryology, when we examine a segmenting egg, of four (or more) segments, we find in like manner, in the majority of cases if not in all, that the same principle is still exemplified. The four segments do not meet in a common centre, but each cell is in contact with two others; and the three, and only three, common boundary walls meet at the normal angle of 120°. A so-called *polar furrow*,[1] the visible edge of a vertical partition-wall, joins (or

[1] It was so termed by Conklin in 1897, in his paper on Crepidula (*J. Morph.* **13**, 1897). It is the *Querfurche* of Rabl (*Morph. Jb.* **5**, 1879); the *Polarfurche* of O. Hertwig

separates) the two triple contacts, precisely as in Fig. 38 (*b*), and so gives rise to a diamond-shaped figure, which was recognised more than a hundred years ago (in a newt or salamander) by Rusconi, and called by him a *tetracitula*.

The Epidermal Layer

While the inner cells of the honeycomb are symmetrically situated, sharing with their neighbours in equally distributed pressures or tensions, and therefore all tending closely to identity of form, the case is obviously different with the cells at the borders of the system. So it is with our froth of soap-bubbles. The bubbles, or cells, in the interior of the mass are all alike in general character, and if they be equal in size are alike in every respect: as we see them in projection their sides are uniformly flattened, and tend to meet at equal angles of 120°. But the bubbles which constitute the outer layer retain their

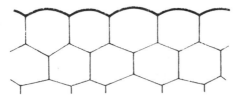

Fig. 39. A froth, with its outer and inner cells or vesicles.

spherical surfaces (just as in the cells of a honeycomb), and these still tend to meet the partition-walls connected with them at constant angles of 120°. This outer layer of bubbles, which forms the surface of our froth, constitutes after a fashion what we should call in botany an 'epidermal' layer. But in our froth of soap-bubbles we have, as a rule, the same kind of contact (that is to say, contact with *air*) both within and without the bubbles; while in our living cell, the outer wall of the epidermal cell is exposed to air on the one side, but is in contact with the protoplasm of the cell on the other: and this involves a difference of tensions, so that the outer walls and their adjacent partitions need no longer meet at precisely equal angles of 120°. Moreover a chemical change, due perhaps to oxidation or possibly also to adsorption, is very apt to affect the external wall and lead to the formation of a 'cuticle'; and this process, as we have seen, is tantamount to a large increase of tension in that

(*Jena Z. Naturw.* **14**, 1880); the *Brechungslinie* of Rauber (Neue Grundlage zur Kenntniss der Zelle, *Morph. Jb.* **8**, 1882); and the *cross-line* of T. H. Morgan (1897). It is carefully discussed by Robert, *op. cit.* pp. 307 *seq.*

outer wall, and will cause the adjacent partitions to impinge upon it at angles more and more nearly approximating to 90°: the bubble-like, or spherical, surfaces of the individual cells being more and more flattened in consequence. Lastly, the chemical changes which affect the outer walls of the superficial cells may extend in greater or less degree to their inner walls also: with the result that these cells will tend to become more or less rectangular throughout, and will cease to dovetail into the interstices of the next subjacent layer. These then are the general characters which we recognise in an epidermis; and we now perceive that its fundamental character simply is that it lies outside, and that its physical characteristics follow, as a matter of course, from the position which it occupies and from the various consequences which that situation entails.

In the young shoot or growing point of a flowering plant botanists (following Hanstein) find three cell-layers, and call them *dermatogen*, *periblem* and *plerome*. The first is an epidermis, such as we have just described. Its cells grow long as the shoot grows long; new partitions cross the lengthening cell and tend to lie at right angles to its hardening walls; and this epidermis, once formed, remains a single superficial layer. The next few layers, the so-called periblem, are compressed and flattened between the epidermis with its tense cuticle and the growing mass within; and under this restraint the cell-layers of the periblem also continue to divide in their own plane or planes. But the cells of the inner mass or plerome, lying in a more homogeneous field, tend to form 'space-filling' polyhedra, twelve- or perhaps fourteen-sided according to the freedom which they enjoy. In a well-known passage Sachs declares that the behaviour of the cells in the growing point is determined not by any specific characters or properties of their own, but by their position and the forces to which they are subject in the system of which they are a part.[1] This was a prescient utterance, and is abundantly confirmed.[2]

Close Packing of Cells

We have hitherto considered our cells, or our bubbles, as lying in a plane of symmetry, and have only considered their appearance as projected on that plane; but we must also begin to consider them as solids, whether they lie in a plane (like the four cells in Fig. 38), or are heaped on one another, like a froth of bubbles or a pile of

[1] *Lectures on the Physiology of the Plant* (Oxford, 1887), p. 460, etc.
[2] Cf. J. H. Priestley in *Biol. Rev.* 3 (1928), 1–20; U. Tetley in *Ann. Bot.* 50 (1936), 522–57; etc.

cannon-balls. We have still much to do with the study of more complex partitioning in a plane, and we have the whole subject to enter on of the solid geometry of bodies in 'close packing', or three-dimensional juxtaposition.

Hexagonal Symmetry

The same principles which account for the development of hexagonal symmetry hold true, as a matter of course, not only of *cells* (in the biological sense), but of any bodies of uniform size and originally circular outline, close-packed in a plane; and hence the hexagonal pattern is of very common occurrence, under widely varying circumstances. The curious reader may consult Sir Thomas Browne's quaint

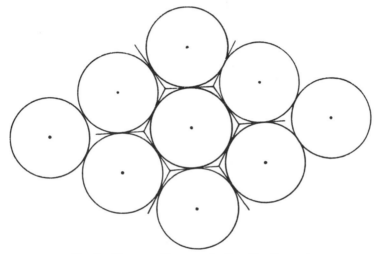

Fig. 40. Diagram of hexagonal cells. After Bonanni.

and beautiful account, in the *Garden of Cyrus*, of hexagonal, and also of quincuncial, symmetry in plants and animals, which 'doth neatly declare how nature Geometrizeth, and observeth order in all things'.

Let us imagine a system of equal cylinders, or equal spheres, in contact with one another in a plane, and represented in section by the equal and contiguous circles of Fig. 40. I borrow my figure from an old Italian naturalist, Bonanni (a contemporary of Borelli, of Ray and Willoughby, and of Martin Lister), who dealt with this matter in a book chiefly devoted to molluscan shells.[1]

[1] A. P. P. Bonanni, *Ricreatione dell'occhio e della mente, nell'Osservatione delle Chiocciole*, Roma, 1681.

It is obvious, as a simple geometrical fact, that each of these co-equal circles is in contact with six others around. Imagine the whole system under some uniform stress—of pressure caused by growth or expansion within the cells, or due to some uniformly applied constricting pressure from without. In these cases the six *points of contact* between the circles in the diagram will be extended into *lines*, representing *surfaces* of contact in the actual spheres or cylinders; and the equal circles of our diagram will be converted into regular and co-equal hexagons. The result is just the same so far as form is concerned—so long as we are concerned only with a morphological result and not with a physiological process—whatever be the force which brings the bodies together. For instance, the cells of a segmenting egg, lying within their vitelline membrane or within some common film or ectoplasm, are pressed together as they grow, and suffer deformation accordingly; their surface tends towards an *area minima*, but we need not even enquire, in the first instance, whether it be surface-tension, mechanical pressure, or what not other physical force, which is the cause of the phenomenon.[1]

Diffusion Models

The production by mutual interaction of polygons, which become regular hexagons when conditions are perfectly symmetrical, is beautifully illustrated by Bénard's *tourbillons cellulaires*, and also in some of Leduc's diffusion experiments. In these latter, a solution of gelatine is allowed to set on a plate of glass, and little drops of weak potassium ferrocyanide are then let fall at regular intervals upon the gelatine. Immediately each little drop becomes the centre of a system of diffusion currents, and the several systems conflict with and repel one another; so that presently each little area becomes the seat of a to-and-fro current system, outwards and back again, until the concentration of the field becomes equalised and the currents cease. When equilibrium is attained, and when the gelatine-layer is allowed to dry, we have an artificial tissue of hexagonal 'cells', which simulate an organic parenchyma very closely; and by varying

[1] The following is one of many curious corollaries to the principle of close-packing here touched upon. A circle surrounded by six similar circles, the whole bounded by a circle of three times the radius of the original one, forms a unit, so to speak, next in order after the circle itself. A round pea or grain of shot will pass through a hole of its own size; but peas or shot will not *run out* of a vessel through a hole less than *three times* their own diameter. There can be no freedom of motion among the close-packed grains when confronted by a smaller orifice. Cf. K. Takahasi, *Sci. Pap. Inst. Phys. Chem. Res. Tokyo*, **26** (1935), 19.

the experiment in ways which Leduc describes, we may imitate various forms of tissue, and produce cells with thick walls or with thin, cells in close contact or with wide intercellular spaces, cells with plane or with curved partitions, and so forth.

James Thomson (Kelvin's elder brother) had observed nearly sixty years ago a curious 'tessellated structure' on a liquid surface, to wit, the soapy water of a wash-tub. The eddies and streaks of swirling water settled down into a cellular configuration, which continued for hours together to alter its details; small areoles disappeared, large ones grew larger, and subdivided into small ones

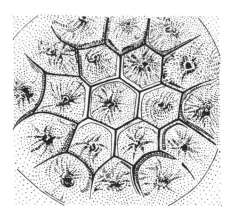

Fig. 41. An 'artificial tissue', formed by coloured drops of sodium chloride solution diffusing in a less dense solution of the same salt. After Leduc.

again. With few and transitory exceptions three partitions and no more met at every node of the meshwork; and (as it seems to me) the subsequent changes were all due to such shifting of the lines as tended to make the three adjacent angles more and more nearly co-equal with one another: the obvious effect of this being to make the pattern more and more regularly hexagonal.[1]

In a not less homely experiment, hot water is poured into a shallow tin and a layer of milk run in below; on blowing gently to cool the water, holes, more or less close-packed and evenly interspaced, appear in the milk. They show how cooling has taken place, so to speak, in spots, and the cooled water has descended in isolated columns.

[1] James Thomson, 'On a Changing Tessellated Structure in Certain Liquids', *Proc. Glasg. Phil. Soc.* (1881–2); *Coll. Papers*, p. 136—a paper with which M. Bénard was not acquainted, but see Bénard's later note in *Ann. Chim.* (December 1911).

Bénard's 'tourbillons cellulaires',[1] set up in a thin liquid layer, are similar to but more elegant than James Thomson's tessellated patterns, and both of them are in their own way still more curious than M. Leduc's; for the latter depend on centres of diffusion artificially inserted into the system and determining the number and position of the 'cells', while in the others the cells make themselves. In Bénard's experiment a thin layer of liquid is warmed in a copper dish. The liquid is under peculiar conditions of instability, for the least fortuitous excess of heat here or there would suffice to start a current, and we should expect the whole system to be highly unstable and unsymmetrical. But if all be kept carefully uniform, small disturbances appear at random all over the system; a current ascends in the centre of each; and a 'steady state', if not a stable equilibrium, is reached in time, when the descending currents, impinging on one another, mark out a 'cellular system'. If we set the fluid gently in motion to begin with, the first 'cell-divisions' will be in the direction of the flow; long tubes appear, or 'vessels', as the botanist would be apt to call them. As the flow slows down new cell-boundaries appear, at right angles to the first and at even distances from one another; parallel rows of cells arise, and this transitory stage of partial equilibrium or imperfect symmetry is such as to remind the botanist of his *cambium* tissues, which are, so to speak, a temporary phase of histological equilibrium. If the impressed motion be not longitudinal but rotary, the first lines of demarcation are spiral curves, followed by orthogonal intersections.

Whether we start with liquid in motion or at rest, symmetry and uniformity are ultimately attained. The cells draw towards uniformity, but four-, five- or seven-sided cells are still to be found among the prevailing hexagons. The larger cells grow less, the smaller enlarge or disappear; where four partition-walls happen to meet, they shift till only three converge; the sides adjust themselves to equal lengths, the angles also to equality. In the final stage the cells are hexagonal prisms of definite dimensions, which depend on temperature and

[1] H. Bénard, 'Les tourbillons cellulaires dans une nappe liquide', *Rev. gén. Sci. pur. appl.* **12** (1900); 1261–71, 1309–28; *Ann. Chim.* (7), **23** (1901), 62–144; *ibid.* (1911). Quincke had seen much the same long before: *Ann. Phys.* **139** (1870), 28. The 'figures of de Heen' are an analogous electrical phenomenon; cf. P. de Heen, 'Les tourbillons et les projections de l'éther', *Bull. Acad. Belg.* (3), **57** (1899), 589; A. Lafay, *Ann. Phys.* (10), **13** (1930), 349–94. These various phenomena, all leading to a pattern of hexagons, have often been studied mathematically: cf. Rayleigh, *Phil. Mag.* **32** (1916), 529–46; *Coll. Papers*, VI, 48; also Ann Pellew and R. V. Southwell, *Proc. Roy. Soc.* A, **176** (1940), 312–43. The hexagonal pattern is a particular case of stability, but not necessarily the simplest; it is only by experiment that we know it to be the permanent condition in an unlimited field.

on the nature and thickness of the liquid layer; molecular forces have not only given us a definite cellular pattern, but also a 'fixed cell-size'.*

The Bee's Cell

The most famous of all hexagonal conformations, and one of the most beautiful, is the bee's cell. As in the basalt or the coral, we have to deal with an assemblage of co-equal cylinders, of circular section, compressed into regular hexagonal prisms; but in this case we have two layers of such cylinders or prisms, one facing one way and one the other, and a new problem arises in connection with their inner ends. We may suppose the original cylinders to have spherical ends,[1] which is their normal and symmetrical way of terminating; then, for closest packing, it is obvious that the end of any one cylinder in the one layer will touch, and fit in between, the ends of three cylinders in the other. It is just as when we pile round-shot in a heap; we begin with three, a fourth fits into its nest between the three others, and the four form a 'tetrad', or regular tetragonal arrangement.

Just as it was obvious, then, that by mutual pressure from the *sides* of six adjacent cells any one cell would be squeezed into a hexagonal prism, so is it also obvious that, by mutual pressure against the *ends* of three opposite neighbours, the end of each and every cell will be compressed into a trihedral pyramid. The three sides of this pyramid are set, *in plane projection*, at co-equal angles of 120° to one another; but the three apical angles (as in the analogous case already described of a system of soap-bubbles) are, by the geometry of the case,[2] co-equal angles of 109° and so many minutes and seconds.

If we experiment, not with cylinders but with spheres, if for instance we pile bread-pills together and then submit the whole to a uniform pressure, as we shall presently find that Buffon did: each ball (like

---

[1] In the combs of certain tropical bees the hexagonal structure is imperfect and the cells are not far removed from cylinders. They are set in tiers, not contiguous but separated by little pillars of wax, and the base of each cell is a portion of a sphere. They differ from the ordinary honeycomb in the same sort of way as the fasciculate from the massive corals. Cf. Leonard Martin, 'Sur les Mélipones de Brésil', *La Nature* (1930), pp. 97–100.

[2] The dihedral angle of 120° is, physically speaking, the essential thing; the Maraldi angle, of 109°, etc., is a geometrical consequence. Cf. G. Cesáro, 'Sur la forme de l'avéole de l'abeille', *Bull. Acad. Belg. Cl. Sci.* (1920), p. 100.

---

* W. Robbins, *Bull. Torrey Bot. Cl.* **79** (1952), 107, showed the presence of hexagons in pure cultures of *Euglena* and T. Morrison at Princeton University found good evidence that this was due to diffusion forces and could be compared to Bénard's experiments.

the seeds in a pomegranate, as Kepler said) will be in contact with *twelve* others—six in its own plane, three below and three above, and under compression it will develop twelve plane surfaces. It will repeat, above and below, the conditions to which the bee's cell is subject at one end only; and, since the sphere is symmetrically situated towards its neighbours on all sides, it follows that the twelve plane sides to which its surface has been reduced will be all similar, equal and similarly situated. Moreover, since we have produced this result by squeezing our original spheres close together, it is evident that the bodies so formed completely fill space. The regular solid which fulfils all these conditions is the *rhombic dodecahedron*. The bee's cell is this figure incompletely formed; it represents, so to speak, one-half of that figure, with its apex and the six adjacent corners proper to the rhombic dodecahedron, but six sides continued, as a hexagonal prism, to an open or unfinished end.[1]

The bee's comb is vertical and the cells nearly horizontal, but sloping slightly downwards from mouth to floor; in each prismatic cell two sides stand vertically, and two corners lie above and below. Thus for every honeycomb or 'section' of honey, there is one and only one 'right way up'; and the work of the hive is so far controlled by gravity. Wasps build the other way, with the cells upright and the combs horizontal; in a hornet's nest, or in that of *Polistes*, the cells stand upright like the wasp's, but their mouths look downwards in the hornet's nest and upwards in the wasp's.

What Jeremy Taylor called 'the discipline of bees and the rare fabric of honeycombs' must have attracted the attention and excited the admiration of mathematicians from time immemorial. 'Ma maison est construite,' says the bee in the *Arabian Nights*, 'selon les lois d'une sévère architecture; et Euclidos lui-même s'instruirait en admirant la géométrie de ses alvéoles.'[2] Ausonius speaks of the *geometrica forma favorum*, and Pliny tells of men who gave a lifetime to its study.

Pappus the Alexandrine has left us an account of its hexagonal plan, and drew from it the conclusion that the bees were endowed with 'a certain geometrical forethought'.[3] 'There being, then, three figures which of themselves can fill up the space round a point, viz.

---

[1] See especially Haüy, the crystallographer: 'Sur le rapport des figures qui existe entre l'avéole des abeilles et le grenat dodécaèdre', *J. hist. naturelle*, **2** (1792), 47.

[2] Ed. Mardrus, xv, 173.

[3] φυσικὴν γεωμετρικὴν πρόνοιαν. Pappus, Bk. v; cf. Heath, *Hist. of Gk. Math.* II, 589. St Basil discusses τὴν γεωμετρίαν τῆς σοφωτάτης μελίσσης: *Hexaem.* VIII, 172 (Migne); Virgil speaks of the *pars divinae mentis* of the bee, and Kepler found the bees *animâ praeditas et geometriae suo modo capaces*.

the triangle, the square and the hexagon, the bees have wisely selected for their structure that which contains most angles, suspecting indeed that it could hold more honey than either of the other two.'[1] Erasmus Bartholin was apparently the first to suggest that the hypothesis of 'economy' was not warranted, and that the hexagonal cell was no more than the necessary result of equal pressures, each bee striving to make its own little circle as large as possible.

The investigation of the ends of the cell was a more difficult matter than that of its sides, and came later. In general terms the arrange-

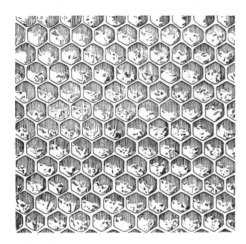 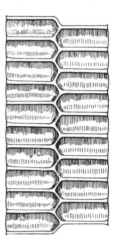

Fig. 42. Portion of a honeycomb. From T. Rayment, *A Cluster of Bees* (The Bulletin, Sydney).

ment was doubtless often studied and described: as for instance, in the *Garden of Cyrus*: 'And the Combes themselves so regularly contrived that their mutual intersections make three Lozenges at the bottom of every Cell; which severally regarded make three Rows of neat Rhomboidall Figures, connected at the angles, and so continue three several chains throughout the whole comb.' Or as Réaumur put it, a little later on: 'trois cellules accolées laissent un vuide pyramidal, précisément semblable à celui de la base d'une autre cellule tournée en sens contraire.'

---

[1] This was according to the 'theorem of Zenodorus'. The use by Pappus of 'economy' as a guiding principle is remarkable. For it means that, like Hero with his mirrors, he had a pretty clear adumbration of that principle of *minima*, which culminated in the *principle of least action*, which guided eighteenth-century physics, was generalised (after Fermat) by Lagrange, inspired Hamilton and Maxwell, and reappears in the latest developments of wave-mechanics.

Kepler had deduced from the space-filling symmetry of the honeycomb that its angles must be those of the rhombic dodecahedron; and Swammerdam also recognised the same geometrical figure in the base of the cell.[1] But Kepler's discovery passed unnoticed, and Maraldi the astronomer, Cassini's nephew, has the credit of ascertaining for the first time the shape of the rhombs and of the solid angle which they bound, while watching the bees in 'les ruches vitrées dans le jardin de M. Cassini attenant l'Observatoire de Paris'.[2] The angles of the rhomb, he tells us, are 110° and 70°: 'Chaque base d'alvéole est formée de trois rombes presque toujours égaux et semblables, qui, *suivant les mesures que nous avons prises*, ont les deux angles obtus chacun de 110 degrés, et par conséquent les deux aigus chacun de 70 degrés.' Further on (p. 312), he observes that on the magnitude of the angles of the three rhombs at the base of the cell depends that of the basal angles of the six trapezia which form its sides; and it occurs to him to ask what must these angles be, if those of the floor and those of the sides be equal one to another. The solution of this problem is that 'les angles aigus des rombes étant de 70 degrés 32 minutes, et les obtus de 109 degrés 28 minutes, ceux des trapèzes qui leur sont contigus doivent être aussi de la même grandeur'. And lastly: 'Il résulte de cette grandeur d'angle non seulement une plus grande facilité et simplicité dans la construction, à cause que par cette manière les abeilles n'employent que deux sortes d'angles, mais il en résulte encore une plus belle simétrie dans la disposition et dans la figure de l'Alvéole.' In short, Maraldi takes the two principles of simplicity and mathematical beauty as his sure and sufficient guides.

The next step was that which had been foreshadowed long before by Pappus. Though Euler had not yet published his famous dissertation on curves *maximi minimive proprietate gaudentes*, the idea of *maxima* and *minima* was in the air as a guiding postulate, an heuristic method, to be used as Maraldi had used his principle of simplicity. So it occurred to Réaumur, as apparently it had not done to Maraldi, that a minimal configuration, and consequent economy of material in the waxen walls of the cell, might be at the root of the matter: and that, just as the close-packed hexagons gave the minimal extent of

---

[1] Kepleri *Opera omnia*, ed. Fritsch (1864), v, 115, 122, 178; vii, p. 719; Swammerdam, *Tractatus de apibus* (observations made in 1673).

[2] 'Obs. sur les abeilles', *Mém. Acad. R. Sci.* (1712), 1731, pp. 297–331. Sir C. Wren had used 'transparent bee-hives' long before; see the letter in which he refers to that pleasant and profitable invention, etc., in S. Hartlib's *Reformed Common-Wealth of Bees* (1655).

boundary in a plane, so the figure determined by Maraldi, namely the rhombic dodecahedron, might be that which employs the minimum of surface for a given content: or which, in other words, should hold the most honey for the least wax. 'Convaincu que les abeilles employent le fond pyramidal qui mérite d'être préféré, j'ai soupçonné que la raison, ou une des raisons, qui les avoit décidées était l'épargne de la cire; qu'entre les cellules de même capacité et à fond pyramidal, celle qui pouvait être faite avec moins de matière ou de circ étoit celle dont chaque rhombe avoit deux angles chacun d'environ 110 degrès, et deux chacun d'environ 70°.' He set the problem to Samuel Koenig, a young Swiss mathematician: Given a hexagonal cell terminated by three similar and equal rhombs, what is the configuration which requires the least quantity of material for its construction? Koenig confirmed Réaumur's conjecture, and gave 109° 26' and 70° 34' as the angles which should fulfil the condition; and Réaumur then sent him the Mémoires de l'Académie for 1712, where Koenig was 'agreeably surprised' to find: 'que les rhombes que sa solution avait déterminé, avait à deux minutes près[1] les angles que M. Maraldi avait trouvés *par des mesures actuelles* à chaque rhombe des cellules d'abeilles.... Un tel accord entre la solution et les mesures actuelles a assurément de quoi surprendre.' Koenig asserted that the bees had solved a problem beyond the reach of the old geometry and requiring the methods of Newton and Leibniz. Whereupon Fontenelle, as Sécretaire Perpétuel, summed up the case in a famous judgment, in which he denied intelligence to the bees, but nevertheless found them blindly using the highest mathematics by divine guidance and command.[2]

When Colin Maclaurin studied the honeycomb in Edinburgh, a few years after Maraldi in Paris, he proceeded to solve the problem without using 'any higher Geometry than was known to the Antients', and he began his account by saying: 'These bases are formed from Three equal Rhombus's, the obtuse angles of which are found to be the doubles of an Angle that often offers itself to mathematicians in

---

[1] The discrepancy was due to a mistake of Koenig's, doubtless misled by his tables, in the determination of $\sqrt{2}$; but Koenig's own paper, sent to Réaumur, remained unpublished and his method of working is unknown. An abridged notice appears in the *Mém. Acad. Sci.* (1739), pp. 30–5.

[2] 'La grande merveille est que la détermination de ces angles passe de beaucoup les forces de la Géométrie commune, et n'appartient qu'aux nouvelles Méthodes fondées sur la Théorie de l'Infini. Mais à la fin les Abeilles en sçauraient trop, et l'excès de leur gloire en est la ruine. Il faut remonter jusqu'à une Intelligence infinie, qui les fait agir aveuglément sous ses ordres, sans leur accorder de ces lumières capables de s'accroître et de se fortifier par elles-mêmes, qui font l'honneur de notre Raison.' *Histoire de l'Académie Royale* (1739), p. 35.

Questions relating to Maxima and Minima.'[1] It was an angle of 109° 28' 16", with its supplement of 70° 31' 44". And this angle of the bee's cell determined by Maraldi, Koenig and Maclaurin in their several ways, this angle which has for its cosine 1/3 and is double of the angle which has for its tangent √2, is on the one hand an angle of the rhombic dodecahedron, and on the other is that very angle of simple tetrahedral symmetry which the soap-films within the tetrahedral cage spontaneously assume, and whose frequent appearance and wide importance we have already touched upon.

That 'the true theoretical angles were 109° 28' and 70° 32', *precisely corresponding with the actual measurement of the bee's cell*', and that the bees had been 'proved to be right and the mathematicians wrong', was long believed by many. Lord Brougham helped notably to spread these and other errors, and his writings on the bee's cell contain, according to Glaisher, 'as striking examples of bad reasoning as are often to be met with in writings relating to mathematical subjects'. The fact is that, were the angles and facets of the honeycomb as sharp and smooth, and as constant and uniform, as those of a quartz-crystal, it would still be a delicate matter to measure the angles within a minute or two of arc, and a technique unknown in Maraldi's day would be required to do it. The minute-hand of a clock (if it move continuously) moves through one degree of arc in ten seconds of time, and through an angle of two minutes in one-third of a second; —and this last is the angle which Maraldi is supposed to have measured. It was eighty years after Maraldi had told Réaumur what the angle was that Boscovich pointed out for the first time that to ascertain the angle to the nearest minute by direct admeasurement of the waxen cell was utterly impossible. Yet Réaumur had certainly believed, and apparently had persuaded Koenig, that Maraldi's determinations, first and last, were the result of measurement; and Fontenelle, the historian of the Academy, epitomising Koenig's paper, speaks of 'les mesures actuelles de M. Maraldi', of the bees being in error to the trifling extent of 2', and of the *grande merveille* of their so nearly solving a problem belonging to the higher geo-

---

[1] Colin Maclaurin, 'On the Bases of the Cells wherein the Bees Deposit their Honey', *Phil. Trans.* **42** (1743), 561–71; also in the Abridgement, **8** (1809), 709–13; it was characteristic of Maclaurin to use geometrical methods for wellnigh everything, even in his book on Fluxions, or in his famous essay on the equilibrium of spinning planets. Cf. also Lhuiller, 'Mémoire sur le minimum du cire des alvéoles des Abeilles, et en particulier sur un *minimum minimorum* relatif à cette matière', *Nouv. Mém. Acad. Berl.* 1781 (1783), pp. 277–300. Cf. Castillon, *ibid.* (commenting on Lhuiller); also Ettore Carruccio, 'Notizie storiche sulla geometria delle api', *Period. Mat.* (4), **16** (1936), 35–54.

metry. Boscovich, in a long-forgotten note, rediscovered by Glaisher, puts the case in a nutshell: 'Mirum sane si Maraldus ex observatione angulum aestimasset intra minuta, quod in tam exigua mole fieri utique non poterat. At is (ut satis patet ex ipsa ejus determinatione) affirmat se invenisse angulos circiter 110° et 70°, nec minuta eruit ex observatione sed ex equalitate angulorum pertinentium ad rhombos et ad trapezia; ad quam habendam Geometria ipsa docuit requiri illa minuta.'[1] Indeed he goes on to say the wonder is that the angles could be measured even within a few *degrees*, variable and irregular as they are seen to be, and as even Réaumur knew well they were.[2] The old misunderstanding was at last explained and corrected by Leslie Ellis; and better still by Glaisher, in a little-known but very beautiful paper.[3] For these two mathematicians showed that, though Maraldi's account of his 'measurements' led to misunderstanding, yet he had really done well and scientifically when he eked out a rough observation by finer theory, and deemed himself entitled thereby to discuss the cell and its angles in the same precise terms that he would use as a mathematician in speaking of its geometrical prototype.

Concerning the subject of wax economy Glaisher sums up the matter as follows: 'As the result of a tolerably careful examination of the whole question, I may be permitted to say that I agree with Lhuiller in believing that the economy of wax has played a very subordinate part in the determination of the form of the cell;* in fact I should not be surprised if it were acknowledged hereafter that the form of the cell had been determined by other considerations, into which the saving of wax did not enter (that is to say did not enter sensibly; of course I do not mean that the amount of wax required was a matter of absolute indifference to the bees). The fact of all the dihedral angles being 120° is, it is not unlikely, the cause that determined the form of the cell.' This last fact, that in such a cell every plane cuts every other plane at an angle of 120°, was known both to

---

[1] In his note *De apium cellulis*, appended to the philosophical poem of Benedict Stay, II, 498–504 (Romae, 1792).

[2] *Op. cit.* v, 382. Several authors recognised that the cells are far from identical, and do no more than approximate to an average or ideal angle: e.g. Swammerdam in the *Biblia Naturae*, II, 379; G. S. Klügel, Grösstes u. Kleinstes, in *Mathem. Wörterb.* (1803); Castillon, *op. cit.*; and especially Jeffries Wyman, 'Notes on the Cells of the Bee', *Proc. Amer. Arts Sci.* 7 (1868).

[3] Leslie Ellis, 'On the Form of Bees' Cells', *Mathematical and other Writings* (1863), p. 353; J. W. L. Glaisher, *Phil. Mag.* (4), **46** (1873), 103–22.

---

* About 2 per cent wax is saved if the ends of the cells are rhombic rather than flat-bottomed.

Klügel and to Boscovich; it is no mere corollary, but the root of the matter. It is, as Glaisher indicates, the fundamental physical principle of construction from which the apical angles of 109° follow as a geometrical corollary. And it is curious indeed to see how the obtuse angle of the rhomb, and its cosine $-\frac{1}{3}$, drew attention all the while; but the dihedral angle of 120° of the rhombohedron, and the inclination of its three short diagonals at 90° to one another, got rare and scanty notice.

Darwin had listened too closely to Brougham and the rest when he spoke of the bee's architecture as 'the most wonderful of known instincts'; and when he declared that 'beyond this stage of perfection in architecture natural selection could not lead; for the comb of the hive-bee, as far as we can see, is *absolutely perfect* in economising labour and wax'.

The minimal properties of the cell and all the geometrical reasoning in the case postulate cell-walls of uniform tenuity and edges which are mathematically straight. But the walls, and still more their edges, are always thickened; the edges are never accurately straight, nor the cells strictly horizontal. The base is always thicker than the side-walls; its solid angles are by no means sharp, but filled up with curving surfaces of wax, after the fashion, but more coarsely, of Plateau's bourrelet. Hence the Maraldi angle is seldom or never attained; the mean value (according to Vogt) is no more than 106·7° for the workers, and 107·3° for the drones. The hexagonal angles of the prism are fairly constant; about 4° is the limit of departure, and about 1·8° the mean error, on either side.

The bee makes no economies; and whatever economies lie in the theoretical construction, the bee's handiwork is not fine nor accurate enough to take advantage of them.[1]

The cells vary little in size, so little that Thévenot, a friend of Swammerdam's, suggested using their dimensions as a modulus or standard of length; but after all, the constancy is not so great as has been supposed. Swammerdam gives measurements which work out at 5·15 mm for the mean diameter of the worker-cells, and 7 mm for those of the drones; Jeffries Wyman found mean values for the worker-cells from 5·1 to 5·2 mm; Vogt, after many careful measurements, found a mean of 5·37 mm for the worker-cells, with an insignificant difference in the various diameters, and a mean of

[1] All this Heinrich Vogt has abundantly shown, in part by making casts of the interior of the cells, as Castellan had done a hundred years before. See his admirable paper on the 'Geometrie und Oekonomie der Bienenzelle', *Festschrift Universität Breslau* (1911), pp. 27–274.

6·9 for the drone-cells, with their horizontal diameter somewhat in excess, and averaging 7·1 mm. A curious attempt has been made of late years by Italian bee-keepers to let the bees work on a larger foundation, and so induce them to build larger cells; and some, but by no means all, assert that the young bees reared in the larger cells are themselves of larger stature.[1]

That the beautiful regularity of the bee's architecture is due to some automatic play of the physical forces, and that it were fantastic to assume (with Pappus and Réaumur) that the bee intentionally seeks for a method of economising wax, is certain; but the precise manner of this automatic action is not so clear. When the hive-bee builds a solitary cell, or a small cluster of cells, as it does for those eggs which are to develop into queens, it makes but a rude construction. The queen-cells are lumps of coarse wax hollowed out and roughly bitten into shape, bearing the marks of the bee's jaws like the marks of a blunt adze on a rough-hewn log.

Omitting the simplest of all cases, when (among some humble-bees) the old cocoons are used to hold honey, the cells built by the 'solitary' wasps and bees are of various kinds. They may be formed by partitioning off little chambers in a hollow stem; they may be rounded or oval capsules, often very neatly constructed out of mud or vegetable fibre or little stones, agglutinated together with a salivary glue; but they show, except for their rounded or tubular form, no mathematical symmetry. The social wasps and many bees build, usually out of vegetable matter chewed into a paste with saliva, very beautiful nests of 'combs'; and the close-set papery cells which constitute these combs are just as regularly hexagonal as are the waxen cells of the hive-bee. But in these cases (or nearly all of them) the cells are in a single row; their sides are regularly hexagonal, but their ends, for want of opponent forces, remain simply spherical (see Fig. 43).

In *Melipona domestica* (of which Darwin epitomises Pierre Huber's description) 'the large waxen honey-cells are nearly spherical, nearly equal in size, and are aggregated into an irregular mass'. But the spherical form is only seen on the outside of the mass; for inwardly each cell is flattened into 'two, three or more flat surfaces, according as the cell adjoins two, three or more other cells. When one cell rests on three other cells, which from the spheres being nearly of the same

---

[1] Cf. (*int. al.*) H. Gontarsi, 'Sammelleistungen von Bienen aus vergrösserten Brutzellen', *Arch. Bienenk.* **16** (1935), 7; A. Ghetti, 'Celli ed api più grandi', *IV Congresso nazion. della S.A.I.* (1935).

size is very frequently and necessarily the case, the three flat surfaces are united into a pyramid; and this pyramid, as Huber has remarked, is manifestly a gross imitation of the three-sided pyramidal base of the cell of the hive-bee'.[1] We had better be content to say that it depends on the same elementary geometry.

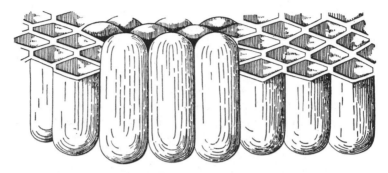

Fig. 43. Brood cells of the bee *Trigona carbonaria* showing spherical caps. From T. Rayment, *A Cluster of Bees* (The Bulletin, Sydney).

The question is, To what particular force are we to ascribe the plane surfaces and definite angles which define the sides of the cell in all these cases, and the ends of the cell in cases where one row meets and opposes another? We have seen that Bartholin suggested, and it is still commonly believed, that this result is due to mere physical pressure, each bee enlarging as much as it can the cell which it is a-building, and nudging its wall outwards till it fills every intervening gap, and presses hard against the similar efforts of its neighbour in the cell next door.[2]

[1] *Origin of Species*, ch. VIII (6th ed. p. 221). The cells of various bees, humble-bees and social wasps have been described and mathematically investigated by K. Müllenhoff, *Pflüg. Arch. ges. Physiol.* 32 (1883), 589; but his many interesting results are too complex to epitomise. For figures of various nests and combs see (e.g.) von Büttel-Reepen, *Biol. Zbl.* 33 (1903), 4, 89, 129, 183.
[2] Darwin had a somewhat similar idea, though he allowed more play to the bee's instinct or conscious intention. Thus, when he noticed certain half-completed cell-walls to be concave on one side and convex on the other, but to become perfectly flat when restored for a short time to the hive, he says: 'It was absolutely impossible, from the extreme thinness of the little plate, that they could have effected this by gnawing away the convex side; and I suspect that the bees in such cases stand on opposite sides and push and bend the ductile and warm wax (which as I have tried is easily done) into its proper intermediate plane, and thus flatten it.' Huber thought the difference in form between the inner and the outer cells a clear proof of intelligence; it is really a direct proof of the contrary. And while cells differ when their situations and circumstances differ, yet over great stretches of comb extreme uniformity, unbroken by any sign of individual differences, is the strikingly mechanical characteristic of the cells.

That the bee, if left to itself, 'works in segments of circles', or in other words builds a rounded and roughly spherical cell, is an old contention[1] which some recent experiments of M. Victor Willem amplify and confirm.[2] M. Willem describes vividly how each cell begins as a little hemispherical basin or 'cuvette', how the workers proceed at first with little apparent order and method, laying on the wax roughly like the mud when a swallow builds; how presently they concentrate their toil, each burying its head in its own cuvette, and slowly scraping, smoothing and ramming home; how those on the other side gradually adjust themselves to their opposite neighbours; and how the rounded ends of the cells fashion themselves into the rhomboidal pyramids, 'à la suite de l'amincissement progressif des cloisons communes, et des pressions antagonistes exercées sur les deux faces de ces cloisons'.

Among other curious and instructive observations, M. Willem has watched the bees at work on the waxen 'foundations' now commonly used, on which a rhomboidal pattern is impressed with a view to starting the work and saving the labour of the bees. The bees (he says) disdain these half-laid foundations of their cells; they hollow out the wax, erase the rhombs, and turn the pyramidal hollows into hemispherical 'cuvettes' in their usual way; and the vertical walls which they raise, more or less on the lines laid down for them, are not hexagonal but cylindrical to begin with. 'La forme plane, en facettes, tant de prismes que des fonds, n'est obtenue que plus tard, progressivement, comme résultat de retouches, d'enlèvements et de pressions exercées sur les cloisons qui s'amincissent, par des groupes d'ouvrières opérant face à face, de manière antagoniste.'

But when all is said and done, it is doubtful whether such *retouches*, *enlèvements* and *pressions antagonistes*, such mechanical forces intermittently exercised, could produce the nearly smooth surfaces, the all but constant angles and the close approach to a minimal configuration which characterise the cell, whether it be constructed by the bee of wax or by the wasp of papery pulp. We have the properties of the material to consider; and it seems much more likely to me that we have to do with a true tension effect: in other words, that the walls

---

[1] It is so stated in the *Penny Cyclopedia* (1835), Art. 'Bees'; and is expounded by Mr G. H. Waterhouse (*Trans. R. Ent. Soc. Lond.* 2 (1864), 115), in an article of which Darwin made good use. Waterhouse showed that when the bees were given a plate of wax, the separate excavations they made therein remained hemispherical, or were built up into cylindrical tubes; but cells in juxtaposition with one another had their party-walls flattened, and their forms more or less prismatic.

[2] Victor Willem, 'L'architecture des abeilles', *Bull. Acad. R. Belg.* (5), **14** (1928), 672–705.

assume their configuration when in a semi-fluid state, while the watery pulp is still liquid or the wax warm under the high temperature of the crowded hive. In the first few cells of a wasp's comb, long before crowding and mutual pressure come into play, we recognise the identical configurations which we have seen exhibited by a group of three or four soap-bubbles, the first three or four cells of a segmenting egg. The direct efforts of the wasp or bee may be supposed to be limited, at this stage, to the making of little hemispherical cups, as thin as the nature of the material permits, and packing these little round cups as close as possible together. It is then conceivable, and indeed probable, that the symmetrical tensions of the semi-fluid films should suffice (however retarded by viscosity) to bring the whole system into equilibrium, that is to say into the configuration which the comb actually assumes.

The remarkable passage in which Buffon discusses the bee's cell and the hexagonal configuration in general is of such historical importance, and tallies so closely with the whole trend of our enquiry, that before we leave the subject I will quote it in full:[1] 'Dirai-je encore un mot: ces cellules des abeilles, tant vantées, tant admirées, me fournissent une preuve de plus contre l'enthousiasme et l'admiration; cette figure, toute géométrique et toute régulière qu'elle nous paraît, et qu'elle est en effet dans la spéculation, n'est ici qu'un résultat mécanique et assez imparfait qui se trouve souvent dans la nature, et que l'on remarque même dans les productions les plus brutes; les cristaux et plusieurs autres pierres, quelques sels, etc., prennent constamment cette figure dans leur formation. Qu'on observe les petites écailles de la peau d'une roussette, on verra qu'elles sont hexagones, parce que chaque écaille croissant en même temps se fait obstacle et tend à occuper le plus d'espace qu'il est possible dans un espace donné: on voit ces mêmes hexagones dans le second estomac des animaux ruminants, on les trouve dans les graines, dans leurs capsules, dans certaines fleurs, etc. Qu'on remplisse un vaisseau de pois, ou plûtot de quelque autre graine cylindrique, et qu'on le ferme exactement après y avoir versé autant d'eau que les intervalles qui restent entre ces graines peuvent en recevoir; qu'on fasse bouillir cette eau, tous ces cylindres deviendront de

---

[1] Buffon, *Histoire naturelle* (Paris, 1753), IV, 99. Bonnet criticised Buffon's explanation, on the ground that his description was incomplete; for Buffon took no account of the Maraldi pyramids. Not a few others discovered impiety in his hypotheses, and some dismissed them with the remark that 'philosophical absurdities are the most difficult to refute'; cf. W. Smellie, *Philosophy of Natural History* (Edinburgh, 1790), p. 424.

colonnes à six pans. On y voit clairement la raison, qui est purement mécanique; chaque graine, dont la figure est cylindrique, tend par son renflement à occuper le plus d'espace possible dans un espace donné, elles deviennent donc toutes nécessairement hexagones par la compression réciproque. Chaque abeille cherche à occuper de même le plus d'espace possible dans un espace donné, il est donc nécessaire aussi, puisque le corps des abeilles est cylindrique, que leurs cellules sont hexagones—par la même raison des obstacles réciproques. On donne plus d'esprit aux mouches dont les ouvrages sont les plus réguliers; les abeilles sont, dit-on, plus ingénieuses que les guêpes, que les frelons, etc., qui savent aussi l'architecture, mais dont les constructions sont plus grossières et plus irrégulières que celles des abeilles: on ne veut pas voir, ou l'on ne se doute pas, que cette régularité, plus ou moins grande, dépend uniquement du nombre et de la figure, et nullement de l'intelligence de ces petites bêtes; plus elles sont nombreuses, plus il y a des forces qui agissent également et s'opposent de même, plus il y a par conséquent de contrainte mécanique, de régularité forcée, et de perfection apparente dans leurs productions.'[1]

The Partitioning of Space

Just as Bonanni and other early writers sought, as we have seen, to explain hexagonal symmetry on mechanical principles, so other early naturalists, relying more or less on the analogy of the bee's cell, endeavoured to explain the cells of vegetable parenchyma; and

[1] Among countless papers on the bee's cell, see John Barclay and others in *Ann. Phil.* **9, 10** (1817); Henry Lord Brougham, in *Dissertations . . . connected with Natural Theology*, app. to Paley's Works (1839), I, 218–368; *C.R. Acad. Sci., Paris*, **46** (1858), 1024–9; *Tracts, Mathematical and Physical* (1860), pp. 103–21, etc.; E. Carruccio, 'Note storiche sulla geometria delle api', *Period. Mat.* (4), **16** (1936), 20 pp.; G. Césaro, 'Sur la forme de l'alvéole des abeilles', *Bull. Acad. R. Belg.* (10 Avril 1929); Sam. Haughton, 'On the Form of the Cells made by Various Wasps and by the Honey-bee', *Proc. Nat. Hist. Soc. Dubl.* **3** (1863), 128–40; *Ann. Mag. Nat. Hist.* (3), **11** (1863), 415–29; A. R. Wallace, 'Remarks on the Foregoing Paper', *ibid.* **12**, 33; J. O. Hennum, *Arch. Math. Vidensk.*, Christiania, **9** (1884), 301; F. Huber, *Nouv. obs. sur les abeilles*, **2** (1814), 475; F. W. Hultmann, *Tidsskr. Math., Uppsala*, **1** (1868), 197; John Hunter, 'Observations on Bees', *Phil. Trans.* (1792), pp. 128–95; Jacob, *Nouv. Ann. Math.* **2** (1843), 160; G. S. Klügel, 'Mathem. Betrachtungen üb. d. kunstreichen Bau. d. Bienenzellen', *Hannoversches Mag.* (1772), pp. 353–68; Léon Lalanne, 'Note sur l'architecture des abeilles', *Ann. Sci. Nat. Zool.* (2), **13** (1840), 358–74; B. Powell, *Proc. Ashmol. Nat. Hist. Soc.* **1** (1844), 10; K. H. Schellbach, *Mathem. Lehrstunde: Lehre v. Grössten u. Kleinsten* (1860), pp. 35–7; Sam. Sharpe, *Phil. Mag.* **4** (1828), 19–21; J. E. Siegwart, 'Die Mathematik im Dienste d. Bienenzucht', *Schw. Bienenzeitung*, **3** (1880); O. Terquem, *Nouv. Ann. Math.* **15** (1856), 176; C. M. Willick, 'On the Angle of Dock-gates and the Bee's Cell', *Phil. Mag.* (4), **18** (1859), 427; *C.R.* **51** (1860), 633; Chauncy Wright, *Proc. Amer. Acad. Arts Sci.* **4** (1860), 432.

to refer them to the rhombic dodecahedron (Fig. 44) or garnet-form, which solid figure, in close-packed association, was believed in their time, and long afterwards, to enclose space with a minimal extent of surface.

Fig. 44. A rhombic dodecahedron.

A few years after the publication of Plateau's book, Lord Kelvin showed, in a short but very beautiful paper,[1] that we must not hastily assume that a close-packed assemblage of rhombic dodecahedra will be the true and general solution of the problem of dividing space with a minimum partitional area, or will be present in a liquid 'foam', in which the general problem is completely and automatically solved. The general mathematical solution of the problem is, that every interface or partition-wall must have constant mean curvature throughout; that where these partitions meet in an edge, they must intersect at angles such that equal forces, in planes perpendicular to the line of intersection, shall balance; that no more than three such interfaces may meet in a line or edge, whence it follows (for symmetry) that the angle of intersection of all surfaces or facets must be 120°; and that neither more nor less than four edges meet in a point or corner. An assemblage of rhombic dodecahedra goes far to meet the case. It fills space; its surfaces or interfaces are planes, and therefore surfaces of constant curvature throughout; and they meet together at angles of 120°. Nevertheless, the proof that the rhombic dodecahedron (which we find exemplified in the bee's cell) is a figure of minimal area is not a comprehensive proof; it is limited to certain conditions, and practically amounts to no more than this, that of the ordinary space-filling solids with all sides plane and similar, this one has the least surface for its solid content.

[1] Sir W. Thomson, 'On the Division of Space with Minimum Partitional Area', *Phil. Mag.* (5), **24** (December 1887), 503–14; cf. *Baltimore Lectures* (1904), p. 615; *Molecular Tactics of a Crystal* (Robert Boyle Lecture, 1894), pp. 21–5.

Lord Kelvin made the remarkable discovery that by means of an assemblage of fourteen-sided figures, or 'tetrakaidekahedra', space is filled and homogeneously partitioned—into equal, similar and similarly situated cells—with an economy of surface in relation to volume even greater than in an assemblage of rhombic dodecahedra.[1]

The tetrakaidekahedron, in its most generalised case, is bounded by three pairs of equal and opposite quadrilateral faces, and four pairs of equal and opposite hexagonal faces, neither the quadrilaterals nor the hexagons being necessarily plane. In its simplest case, with all its facets plane and equilateral, it is Kelvin's 'ortho-

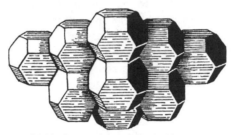

Fig. 45. A set of 14-hedra, to show close-packing. From F. T. Lewis.

tetrakaidekahedron'; and also (though Kelvin was unaware of the fact) one of the thirteen semi-regular and isogonal polyhedra, or 'Archimedean bodies'. In a particular case, the quadrilaterals are plane surfaces with curved edges, but the hexagons are slightly curved 'anticlastic' surfaces; and these latter have at every point equal and opposite curvatures, and are surfaces of minimal curvature for a boundary of six curved edges. This figure has the remarkable property that, like the plane rhombic dodecahedron, it so partitions space that three faces meeting in an edge do so everywhere at co-equal angles of 120°; and, unlike the rhombic dodecahedron, four edges meet in each point or corner at co-equal angles of 109° 28'.

We may take it as certain that, in a homogeneous system of fluid films like the interior of a froth of soap-bubbles, where the films are perfectly free to glide or turn over one another and are of approximately co-equal size, the mass is actually divided into cells of this remarkable conformation: and the possibility of such a configuration being present even in the cells of an ordinary vegetable parenchyma was suggested in the first edition of this book. It is all a question

[1] First Kelvin described the plane-faced tetrakaidekahedron; later he showed how that figure must have its faces warped and edges curved to fulfil all the conditions of minimal area.

of *restraint*, of degrees of mobility or fluidity. If we squeeze a mass of clay pellets together, like Buffon's peas, they come out, or all the inner ones do, in neat garnet-shape, or rhombic dodecahedra. But a young student once showed me (in Yale)* that if you wet these clay pellets thoroughly, so that they slide easily on one another and so acquire a sort of pseudo-fluidity in the mass, they no longer come out as regular dodecahedra, but with square and hexagonal facets recognisable as those of ill-formed or half-formed tetrakaidekahedra.

Dr F. T. Lewis† has made a long and careful study of various vegetable parenchymas, by simple maceration, wax-plate reconstruction and otherwise, and has succeeded in showing that the tetrakaidekahedral form is closely approached, or even attained, in certain simple and homogeneous tissues. After reconstructing a large model of the cells of elder-pith, he finds that the fourteen-sided figure clearly manifests itself as the characteristic or typical form to which the cells approximate, in spite of repeated cell-divisions and consequent inequalities of size. Counting in a hundred cells the number of contacts which each made with its neighbours, that is to say the total number both of actual and potential facets, Lewis found that 74 per cent of the cells were either 12, 13, 14, 15 or 16-sided, 56 per cent either 13, 14 or 15-sided, and that the average number of facets or contacts was, in this instance, just 13·96. These figures indicate the general symmetry of the cells, their departure from the dodecahedral, and their tendency towards the tetrakaidekahedral, form.[1]

But after all, the geometry of the 14-hedron, displayed to perfection by our soap-films in the twinkling of an eye, is only roughly developed in an organic structure, even one so delicate as elder-pith; the conditions are no longer simple, for friction, viscosity and solidification have vastly complicated the case. We get a curious and an unexpected variant of the same phenomenon in the microscopic foam-like structure assumed as molten metal cools. If these foam-cells were again

[1] F. T. Lewis, 'The Typical Shape of Polyhedral Cells, in Vegetable Parenchyma, and the Restoration of that Shape Following Cell-division', *Proc. Amer. Acad. Arts Sci.* **58** (1923), 537–52, and other papers. See also (*int. al.*) J. W. Marvin, 'The Aggregation of Orthis-tetrakaidekahedra', *Science*, **83** (1936), 188; E. B. Matzke, 'An Analysis of the Orthotetrakaidekahedron', *Bull. Torrey Bot. Cl.* **54** (1927), 341–8. Professor van Iterson of Delft tells me that *Asparagus sprengeri* (a common greenhouse plant) is a good subject for showing the 14-hedral cells.

* This 'young student' is the well-known protozoologist Dr Vance Tartar of the University of Washington, Seattle, U.S.A.
† It is of interest to note that Lewis stated explicitly that it was the first edition of this book that led him to these investigations. See E. Mayer, *Anat. Rec.* **85** (1943), 116.

14-hedra, their facets would all be either squares or hexagons; but pentagonal facets are commoner than either, and the cells often approach closely to the form of a regular *pentagonal* dodecahedron! The edges of this figure meet at angles of 108°, not far from the characteristic Maraldi angle of 109° 28′; and the faces meet at an angle not far removed from 120°. A slight curvature of the sides is enough to turn our pentagonal dodecahedron into a possible figure of equilibrium for a foam-cell. We cannot close-pack pentagonal dodecahedra, whether equal or unequal, so as to fill space; but still the figure may be, and seems to be, common, interspersed among the polyhedra of various shapes and sizes which are packed together in a metallic foam.[1]

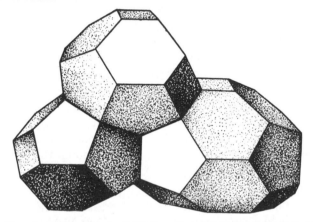

Fig. 46. Three irregular polyhedra (14-hedron above, 13-hedron below, left; 15-hedron below, right). From F. T. Lewis, *Amer. J. Botany*, **30** (1943), 77.

A somewhat similar result, and a curious one, was found by Mr J. W. Marvin, who compressed leaden small-shot in a steel cylinder, as Buffon compressed his peas; but this time the pressure on the plunger ran from 1000 to 35,000 lb, or nearly twenty tons to the square inch. When the shot was introduced carefully, so as to lie in ordinary close packing, the result was an assemblage of regular rhombic dodecahedra, as might be expected and as Buffon had found. But the result was very different when the shot was poured at random into the cylinder, for the average number of facets on each grain now varied with the pressure, from about 8·5 at 1000 lb to 12·9 at 10,000 lb, and to no less than 14·16 facets or contacts

[1] Cf. Cecil H. Desch, 'The Solidification of Metals from the Liquid State', *J. Inst. Met.* **22** (1919), 247.

after all interstices were eliminated, which took the full pressure of 35,000 lb to do. An average of just over fourteen facets might seem to indicate a tendency to the production of tetrakaidekahedra, just as in the froth of soap-bubbles; but this is not so. The squeezed grains are irregular in shape, and pentagonal facets are much the commonest, just as we found them to be in the microscopic structure of a once-molten metal. At first sight it might seem that, though the experiment has something to teach us about random packing in a limited space, it has no biological significance; but it is curious to find that the pith-cells of Eupatorium have a similar average configuration, with the same predominance of pentagonal facets.[1]

We learn, in short, from Lewis and from Marvin that the mechanical result of mutual pressure, even in an assemblage of co-equal spheres, is more varied and more complex than we had supposed. The two simple and homogeneous configurations—the rhombododecahedral and tetrakaidekahedral assemblages—are easily and commonly produced, the one by the compression of solid spheres in ordinary close-packing, the other when a liquid system of spheres or bubbles is free to slide and glide into a packing which is closer still. Between these two configurations there is no other symmetrical or homogeneous arrangement possible; but random packing and degrees of compression leave their random effects, among which are traces here and there of regular shape and symmetry.

There are, in this discussion of the close-packing of cells within a tissue, a number of points that need further clarification as a result of the careful work of E. B. Matzke and his collaborators. This is all neatly summarised by Matzke in a short paper* in which he shows that there have been a series of misconceptions or errors and that, as so often is the case, they seem to have been carefully preserved in the literature.

The first of these are the 'peas of Buffon' to which D'Arcy Thompson refers. By a little detective work Matzke finds that these peas were first described by Stephen Hales in his *Vegetable Staticks* in 1727. 'I compressed several fresh parcels of Pease in the same Pot with a force equal to 1600, 800 and 400 pounds; in which Experiments, tho' the Pease dilated, yet they did not raise the lever, because what they increased in bulk was, by the great incumbent weight, pressed into the interstices of the Pease, which they adequately filled up, being thereby formed into pretty regular Dodecahedra.' Now Buffon, in 1753 in his *Histoire Naturelle*, makes a somewhat similar statement but fails to refer to Hales. This is unfortunate

---

[1] J. W. Marvin, 'The Shape of Compressed Lead-shot, etc.', *Amer. J. Bot.* 26 (1939), 280–8; 'Cell-shape Studies in the Pith of Eupatorium', *ibid.* pp. 487–504.

* *Torreya*, 72 (1950), 222.

simply because, as Matzke points out, one of Buffon's first scholarly undertakings as a young man of 28 (in 1735) was the translation of Hales's book into French. Therefore the 'peas of Buffon' should be the 'peas of Hales'.

To make matters worse, when Matzke carefully repeated the experiment the peas failed to produce 'pretty regular dodecahedra' but only showed a few suggestions of rhombic faces, and this perhaps is to be expected since peas are so irregular and possess little homogeneity in their internal structure.

The next step, as D'Arcy Thompson describes, was for Matzke and his collaborator Marvin to compress uniform lead-shot, and the rhombic dodecahedra appeared only if the shot was regularly stacked like the traditional cannon-balls. Otherwise they produced irregular figures which merely averaged about 14 faces, a result similar to that found by Lewis, and later others, for the cells in living tissues.

All that remains then for Kelvin's ideal 14-hedra is the frictionless world of soap-bubbles. But here again expectations do not fit with facts. Matzke devised a system of making uniform soap-bubbles and analysed 600 of the internal bubbles in a froth, only to find that, like living cells and lead-shot, they averaged about 14 faces and the individual bubbles tended to be irregular.

There is a danger in making hypotheses that can be tested, but even with the new information before us the problem has not lost its interest. The fact that in all these instances the 'cells' are irregular, may indicate that in each case there are a set of forces acting that will automatically frustrate any perfect and regular configuration, and that for each these forces may well differ. The fact that in all these instances the 'cells' average approximately 14 faces may indicate that along with the forces that promote irregularity, there is a general tendency towards a minimum configuration, which is in essential agreement with the final conclusion of D'Arcy Thompson.

On the Form and Branching of Blood Vessels

Passing to what may seem a very different subject, we may investigate a number of interesting points in connection with the form and structure of the blood-vessels.

We know that the fluid pressure (P) within the vessel is balanced by (1) the tension (T) of the wall, divided by the radius of curvature, and (2) the external pressure (p_n), normal to the wall: according to our formula

$$P = p_n + T(1/r + 1/r').$$

If we neglect the external pressure, that is to say any support which may be given to the vessel by the surrounding tissues, and if we deal only with a cylindrical vein or artery, this formula becomes simplified to the form $P = T/R$. That is to say, under constant pressure, the tension varies as the radius. But the tension, per unit area of the vessel, depends upon the thickness of the wall, that is to say on the amount of membranous and especially of muscular tissue of which it is composed. Therefore, so long as the pressure is constant, the thickness of the wall should vary as the radius, or as the diameter, of the blood-vessel.

But it is not the case that the pressure is constant, for it gradually falls off, by loss through friction, as we pass from the large arteries to the small; and accordingly we find that while, for a time, the cross-sections of the larger and smaller vessels are symmetrical figures, with the wall-thickness proportional to the size of the tube, this proportion is gradually lost, and the walls of the small arteries, and still more of the capillaries, become exceedingly thin, and more so than in strict proportion to the narrowing of the tube.

In the case of the heart we have, within each of its cavities, a pressure which, at any given moment, is constant over the whole wall-area, but the thickness of the wall varies very considerably. For instance, in the left ventricle the apex is by much the thinnest portion, as it is also that with the greatest curvature. We may assume, therefore (or at least suspect), that the formula, $t(1/r + 1/r') = C$, holds good; that is to say, that the thickness (t) of the wall varies inversely as the mean curvature. This may be tested experimentally, by dilating a heart with alcohol under a known pressure, and then measuring the thickness of the walls in various parts after the whole organ has become hardened. By this means it is found that, for each of the cavities, the law holds good with great accuracy.[1] Moreover, if we begin by dilating the right ventricle and then dilate the left in like manner, until the whole heart is equally and symmetrically dilated, we find (1) that we have had to use a pressure in the left ventricle from six to seven times as great as in the right ventricle, and (2) that the thickness of the walls is just in the same proportion.[2]

Many problems of a hydrodynamical kind arise in connection with the flow of blood through the blood-vessels; and while these are of primary importance to the physiologist they interest the morpholo-

[1] R. H. Woods, 'On a Physical Theorem Applied to Tense Membranes', *J. Anat., Lond.* **26** (1892), 362–71. A similar investigation of the tensions in the uterine wall, and of the varying thickness of its muscles, was attempted by Haughton in his *Animal Mechanics* (1873), pp. 151–8.

[2] This corresponds with a determination of the normal pressures (in systole) by Knohl, as being in the ratio of $1:6\cdot8$.

gist in so far as they bear on questions of structure and form. As an example of such mechanical problems we may take the conditions which go to determine the manner of branching of an artery, or the angle at which its branches are given off; for, as John Hunter said,[1] 'To keep up a circulation sufficient for the part, and no more, Nature has varied the angle of the origin of the arteries accordingly'. This is a vastly important theme, for the theorem which John Hunter has set forth in these simple words is no other than that 'principle of minimal work' which is fundamental in physiology, and which some have deemed the very criterion of 'organisation'.[2] For the principle of Lagrange, the 'principle of virtual work', is the key to physiological equilibrium, and physiology itself has been called a problem in maxima and minima.[3]

This principle, overflowing into morphology, helps to bring the morphological and the physiological concepts together. We have dealt with problems of maxima and minima in many simple configurations, where form alone seemed to be in question; and we meet with the same principle again wherever work has to be done and mechanism is at hand to do it. That this mechanism is the best possible under all the circumstances of the case, that its work is done with a maximum of efficiency and at a minimum of cost, may not always lie within our range of quantitative demonstration, but to believe it to be so is part of our common faith in the perfection of Nature's handiwork. All the experience and the very instinct of the physiologist tells him it is true; he comes to use it as a postulate, or *methodus inveniendi*, and it does not lead him astray. The dicovery of the circulation of the blood was implicit in, or followed quickly after, the recognition of the fact that the valves of heart and veins are adapted to a one-way circulation; and we may begin likewise by assuming a perfect fitness or adaptation in all the minor details of the circulation.

As part of our concept of organisation we assume that the cost of operating a physiological system is a minimum, what we mean by *cost* being measurable in calories and ergs, units whose dimensions are equivalent to those of *work*. The circulation teems with illustra-

---

[1] *Essays*, edited by Owen (1861), I, 134. The subject greatly interested Keats. See his *Notebook*, edited by M. B. Forman (1932), p. 7; and cf. *Keats as a Medical Student*, by Sir Wm Hale-White, in *Guy's Hosp. Rep.* 73 (1925), 249–62.

[2] Cf. Cecil D. Murray, 'The Physiological Principle of Minimal Work, in the Vascular System, and the Cost of Blood-volume', *Proc. Nat. Acad. Sci., Wash.* 12 (1926), 207–14; 'The Angle of Branching of the Arteries', *J. Gen. Physiol.* 19 (1926), 835–41; 'On the Branching-angles of Trees', *ibid.* 10 (1927), 725.

[3] By Dr F. H. Pike, quoted by C. D. Murray.

tions of this great and cardinal principle. 'To keep up a circulation sufficient for the part and no more' Nature has not only varied the angle of branching of the blood-vessels to suit her purpose, she has regulated the dimensions of every branch and stem and twig and capillary; the normal operation of the heart is perfection itself, even the amount of oxygen which enters and leaves the capillaries is such that the work involved in its exchange and transport is a minimum. In short, oxygen transport is the main object of the circulation, and it seems that through all the trials and errors of growth and evolution an efficient mode of transport has been attained. To prove that it is the very best of all possible modes of transport may be beyond our powers and beyond our needs; but to assume that it is *perfectly economical* is a sound working hypothesis.[1] And by this working hypothesis we seek to understand the form and dimensions of this structure or that, in terms of the work which it has to do.

The general principle, then, is that the form and arrangement of the blood-vessels is such that the circulation proceeds with a minimum of effort, and with a minimum of wall-surface, the latter condition leading to a minimum of friction and being therefore included in the first. What, then, should be the angle of branching, such that there shall be the least possible loss of energy in the course of the circulation? In order to solve this problem in any particular case we should obviously require to know (1) how the loss of energy depends upon the distance travelled, and (2) how the loss of energy varies with the diameter of the vessel. The loss of energy is evidently greater in a narrow tube than in a wide one, and greater, obviously, in a long journey than a short. If the large artery, *AB*, gives off a comparatively narrow branch leading to *P* (such as *CP*, or *DP*), the route *ACP* is evidently shorter than *ADP*, but on the other hand, by

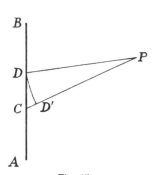

Fig. 47.

[1] Cf. A. W. Volkmann, *Die Haemodynamik nach Versuchen* (Leipzig, 1850) (a work of great originality); G. Schwalbe, 'Ueber...die Gestaltung des Arterien-systems', *Jena. Z. Naturw.* **12** (1878), 267; W. Hess, 'Eine mechanisch bedingte Gesetzmässigkeit im Bau des Blutgefäßsystems', *Arch. Entw. Mech.* **16** (1903), 632; 'Ueber die peripherische Regulierung der Blutzirkulation', *Pflüg. Arch. ges. Physiol.* **168** (1917), 439–90; R. Thoma, 'Die mittlere Durchflussmengen der Arterien des Menschen als Funktion des Gefässradius', *ibid.* **189**, 282–310, **193**, 385–406 (1921–2); E. Blum, 'Querschnittsbeziehungen zwischen Stamm u. Ästen im Arteriensystem', *ibid.* **175** (1919), 1–19.

the latter path, the blood has tarried longer in the wide vessel *AB*, and has had a shorter course in the narrow branch. The relative advantage of the two paths will depend on the loss of energy in the portion *CD*, as compared with that in the alternative portion *CD'*, the one being short and narrow, the other long and wide. If we ask, then, which factor is the more important, length or width, we may safely take it that the question is one of degree; and that the factor of width will become the more important of the two wherever artery and branch are markedly unequal in size. In other words, it would seem that for small branches a large angle of bifurcation, and for large branches a small one, is always the better. Roux has laid down certain rules in regard to the branching of arteries, which correspond with the general conclusions which we have just arrived at. The most important of these are as follows: (1) If an artery bifurcates into two equal branches, these branches come off at equal angles to the main stem. (2) If one of the two branches be smaller than the other, then the main branch, or continuation of the original artery, makes with the latter a smaller angle than does the smaller or 'lateral' branch. And (3) all branches which are so small that they scarcely seem to weaken or diminish the main stem come off from it at a large angle, from about 70° to 90°.

After thus dealing with the most suitable angle of branching, we have still to consider the appropriate cross-section of the branches compared with the main trunk, for instance in the special case where a main artery bifurcates into two. That the sectional area of the two branches may together equal the area of the parent trunk, it is (of course) only necessary that the diameters of trunk and branch should be as $\sqrt{2}:1$, or (say) as 14:10, or (still more roughly) as 10:7; and in the great vessels, this simple ratio comes very nearly true. We have, for instance, the following measurements of the common iliac arteries, into which the abdominal aorta subdivides:

Internal diameter of abdominal arteries (mm)[1]

Aorta abdom.	15·2	12·0	14·1	13·9
Iliaca comm. d.	10·8	8·8	10·4	8·6
Iliaca comm. s.	10·7	8·6	9·5	10·0
Mean of do.	10·8	8·7	10·0	9·3
Ratio	71	72	69	67 p.c.
			Av. 70 p.c.	

But the increasing surface of the branches soon means increased friction, and a slower pace of the blood travelling through; and therefore the branches must be more capacious than at first appears.

[1] From R. Thoma, *op. cit.* p. 388.

129

It becomes a question not of capacity but of resistance; and in general terms the answer is that the ratio of resistance to cross-section shall be equal in every part of the system, before and after bifurcation, as a condition of least possible resistance in the whole system; the total cross-section of the branches, therefore, must be greater than that of the trunk in proportion to the increased resistance.

An approximate result, familiar to students of hydrodynamics, is that the resistance is a minimum, and the condition an optimum, when the cross-section of the main stem is to the sum of the cross-sections of the branches as $1:\sqrt[3]{2}$, or $1:1\cdot26$. Accordingly, in the case of a blood-vessel bifurcating into two equal branches, the diameter of each should be to that of the main stem (approximately) as

$$\sqrt{\frac{1\cdot26}{2}}:1 \quad \text{or (say) } 8:10.$$

While these statements are so far true, and while they undoubtedly cover a great number of observed facts, yet it is plain that, as in all such cases, we must regard them not as a complete explanation, but as *factors* in a complicated phenomenon: not forgetting that (as one of the most learned of all students of the heart and arteries, Dr Thomas Young, said in his Croonian lecture)[1] all such questions as these, and all matters connected with the muscular and elastic powers of the blood-vessels, 'belong to the most refined departments of hydraulics'; and Euler himself had commented on the 'insuperable difficulties' of this sort of problem.[2] Some other explanation must be sought in order to account for a phenomenon which particularly impressed John Hunter's mind, namely the gradually altering angle at which the successive intercostal arteries are given off from the thoracic aorta: the special interest of this case arising from the regularity and symmetry of the series, for 'there is not another set of arteries in the body whose origins are so much the same, whose

---

[1] On the functions of the heart and arteries, *Phil. Trans.* (1809), pp. 1–31, cf. (1808), pp. 164–86; *Collected Works*, I (1855), 511–34. The same lesson is conveyed by all such work as that of Volkmann, E. H. Weber and Poiseuille. Cf. Stephen Hales's *Statical Essays*, II, *Introduction*: 'Especially considering that they [i.e. animal Bodies] are in a manner framed of one continued Maze of innumerable Canals, in which Fluids are incessantly circulating, some with great Force and Rapidity, others with very different Degrees of rebated Velocity: Hence, etc.' Even Leonardo had brought his knowledge of hydrodynamics to bear on the valves of the heart and the vortex-like eddies of the blood. Cf. J. Playfair McMurrich, *L. da Vinci, the Anatomist* (1930), p. 165; etc. How complicated the physiological aspect of the case becomes may be judged by Thoma's papers quoted above.

[2] In a tract entitled *Principia pro motu sanguinis per arterias determinando* (*Op. posth.* 11 (1862), 814–23).

offices are so much the same, whose distances from their origin to the place of use, and whose uses [? sizes][1] are so much the same'.

To conclude, we may now approach the question of economical size of the blood-vessels in a broader way. They must not be too small, or the work of driving blood through them will be too great; they must not be too large, or they will hold more blood than is needed—and blood is a costly thing. We rely once more on Poiseuille's Law,[2] which tells us the amount of work done in causing so much fluid to flow through a tube against resistance, the said resistance being measured by the viscosity of the fluid, the coefficient of friction and the dimensions of the tube; but we have also to account for the blood itself, whose maintenance requires a share of the bodily fuel, and whose cost per c.c. may (in theory at least) be expressed in calories, or in ergs per day. The total cost, then, of operating a given section of artery will be measured by (1) the work done in overcoming its resistance, and (2) the work done in providing blood to fill it; we have come again to a differential equation, leading to an equation of maximal efficiency. The general result is as follows:[3] it can be made a quantitative one by introducing known experimental values. Were blood a cheaper thing than it is we might expect all arteries to be uniformly larger than they are, for thereby the burden on the heart (the flow remaining equal) would be greatly reduced— thus if the blood-vessels were doubled in diameter, and their volume thereby quadrupled, the work of the heart would be reduced to one-sixteenth. On the other hand, were blood a scarcer and still costlier fluid, narrower blood-vessels would hold the available supply; but a larger and stronger heart would be needed to overcome the increased resistance.

[1] 'Sizes' is Owen's editorial emendation, which seems amply justified.
[2] Owing to faulty determination of the fall of pressure in the capillaries, Poiseuille's equation used to be deemed inapplicable to them; but Krogh's recent work removes, or tends to remove, the inconsistency (*Anatomy and Physiology of the Capillaries*, Yale University Press, 1922).
[3] Cf. C. D. Murray, *op. cit.* p. 211.

CHAPTER V

ON SPICULES AND SPICULAR SKELETONS

In this chapter and the following ones D'Arcy Thompson is struggling against the notion that all form can simply be explained by heredity, and that therefore changes in form inevitably map out phylogenetic relations. Instead he repeatedly suggests that physical forces (such as those which produce the variations of shapes of snow-flakes) are of prime importance and relationships of shape may not justify any family tree or a sequence in time, but simply show mathematical kinship. Today we are inclined to combine the two and say that the genes, the units of heredity, do control shapes, but that the activities of genes are constrained by the physico-chemical properties of the chemical substances and the configuration of these substances present in the organism. This does not touch upon the question of whether or not all the shapes produced are adaptively signifi-cant. D'Arcy Thompson's strong arguments that they are not is a reaction against those who see a selective advantage to all structures. But this issue cannot be resolved without an ecological study of each example, a Gar-gantuan task that is unlikely ever to be achieved. All we can say at the moment is that there is no *a priori* reason why some structures, which have been initiated by mutation and formed within the confines of physico-chemical laws, should not utterly lack any adaptive significance, and yet remain fixed in the population. It might even be argued that a particular gene-change produced other effects that were adaptively significant and these less obvious gene-effects elicited the selection pressure which pre-served the adaptively inert structure in the population.

But with this emphasis on the importance of physical forces, there would seem, at first glance, to be a curious conflict in this chapter. The most original and effective point he makes is that the forces of crystallisation are insufficient to account for the shapes of spicules and skeletons. This does not mean, however, that he has abandoned any hope of a physical explanation; instead he makes a first step towards such an explanation in terms of surface deposition of materials between cells or the bubbles of a frothy protoplast, and concludes that the problem is hardly solved, for 'Nature keeps some of her secrets longer than others'.

The Problem

The deposition of inorganic material in the living body, usually in the form of calcium salts or of silica, is a common phenomenon. It begins by the appearance of small isolated particles, crystalline or non-crystalline, whose form has little relation or none to the structure of the organism; it culminates in the complex skeletons of the vertebrate animals, in the massive skeletons of the corals, or in the polished, sculptured and mathematically regular molluscan shells. Even among very simple organisms, such as diatoms, radiolarians, foraminifera or sponges, the skeleton displays extraordinary variety and beauty, whether by reason of the intrinsic form of its elementary constituents or the geometric symmetry with which these are interconnected and arranged.

With regard to the form of these various structures (and this is all that immediately concerns us here), we have to do with two distinct problems, which merge with one another though they are theoretically distinct. For the form of the spicule or other skeletal element may depend solely on its chemical nature, as for instance, to take a simple but not the only case, when it is purely crystalline; or the inorganic material may be laid down in conformity with the shapes assumed by cells, tissues or organs, and so be, as it were, moulded to the living organism; and there may well be intermediate stages in which both phenomena are simultaneously at work, the molecular forces playing their part in conjunction with the other forces inherent in the system.

So far as the problem is a purely chemical one we must deal with it very briefly indeed: all the more because special investigations regarding it have as yet been few, and even the main facts of the case are very imperfectly known. This at least is clear, that the phenomena with which we are about to deal go deep into the subject of colloid chemistry, and especially that part of the science which deals with colloids in connection with surface phenomena. It is to the special student of the chemistry and physics of the colloids that we must look for the elucidation of our problem.[1]

In the first and simplest part of our subject, the essential problem is the problem of crystallisation in presence of colloids. In the cells of plants true crystals are found in comparative abundance, and consist, in the majority of cases, of calcium oxalate. In the stem and root of the rhubarb for instance, in the leaf-stalk of *Begonia* and in countless other cases, sometimes within the cell, sometimes in the substance of the cell-wall, we find large and well-formed crystals of

---

[1] There is much information regarding the chemical composition and mineralogical structure of shells and other organic products in H. C. Sorby's Presidential Address to the Geological Society (*Proc. Geol. Soc.* 1879, pp. 56–93); but Sorby failed to recognise that association with 'organic' matter, or with colloid matter whether living or dead, introduced a new series of purely physical phenomena.

this salt; their varieties of form, which are extremely numerous, are simply the crystalline forms proper to the salt itself, and belong to the two systems, cubic and monoclinic, in one or other of which, according to the amount of water of crystallisation, this salt is known to crystallise. When calcium oxalate crystallises according to the latter system (as it does when its molecule is combined with two molecules of water), the microscopic crystals have the form of fine needles, or 'raphides'; these are very common in plants, and may be artificially produced when the salt is crystallised out in presence of glucose or of dextrin.[1]

Calcium carbonate, on the other hand, when it occurs in plant-cells, as it does abundantly (for instance in the 'cystoliths' of the Urticaceae and Acanthaceae, and in great quantities in *Melobesia* and the other calcareous or 'stony' algae), appears in the form of fine rounded granules, whose inherent crystalline structure is only revealed (like that of a molluscan shell) under polarised light. Among animals, a skeleton of carbonate of lime occurs under a multitude of forms, of which we need only mention a few of the most conspicuous. The spicules of the calcareous sponges are triradiate, occasionally quadriradiate, bodies, with pointed rays, not crystalline in outward form but with a definitely crystalline internal structure; we shall return again to these, and find for them what would seem to be a satisfactory explanation of their form. Among the Alcyonarian zoophytes we have a great variety of spicules,[2] which are sometimes straight and slender rods, sometimes flattened and more or less striated plates, and still more often disorderly aggregations of micro-crystals, in the form of rounded or branched concretions with rough or knobby surfaces.[3] A third type, presented by several very different things, such as a pearl or the ear-bone of a bony fish, consists of a more or less rounded body, sometimes spherical, sometimes flattened, in which the calcareous matter is laid down in concentric zones, denser and clearer layers alternating with one another. In the development of the molluscan shell and in the calcification of a bird's egg or a crab's shell, small spheroidal bodies with similar concentric striation make their appearance; but instead of remaining separate they become crowded together, and in doing so are apt to form a pattern of

---

[1] Julien Vesque, 'Sur la production artificielle de cristaux d'oxalate de chaux semblables à ceux qui se forment dans les plantes'. *Ann. Sci. Nat. (Bot.)* (5), **19** (1874), 300–13.

[2] Cf. Kölliker, *Icones Histologicae* (1864), p. 119, etc.

[3] In rare cases, these show a single optic axis and behave as individual crystals: W. J. Schmidt, *Arch. Entw. Mech.* **51** (1922), 509–51.

hexagons. In some cases the carbonate of lime, on being dissolved away by acid, leaves behind it a certain small amount of organic residue; in many cases other salts, such as phosphates of lime, ammonia or magnesia, are present in small quantities; and in most cases, if not all, the developing spicule or concretion is somehow so associated with living cells that we are apt to take it for granted that it owes its form to the constructive or plastic agency of these.

The appearance of direct association with living cells, however, is apt to be fallacious; for the actual *precipitation* takes place, as a rule, not in actively living, but in dead or at least inactive tissue;[1] that is to say in the 'formed material' or matrix which accumulates round the living cells, or in the interspaces between these latter, or, as often happens, in the cell-wall or cell-membrane rather than within the substance of the protoplasm itself. We need not go the length of asserting that this is a rule without exception; but, so far as it goes, it is of great importance and to its consideration we shall presently return.[2]

Cognate with this is the fact that, at least in some cases, the organism can go on, in apparently unimpaired health, when stinted or even wholly deprived of the material of which it is wont to make its spicules or its shell. Thus the eggs of sea-urchins reared in lime-free water develop, in apparent health and comfort, into larvae which lack the usual skeleton of calcareous rods: and in which, accordingly, the long arms of the Pluteus larva, which the rods should support and extend, are entirely absent.[3] Again, when foraminifera are kept for generations in water from which they gradually exhaust the lime, their shells grow hyaline and transparent, and dwindle to a mere chitinous pellicle; on the other hand, in the presence of excess of lime their shells become much altered, are strengthened with various ridges or 'ornaments', and come to resemble other varieties and even 'species'.[4]

The crucial experiment, then, is to attempt the formation of similar

[1] In an interesting paper by Robert Irvine and Sims Woodhead on the 'Secretion of Carbonate of Lime by Animals' (*Proc. Roy. Soc. Edinb.* **15**, 308–16; **16**, 324–51, 1889–90) it is asserted (p. 351) that 'lime salts, of whatever form, are deposited *only* in vitally inactive tissue'.

[2] The tube of *Teredo* shows no trace of organic matter, but consists of irregular prismatic crystals: the whole structure 'being identical with that of small veins of calcite, such as are seen in thin sections of rocks' (Sorby, *Proc. Geol. Soc.* 1879, p. 58). This, then, would seem to be a somewhat exceptional case of a shell laid down completely outside of the animal's external layer of organic substance.

[3] Cf. Pouchet and Chabry, *C.R. Soc. Biol., Paris* (9), **1** (1889), 17–20; *C.R. Acad. Sci. Paris*, **108** (1889), 196–8.

[4] Cf. Heron-Allen, *Phil. Trans.* B, **206** (1915), 262.

spicules or concretions apart from the living organism. But however feasible the attempt may be in theory, we must be prepared to encounter many difficulties; and to realise that, though the reactions involved may be well within the range of physical chemistry, yet the actual conditions of the case may be so complex, subtle and delicate that only now and then, and only in the simplest of cases, has it been found possible to imitate the natural objects successfully. Such an attempt is part of that wide field of enquiry through which Stéphane Leduc and other workers have sought to produce, by synthetic means, forms similar to those of living things; but it is a circumscribed and well-defined part of that wider investigation.[1] When by chemical or physical experiment we obtain configurations similar, for instance, to the phenomena of nuclear division, or conformations similar to a pattern of hexagonal cells, or a group of vesicles which resemble some particular tissue or cell-aggregate, we indeed prove what is the main object of this book to illustrate, namely, that the physical forces are capable of producing particular organic forms. But it is by no means always that we can feel perfectly assured that the physical forces which we deal with in our experiment are identical with, and not merely analogous to, the physical forces which, at work in nature, are bringing about the result which we have succeeded in imitating. In the present case, however, our enquiry is restricted and apparently simplified; we are seeking in the first instance to obtain by purely chemical means a purely chemical result, and there is little room for ambiguity in our interpretation of the experiment.

Molecular Asymmetry

When we find ourselves investigating the forms assumed by chemical compounds under the peculiar circumstances of association with a living body, and when we find these forms to be characteristic or

[1] Leduc's artificial growths were mostly obtained by introducing salts of the heavy metals or alkaline earths into solutions which form with them a 'precipitation-membrane'—as when we introduce copper sulphate into a ferrocyanide solution. See his *Mechanism of Life* (1911), ch. x, for copious references to other works on the 'artificial production of organic forms'. Closely related to Leduc's experiments are those of Denis Monnier and Carl Vogt, 'Sur la fabrication artificielle des formes des éléments organiques', *J. Anat., Paris*, **18** (1882), 117–23; cf. Moritz Traube, 'Zur Geschichte der mechanischen Theorie des Wachstums der organischen Zelle', *Botan. Ztg.* **36** (1878). Cf. also A. L. Herrera, 'Sur les phénomènes de vie apparente observés dans les émulsions de carbonate de chaux dans la silice gélatineuse', *Mem. Soc. Alzate, Mexico*, **26** (1908); *Los Protobios*, *Boll. Dir. Estud. Biolog., Mexico*, **1**, 607–31, and other papers. Also (*int. al.*) R. S. Lillie and E. N. Johnston, 'Precipitation-structures Simulating Organic Growth', *Biol. Bull.* **23** (1917), 135; **26** (1919), 225–72; *Sci. Mon.* (February 1922), p. 125; H. W. Morse, C. H. Warren and J. D. H. Donnay, 'Artificial Spherulites, etc.', *Amer. J. Sci.* (5), **23** (1932), 421–39.

recognisable, and somehow different from those which the same substance is wont to assume under other circumstances, an analogy, captivating though perhaps remote, presents itself to our minds between this subject of ours and certain synthetic problems of the organic chemist. There is doubtless an essential difference, as well as a difference of scale, between the visible form of a spicule or concretion and the hypothetical form of an individual molecule. But molecular form is a very important concept; and the chemist has not only succeeded, since the days of Wöhler, in synthetising many substances which are characteristically associated with living matter, but his task has included the attempt to account for the molecular *forms* of certain 'asymmetric' substances—glucose, malic acid and many more—as they occur in Nature. These are bodies which, when artificially synthetised, have no optical activity, but which, as we actually find them in organisms, turn (when *in solution*) the plane of polarised light in one direction rather than the other; thus dextroglucose and laevomalic acid are common products of plant metabolism, but dextromalic acid and laevoglucose do not occur in Nature at all. The optical activity of these bodies depends, as Pasteur showed eighty years ago,[1] upon the form, right-handed or left-handed, of their molecules, which molecular asymmetry further gives rise to a corresponding right- or left-handedness (or enantiomorphism) in the crystalline aggregates. It is a distinct problem in organic or physiological chemistry, and by no means without its interest for the morphologist, to discover how it is that Nature, for each particular substance, habitually builds up, or at least selects, its molecules in a one-sided fashion, right-handed or left-handed as the case may be. It will serve us no better to assert that this phenomenon has its origin in 'fortuity' than to repeat the Abbé Galiani's saying, '*les dés de la nature sont pipés*'.

The problem is not so closely related to our immediate subject that we need discuss it at length; but it has its relation, such as it is, to the general question of *form* in relation to vital phenomena, and it has its historic interest as a theme of long-continued discussion. According to Pasteur, there lay in the molecular asymmetry of the natural bodies and their symmetry when artificially produced, one of the most deep-seated differences between vital and non-vital phenomena: he went further, and declared that 'this was perhaps the

[1] Lectures on the molecular asymmetry of natural organic compounds, *Chemical Soc. of Paris* (1860); also in Ostwald's *Klassiker exact. Wiss.* no. 28, and in *Alembic Club Reprints*, no. 14 (Edinburgh, 1897); cf. G. M. Richardson, *Foundations of Stereochemistry* (New York, 1901).

only well-marked line of demarcation that can at present [1860] be drawn between the chemistry of dead and of living matter'. Nearly forty years afterwards the same theme was pursued and elaborated by Japp in a celebrated lecture,[1] and the distinction still has its weight, I believe, in the minds of many chemists. 'We arrive at the conclusion', said Professor Japp, 'that the production of single asymmetric compounds, or their isolation from the mixture of their enantiomorphs, is, as Pasteur firmly held, the prerogative of life. Only the living organism, or the living intelligence with its conception of asymmetry, can produce this result. Only asymmetry can beget asymmetry.' In these last words (which, so far as the chemist and the biologist are concerned, we may acknowledge to be true) lies the crux of the difficulty.

Sponge Spicules: Triradiate Forms

The spicules of the calcareous sponges are commonly triradiate, and the three radii are usually inclined to one another at *nearly* equal angles; in certain cases, two of the three rays are nearly in a straight line, and at right angles to the third.[2]* They are not always in a plane, but are often inclined to one another in a trihedral angle, not easy of precise measurement under the microscope. The three rays are often supplemented by a fourth, which is set tetrahedrally, making *nearly* co-equal angles with the other three. The calcareous spicule consists mainly of carbonate of lime in the form of calcite, with (according to von Ebner) some admixture of soda and magnesia, of sulphates and of water. According to the same writer there is no organic matter in the spicule, either in the form of an axial filament or otherwise, and the appearance of stratification, often simulating the presence of an axial fibre, is due to 'mixed crystallisation' of the various constituents. The spicule is a true crystal, and therefore its existence and its form are *primarily* due to the molecular forces of crystallisation; moreover it is a single crystal and not a group of crystals, as is seen by its behaviour in polarised light. But its axes are not crystalline axes, its angles are variable and indefinite, and its

---

[1] F. R. Japp, 'Stereochemistry and Vitalism', *Brit. Ass. Rep.* (Bristol, 1898), p. 813; cf. also a voluminous discussion in *Nature, Lond.* 1898–9.

[2] For very numerous illustrations of the triradiate and quadriradiate spicules of the calcareous sponges, see (*int. al.*), papers by Dendy (*Quart. J. Micr. Sci.* 35, 1893), Minchin (*Proc. Zool. Soc. Lond.* 1904), Jenkin (*Proc. Zool. Soc. Lond.* 1908), etc.

---

* There has been a recent series of valuable contributions by W. C. Jones on spicule formation in sponges. *Quart. J. Micr. Sci.* 95 (1954), 33, 191; 96 (1955), 129, 411; 99 (1958), 263.

form neither agrees with, nor in any way resembles, any one of the countless, polymorphic forms in which calcite is capable of crystallising. It is as though it were carved out of a solid crystal; it is, in fact, a crystal under restraint, a crystal growing, as it were, in an artificial mould, and this mould is constituted by the surrounding cells or structural vesicles of the sponge.

We have already studied in an elementary way, but enough for our purpose, the manner in which three, four or more cells, or bubbles, meet together under the influence of surface-tension, in configurations geometrically similar to what may be brought about by a uniform distribution of mechanical pressure. And we know that surface-energy leads to the *adsorption* of certain chemical substances, first at the corners, then at the edges, lastly in the partition-walls, of such an assemblage of cells. A spicule formed in the interior of such a mass, starting at a corner where four cells meet and extending along the adjacent edges, would then (in theory) have the characteristic form which the geometry of the bee's cell has taught us, of four rays radiating from a point, and set at co-equal angles to one another of 109°, approximately. Precisely such 'tetractinellid' spicules are often formed.

But when we confine ourselves to a plane assemblage of cells, or to the outer surface of a mass, we need only deal with the simpler geometry of the hexagon. In such a plane assemblage we find the cells meeting one another in threes; when the cells are uniform in size the partitions are straight lines, and combine to form regular hexagons; but when the cells are unequal, the partitions tend to be curved, and to combine to form other and less regular polygons. Accordingly, a skeletal secretion originating in a layer or surface of cells will begin at the corners and extend to the edges of the cells, and will thus take the form of triradiate spicules, whose rays (in a typical case) will be set at co-equal angles of 120° (Fig. 48, *f*). This latter condition of inequality will be open to modification in various ways. It will be modified by any inequality in the specific tensions of adjacent cells; as a special case, it will be apt to be greatly modified at the surface of the system, where a spicule happens to be formed in a plane perpendicular to the cell-layer, so that one of its three rays lies between two adjacent cells and the other two are associated with the surface of contact between the cells and the surrounding medium; in such a case (as in the cases considered in connection with the forms of the cells themselves on p. 101), we shall tend to obtain a spicule with two equal angles and one unequal (Fig. 48, *a*, *c*); in the last case, the two outer, or superficial rays, will tend to be markedly curved. Again, the equiangular condition will be departed from, and more or less curvature will be imparted to the rays, wherever the cells of the system cease to be uniform in size, and when the hexagonal symmetry of the system is lost accordingly. Lastly, although we speak of the rays as meeting at certain definite angles, this statement applies to their *axes* rather than to the rays themselves. For if the triradiate spicule be developed in the *interspace* between three juxta-

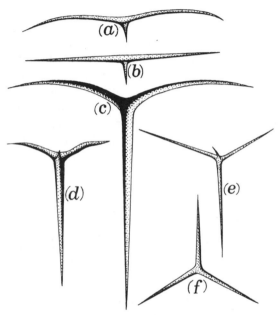

Fig. 48. Spicules of *Grantia* and other calcareous sponges.
After Haeckel.

posed cells it is obvious that its sides will tend to be concave, because the space between three contiguous equal circles is an equilateral, curvilinear triangle; and even if our spicule be deposited, not in the space between our three cells, but in the mere thickness of an intervening wall, then we may recollect that the several partitions never actually meet at sharp angles, but the angle of contact is always bridged over by an accumulation of material (varying in amount according to its fluidity) whose boundary takes the form of a circular arc, and which constitutes the 'bourrelet' of Plateau. In any sample of the triradiate spicules of *Grantia*, or in any series of careful draw-

140

ings, such as Haeckel's, we shall find all these various configurations severally and completely illustrated.

The tetrahedral, or rather tetractinellid, spicule needs no further explanation in detail (Fig. 48, *d, e*). For just as a triradiate spicule corresponds to the case of three cells in mutual contact, so does the four-rayed spicule to that of a solid aggregate of four cells: these latter tending to meet one another in a tetrahedral system, showing four edges, at each of which three facets or partitions meet, their edges being inclined to one another at equal angles of about 109°—the 'Maraldi' angle. And even in the case of a single layer, or superficial layer, of cells, if the skeleton originate in connection with all the edges of mutual contact, we shall (in complete and typical cases) have a four-rayed spicule, of which one straight limb will correspond to the line of junction between the three cells, and the other three limbs (which will then be curved limbs) will correspond to the three edges where the three cells meet in pairs on the surface of the system.

Mechanics of Spicule Formation

But if such a physical explanation of the forms of our spicules is to be accepted, we must seek for some physical agency to explain the presence of the solid material just at the junctions or interfaces of the cells, and for the forces by which it is confined to, and moulded to the form of, these intercellular or interfacial contacts. We owe to Dreyer the physical or mechanical theory of spicular conformation which I have just described—a theory which ultimately rests on the form assumed, under surface-tension, by an aggregation of cells or vesicles. But this fundamental point being granted, we have still several possible alternatives by which to explain the details of the phenomenon.

Dreyer, if I understand him aright, was content to assume that the solid material, secreted or excreted by the organism, accumulates in the interstices between the cells, and is there subjected to mechanical pressure or constraint as the cells get crowded together by their own growth and that of the system generally. As far as the general form of the spicules goes such explanation is not inadequate, though under it we might have to renounce some of our assumptions as to what takes place at the surface of the system. But where a few years ago the concept of secretion seemed precise enough, we turn nowadays to the phenomenon of adsorption as a further stage towards the elucidation of our facts, and here we have a case in point. In the

141

tissues of our sponge, wherever two cells meet, there we have a definite *surface* of contact, and there accordingly we have a manifestation of surface-energy; and the concentration of surface-energy will tend to be a maximum at the *lines* or *edges* whereby such surfaces are conjoined. Of the micro-chemistry of the sponge-cells our ignorance is great; but (without venturing on any hypothesis involving the chemical details of the process) we may safely assert that there is an inherent probability that certain substances will tend to be concentrated and ultimately deposited just in these lines of intercellular contact and conjunction. In other words, adsorptive concentration, under osmotic pressure, at and in the surface-film which bounds contiguous cells, and especially in the *edges* where these films meet and intersect, emerges as an alternative (and, as it seems to me, a highly preferable alternative) to Dreyer's conception of an accumulation under mechanical pressure in the vacant spaces left between one cell and another.

But a purely chemical, or purely molecular, adsorption is not the only form of the hypothesis on which we may rely. For from the purely physical point of view, angles and edges of contact between adjacent cells will be *loci* in the field of distribution of surface-energy, and any material particles whatsoever will tend to undergo a diminution of freedom on entering one of those boundary regions. Let us imagine a couple of soap bubbles in contact with one another; over the surface of each bubble tiny bubbles and droplets glide in every direction; but as soon as these find their way into the groove or re-entrant angle between the two bubbles, there their freedom of movement is so far restrained, and out of that groove they have little tendency, or little freedom, to emerge. A cognate phenomenon is to be witnessed in microscopic sections of steel or other metals. Here, together with its crystalline structure, the metal develops a cellular structure by reason of its lack of homogeneity; for in the molten state one constituent tends to separate out into drops, while the other spreads over these and forms a filmy reticulum between— the *disperse* phase and the *continuous* phase of the colloid chemists. In a polished section we easily observe that the little particles of graphite and other foreign bodies common in the matrix have tended to aggregate themselves in the walls and at the angles of the polygonal cells—this being a direct result of the diminished freedom which they undergo on entering one of these boundary regions. And the same phenomenon is turned to account in the various 'separation-processes' in which metallic particles are caught up in the interstices of a froth, that is to say in the walls of the *foam-cells* or *Schaum-kammern*.[1]

It is by a combination of these two principles, chemical adsorption on the one hand and physical quasi-adsorption or concentration of grosser particles on the other, that I conceive the substance of the sponge-spicule to be concentrated and aggregated at the cell boundaries; and the forms of the triradiate and tetractinellid spicules are in precise conformity with this hypothesis. A few general matters, and a few particular cases, remain to be considered. It matters little or not at all for the phenomenon in question, what is the histological nature or 'grade' of the vesicular structures on which it depends. In some cases (apart from sponges), they may be no more than little alveoli of an intracellular protoplasmic network, and this would seem to be the case at least in the protozoan *Entosolenia aspera*, within the vesicular protoplasm of whose single cell Möbius has described tiny spicules in the shape of little tetrahedra with sunken or concave sides. It is probably the case also in the small beginnings of Echinoderm spicules, which are likewise intracellular and are of similar shape. Among the sponges we have many varying conditions. In some cases there is reason to believe that the spicule is formed at the boundaries of true cells or histological units; but in the case of the larger triradiate or tetractinellid spicules they far surpass in size the actual 'cells'. We find them lying, regularly and symmetrically arranged, between the 'pore-canals' or 'ciliated chambers', and it is in conformity with the shape and arrangement of these large rounded or spheroidal structures that their shape is assumed.

Again, it is not at variance with our hypothesis to find that, in the adult sponge, the larger spicules may greatly outgrow the bounds not only of actual cells but also of the ciliated chambers, and may even appear to project freely from the surface of the sponge. For we have already seen that the spicule is capable of growing, without marked change of form, by further deposition, or crystallisation, of layer upon layer of calcareous molecules, even in an artificial solution; and we are entitled to believe that the same process may be carried on in the tissues of the sponge, without greatly altering the symmetry of the spicule, long after it has established its characteristic but non-crystalline form of a system of slender trihedral or tetrahedral rays.

[1] The crystalline composition of iron was recognised by Hooke in the *Micrographia* (1665); and the cellular or polyhedral structure of the metal was clearly recognised by Réaumur, in his *Art de convertir le fer forgé en acier* (1722).

Neither is it of great importance to our hypothesis whether the rayed spicule necessarily arises as a single structure, or does so from separate minute centres of aggregation. Minchin has shown that, in some cases at least, the latter is the case; the spicule begins, he tells us, as three tiny rods, separate from one another, each developed in the interspace between two sister-cells, which are themselves the results of the division of one of a little trio of cells; and the little rods meet and fuse together while still very minute, when the whole spicule is only about $\frac{1}{200}$ of a millimetre long. At this stage, it is interesting to learn that the spicule is non-crystalline; but the new accretions of calcareous matter are soon deposited in crystalline form.

This observation threw difficulties in the way of former mechanical theories of the conformation of the spicule, and was quite at variance with Dreyer's theory, according to which the spicule was bound to begin from a central nucleus coinciding with the meeting-place of three contiguous cells, or rather the interspace between them. But the difficulty is removed when we import the concept of adsorption; for by this agency it is natural enough, or conceivable enough, that deposition should go on at separate parts of a common system of surfaces; and if the cells tend to meet one another by their interfaces before these interfaces extend to the angles and so complete the polygonal cell, it is again only natural that the spicule should first arise in the form of separate and detached limbs or rays.

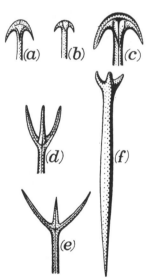

Among the 'tetractinellid' sponges, whose spicules are composed of amorphous silica or opal, all or most of the above-described main types of spicule occur, and, as the name of the group implies, the four-rayed, tetrahedral spicules are especially represented. A somewhat frequent type of spicule is one in which one of the four rays is greatly developed, and the other three constitute small prongs diverging at equal angles from the main or axial ray. In all probability, as Dreyer suggests, we have here had to do with a group of four vesicles, of which three were large and co-equal, while a fourth

Fig. 49. Spicules of tetractinellid sponges (after Sollas). (a–e) anatriaenes; (d–f) protriaenes.

and very much smaller one lay above and between the other three. In certain cases where we have likewise one large and three much smaller rays, the latter are recurved, as in Fig. 49, *a–c*. This type, save for the constancy of the number of rays and the limitation of the terminal ones to three, and save also for the more important difference that they occur only at one and not at both ends of the long axis, is similar to the type which we have explained as being probably developed within an oval cell, by whose walls its branches have been cabined and confined. But it is more probable that we have here to do with a spicule developed in the midst of a group of three co-equal and more or less elongated or cylindrical cells or vesicles, the long axial ray corresponding to their common edge or line of contact, and the three short rays having each lain in the surface furrow between two out of the three adjacent cells.

Holothurian Spicules

Just as in the case of the little S-shaped spicules formed within the bounds of a single cell, so also in the case of the larger tetractinellid types do we find the same configurations reproduced among the holothuroids as we have dealt with in the sponges. The holothurian spicules are a little less neatly formed, a little rougher, than the sponge-spicules, and certain forms occur among the former group which do not present themselves among the latter; but for the most part a community of type is obvious and striking (Fig. 50).

The very peculiar spicules of the holothurian *Synapta*, where a tiny anchor is pivoted or hinged on a perforated plate, are a puzzle indeed; but we may at least solve part of the riddle. How the hinge is formed, I do not know; the anchor gets its shape, perhaps, in some such way as we have supposed the 'amphidiscs' of *Hyalonema* to acquire their reflexed spokes, but the perforated plate is more comprehensible. Each plate starts in a little clump of cells in whose boundary-walls calcareous matter is deposited, doubtless by adsorption, the holes in the finished plate thus corresponding to the cells which formed it. Close-packing leads to an arrangement of six cells round a central one, and the normal pattern of the plate displays this hexagonal configuration. The calcareous plate begins as a little rod whose ends fork, and then fork again: in the same inevitable trinodal pattern which includes the 'polar furrow' of the embryologists. The anchor had been first formed, and the little plate is added on beneath it. The first spicular rudiment of the plate may lie parallel

145

to the stock of the anchor or it may lie athwart it.[1] From the physical point of view it would seem to be a mere matter of chance which way the cluster of cells happens to lie; but this difference of direction will cause a certain difference in the symmetry of the resulting plate. It is this very difference which systematic zoologists at one time seized upon to distinguish *S. Buskii* from our two commoner 'species'. The two latter (*S. inhaerens* and *S. digitata*) are mainly distinguished

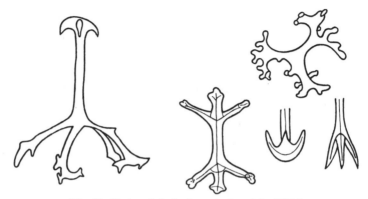

Fig. 50. Various holothurian spicules. After Théel.

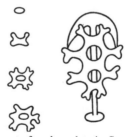

Fig. 51. Development of anchor-plate in *Synapta*. After Semon.

from one another by the number of holes in the plate, that is to say, by the average number of cells in the little cluster of which the plate or spicule was formed. In many or perhaps most other holothurians the spicules consist of little perforated plates or baskets, developed in the same way, about cells or vesicles more or less close-packed, and therefore more or less symmetrically arranged (Fig. 51).

[1] Cf. S. Becher, 'Nicht-funktionelle Korrelation in der Bildung selbständiger Skelet-elemente', *Zool. Jb. Zool. Physiol.* **31** (1912), 1–189; Hedwiga Wilhelmi, 'Skelet-bildung der füsslosen Holothurien', *ibid.* **37** (1920), 493–547; *Arch. Entw. Mech.* **46** (1920), 210–58. See also W. Woodland, 'Studies in Spicule-formation', *Quart. J. Micr. Sci.* **49** (1906), 535–59; **51** (1907), 483–509, and R. Semon, 'Naturgeschichte der Synaptiden', *Mitt. Zool. Sta. Neapel*, **7** (1886), 272–99. On the common species of *Synapta*, see Koehler, *Faune de France, Echinodermes* (1921), pp. 188–9.

Hexactinellid Sponge Spicules

The six-rayed siliceous spicules of the hexactinellid sponges, while they are perhaps the most regular and beautifully formed spicules to be found within the entire group, have been found very difficult to explain, and Dreyer has confessed his complete inability to account for their conformation.[1] But, though it may only be throwing the difficulty a little further back, we may so far account for them by considering that the cells or vesicles by which they are conformed are not arranged in what is known as 'closest packing', but in linear series; so that in their arrangement, and by their mutual compression,

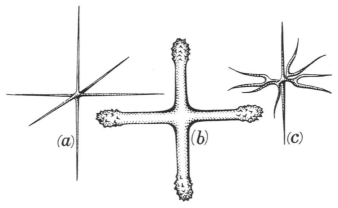

Fig. 52. Spicules of hexactinellid sponges. After F. E. Schultze.

we tend to get a pattern not of hexagons but of squares; or, looking to the solid, not of dodecahedra but of cubes or parallelepipeda. This indeed appears to be the case, not with the individual cells (in the histological sense), but with the larger units or vesicles, which make up the body of the hexactinellid. And this being so, the spicules formed between the linear, or cubical series of vesicles, will have the same tendency towards a 'hexactinellid' shape, corresponding to the angles and adjacent edges of a system of cubes, as in our former case they had to a triradiate or a tetractinellid form, when developed in connection with the angles and edges of a system of hexagons, or a system of rhombic dodecahedra.

[1] Cf. Albr. Schwan, 'Ueber die Funktion des Hexactinellidenskelets, u. seine Vergleichbarkeit mit dem Radiolarienskelet', *Zool. Jb. Zool. Physiol.* 33 (1913), 603–16; cf. V. Hacker, 'Bericht über d. Tripyleenausbeute d. d. Tiefsee-Exped.' *Verh. zool.-bot. Ges.* (1904).

However the hexactinellid spicules be arranged (and this is none too easy to determine) in relation to the tissues and chambers of the sponge, it is at least clear that, whether they lie separate or be fused together in a composite skeleton, they effect a symmetrical partitioning of space according to the cubical system, in contrast to that closer packing which is represented and effected by the tetrahedral system.[1]

Histologically, the case is illustrated by a well-known phenomenon in embryology. In the segmenting ovum, there is a tendency for the cells to be budded off in linear series; and so they often remain, in rows side by side, at least for a considerable time and during the course of several consecutive cell divisions. Such an arrangement constitutes what the embryologists call the 'radial type' of segmentation.[2] But in what is described as the 'spiral type' of segmentation, it is stated that, as soon as the first horizontal furrow has divided the cells into an upper and a lower layer, those of 'the upper layer are shifted in respect to the lower layer, by means of a rotation about the vertical axis'.[3] It is, of course, evident that the whole process is merely that which is familiar to physicists as 'close packing'. It is a very simple case of what Lord Kelvin used to call 'a problem in tactics'. It is a mere question of the rigidity of the system, of the freedom of movement on the part of its constituent cells, whether or at what stage this tendency to slip into the closest propinquity, or position of *minimum potential*, will be found to manifest itself.

Natural Selection and Spicules

This question of the origin and causation of the forms of sponge-spicules, with which we have now sought to deal, is all the more important and all the more interesting because it has been discussed time and again, from points of view which are characteristic of very different schools of thought in biology. Haeckel found in the form of the sponge-spicule a typical illustration of his theory of 'bio-crystallisation'; he considered that these 'biocrystals' represented something midway—*ein Mittelding*—between an inorganic crystal and an organic secretion; that there was a 'compromise between the

[1] *Chall. Rep., Hexactinellida*; pls. xvi, liii, lxxvi, lxxxviii.
[2] See, for instance, the figures of the segmenting egg of *Synapta* (after Selenka), in Korschelt and Heider's *Vergleichende Entwicklungsgeschichte*. On the spiral type of segmentation as a secondary derivative, due to mechanical causes, of the 'radial' type of segmentation, see E. B. Wilson, 'Cell-lineage of *Nereis*', *J. Morph.* **6** (1892), 450.
[3] Korschelt and Heider, p. 16.

crystallising efforts of the calcium carbonate and the formative activity of the fused cells of the syncytium'; and that the semi-crystalline secretions of calcium carbonate 'were utilised by natural selection as "spicules" for building up a skeleton, and afterwards, by the interaction of adaptation and heredity, became modified in form, and differentiated in a vast variety of ways, in the struggle for existence'.[1] What Haeckel precisely meant by these words is not clear to me.

F. E. Schultze, perceiving that identical forms of spicule were developed whether the material were crystalline or non-crystalline, abandoned all theories based upon crystallisation; he simply saw in the form and arrangement of the spicules something which was 'best fitted' for its purpose, that is to say for the support and strengthening of the porous walls of the sponge, and finding clear evidence of 'utility' in the specific characters of these skeletal elements, had no difficulty in ascribing them to natural selection.

Sollas and Dreyer, as we have seen, introduced in various ways the conception of physical causation—as indeed Haeckel himself had done in regard to one particular, when he supposed the *position* of the spicules to be due to the constant passage of the water-currents; though even here, by the way, if I understand Haeckel aright, he was thinking not of a direct or immediate physical causation, but rather of one manifesting itself through the agency of natural selection.[2] Sollas laid stress upon the 'path of least resistance' as determining the direction of growth; while Dreyer dealt in greater detail with the tensions and pressures to which the growing spicule was exposed, amid the alveolar or vesicular structure which was represented alike by the chambers of the sponge, by the constituent cells, or by the minute structure of the intracellular protoplasm. But neither of these writers, so far as I can discover, was inclined to doubt for a moment the received canon of biology, which sees in such structures as these the characteristics of true organic *species*, the in-

---

[1] 'Hierbei nahm der kohlensäure Kalk eine halb-krystallinische Beschaffenheit an, und gestaltete sich unter Aufnahme von Krystallwasser und in Verbindung mit einer geringen Quantität von organischer Substanz zu jenen individuellen, festen Körpern, welche durch die natürliche Züchtung als *Spicula* zur Skeletbildung benützt, und späterhin durch die Wechselwirkung von Anpassung und Verebung im Kampfe ums Dasein auf das Vielfältigste umgebildet und differenziert wurden.' *Die Kalkschwämme* (1872), I, 377; cf. also pp. 482, 483.

[2] *Op. cit.* p. 483. 'Die geordnete, oft so sehr regelmässige und zierliche Zusammensetzung des Skeletsystems ist zum grössten Theile unmittelbares Product der Wasserströmung; die charakteristische Lagerung der Spicula ist von der constanten Richtung des Wasserstroms hervorgebracht; zum kleinsten Theile ist sie die Folge von Anpassungen an untergeordnete äussere Existenzbedingungen.'

dications of blood-relationship and family likeness, and the evidence by which evolutionary descent throughout geologic time may be deduced or deciphered.

Minchin, in a well-known paper,[1] took sides with F. E. Schultze, and gave his reasons for dissenting from such mechanical theories as those of Sollas and of Dreyer. For example, after pointing out that all protoplasm contains a number of 'granules' or microsomes, contained in an alveolar framework and lodged at the nodes of a reticulum, he argued that these also ought to acquire a form such as the spicules possess, if it were the case that these latter owed their form to their similar or identical position. 'If vesicular tension cannot in any other instance cause the granules at the nodes to assume a tetraxon form, why should it do so for the sclerites?' The answer is not far to seek. If the force which the 'mechanical' hypothesis has in view were simply that of *mechanical pressure*, as between solid bodies, then indeed we should expect that *any* substances lying between the impinging spheres would tend to assume the quadriradiate or 'tetraxon' form; but this conclusion does not follow at all, in so far as it is to *surface-energy* that we ascribe the phenomenon. Here the specific nature of the substance makes all the difference. We cannot argue from one substance to another; adsorptive attraction shows its effect on one and not on another; and we have no reason to be surprised if we find that the little granules of protoplasmic material, which as they lie bathed in the more fluid protoplasm have (presumably, and as their shape indicates) a strong surface-tension of their own, behave towards the adjacent vesicles in a very different fashion to the incipient aggregations of calcareous or siliceous matter in a colloid medium. 'The ontogeny of the spicules,' says Professor Minchin, 'points clearly to their regular form being a *phylogenetic adaptation, which has become fixed and handed on by heredity, appearing in the ontogeny as a prophetic adaptation.*' And again, 'The forms of the spicules are the result of adaptation to the requirements of the sponge as a whole, produced by *the action of natural selection upon variation in every direction*'. It would scarcely be possible to illustrate more briefly and more cogently than by these few words (or the similar words of Haeckel quoted on p. 149), the fundamental difference between the Darwinian conception of the causation and determination of Form, and that which is based on, and characteristic of, the physical sciences.

[1] 'Materials for a Monograph of the Ascones', *Quart. J. Micr. Sci.* **40** (1898), 469–587.

Last of all, Dendy took a middle course. While admitting that the majority of sponge-spicules are 'the outcome of conditions which are in large part purely physical', he still saw in them 'a very high taxonomic value', as 'indications of phylogenetic history' all on the ground that 'it seems impossible to account in any other way for the fact that we can actually arrange the different forms in such well-graduated series'. At the same time he believed that 'the vast majority of spicule-characters appear to be non-adaptive', 'that no one form of spicule has, as a rule, any greater survival-value than another', and that 'the natural selection of favourable varieties can have had very little to do with the matter'.[1]

The quest after lines and evidences of descent dominated morphology for many years, and preoccupied the minds of two or three generations of naturalists. We find it easier to see than they did that a graduated or consecutive series of forms may be based on physical causes, that forms mathematically akin may belong to organisms biologically remote, and that, in general, mere formal likeness may be a fallacious guide to evolution in time and to relationship by descent and heredity.

The Skeletons of Radiolaria

If I have dealt comparatively briefly with the inorganic skeletons of sponges, in spite of the interest of the subject from the physical point of view, it has been owing to several reasons. In the first place, though the general trend of the phenomena is clear, it must be admitted that many points are obscure, and could only be discussed at the cost of a long argument. In the second place, the physical theory is too often (as I have shown) in conflict with the accounts given by embryologists of the development of the spicules, and with the current biological theories which their descriptions embody; it is beyond our scope to deal with such descriptions in detail. Lastly, we find ourselves able to illustrate the same physical principles with greater clearness and greater certitude in another group of animals, namely the Radiolaria.

The group of microscopic organisms known as the Radiolaria is extraordinarily rich in diverse forms or 'species'. I do not know how many of such species have been described and defined by

---

[1] Cf. A. Dendy, 'The Tetraxonid Sponge-spicule: a Study in Evolution', *Acta Zool.* (1921), pp. 136, 146, etc. Cf. also 'Bye-products of Organic Evolution', *J. Quekett Micr. Cl.* 12 (1913), 65–82.

naturalists, but some fifty years ago the number was said to be over four thousand, arranged in more than seven hundred genera;[1] of late years there has been a tendency to reduce the number. But apart from the extraordinary multiplicity of forms among the Radiolaria, there are certain features in this multiplicity which arrest our attention. Their distribution in space is curious and vague; many species are found all over the world, or at least every here and there, with no evidence of specific limitations of geographical habitat; some occur in the neighbourhood of the two poles, some are confined to warm and others to cold currents of the ocean. In time their distribution is not less vague: so much so that it has been asserted of them that 'from the Cambrian age downwards, the families and even genera appear identical with those now living'. Lastly, except perhaps in the case of a few large 'colonial forms', we seldom if ever find, as is usual in most animals, a local predominance of one particular species. On the contrary, in a little pinch of deep-sea mud or of some fossil 'radiolarian earth', we shall probably find scores, and it may be even hundreds, of different forms. Moreover, the radiolarian skeletons are of quite extraordinary delicacy and complexity, in spite of their minuteness and the comparative simplicity of the 'unicellular' organisms within which they grow; and these complex conformations have a wonderful and unusual appearance of geometric regularity.

Analogy with Snow-Crystals

All these general considerations seem such as to prepare us for some physical hypothesis of causation. The little skeletons remind us of such things as snow-crystals (themselves almost endless in their diversity), rather than of a collection of animals, constructed in accordance with functional needs and distributed in accordance with their fitness for particular situations. Nevertheless, great efforts have been made to attach 'a biological meaning' to these elaborate structures, and 'to justify the hope that in time their utilitarian character will be more completely recognised'.[2]

As Ernst Haeckel described and figured many hundred 'species' of radiolarian skeletons, so have the physicists depicted snow-crystals

[1] Haeckel, in his *Challenger Monogr.* (1887), p. clxxxviii, estimated the number of known forms at 4314 species, included in 739 genera. Of these, 3508 species were described for the first time in that work.

[2] Cf. Gamble, *Radiolaria* (Lankester's *Treatise on Zoology*), I (1909), 131. Cf. also papers by V. Häcker, in *Jena Z. Naturw.* 39 (1905), 581; *Z. wiss. Zool.* 83 (1905), 336; *Arch. Protistenk.* 9 (1907), 139; etc.

in several thousand different forms.[1] These owe their multitudinous variety to symmetrical repetitions of one simple crystalline form—a beautiful illustration of Plato's *One among the Many*, τὸ ἕν παρα τὰ πολλά. On the other hand, the radiolarian skeleton rings its endless changes on combinations of certain facets, corners and edges within a filmy and bubbly mass. The broad difference between the two is very plain and instructive.

Kepler studied the snowflake with care and insight, though he said that to care for such a trifle was like Socrates measuring the hop of a flea. The first drawings I know are by Dominic Cassini; and if that great astronomer was content with them they show how the physical sciences lagged behind astronomy. They date from the time

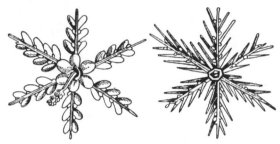

Fig. 53. Snow-crystals, or 'snow-flowers'. From Dominic Cassini (*c.* 1600).

when Maraldi, Cassini's nephew, was studying the bee's cell; and they show once more how very rough his measurements of the honeycomb are bound to have been.

Crystals lie outside the province of this book; yet snow-crystals, and all the rest besides, have much to teach us about the variety, the beauty and the very nature of form. To begin with, the snow-crystal is a regular hexagonal plate or thin prism; that is to say, it shows hexagonal faces above and below, with edges set at co-equal angles of 120°. Ringing her changes on this fundamental form, Nature superadds to the primary hexagon endless combinations of similar plates or prisms, all with identical angles but varying lengths of side; and she repeats, with an exquisite symmetry, about all three axes of the hexagon, whatsoever she may have done for the adornment and elaboration of one. These snow-crystals seem (as Tutton says)

[1] See Kepler, *De nive sexangula* (1611), *Opera*, ed. Fritsch, VII, 715–30; Erasmus Bartholin, *De figura nivis*, *Diss.* (Hafniae, 1661); Dom. Cassini, 'Obs. de la figure de la neige' (Abstr.), *Mém. Acad. R. Sci.* (1666–99), x, 1730; J. C. Wilcke, 'Om de naturliga snö-figurers', *K. V. Akad. Handl.* **22** (1761). Also numerous modern references.

153

to give visible proof of the space-lattice on which their structure is framed.

The beauty of a snow-crystal depends on its mathematical regularity and symmetry; but somehow the association of many variants of a single type, all related but no two the same, vastly increases our pleasure and admiration. Such is the peculiar beauty which a Japanese artist sees in a bed of rushes or a clump of bamboos, especially when the wind's ablowing; and such (as we saw before) is the phase-beauty of a flowering spray when it shows every gradation from opening bud to fading flower.

Fig. 54. Snow-crystals. From Bentley and Humphreys, 1931.

The snow-crystal is further complicated, and its beauty is notably enhanced, by minute occluded bubbles of air or drops of water, whose symmetrical form and arrangement are very curious and not always easy to explain.[1] Lastly, we are apt to see our snow-crystals after a slight thaw has rounded their sharp edges, and has heightened their beauty by softening their contours.

Construction of Radiolaria

In the majority of cases, the skeleton of the Radiolaria is composed, like that of so many sponges, of silica; in one large family, the Acantharia, and perhaps in some others, it is made of a very unusual constituent, namely strontium sulphate.[2] There is no important morphological character in which the shells made of these two constituents differ from one another; and in no case can the chemical properties of these inorganic materials be said to influence the form

[1] We may find some suggestive analogies to these occlusions in Emil Hatschek's paper, 'Gestalt und Orientierung von Gasblasen in Gelen', *Kolloid. Z.* **20** (1914), 226–34.
[2] Bütschli, 'Ueber die chemische Natur der Skeletsubstanz der Acantharia', *Zool. Anz.* **30** (1906), 784.

of the complex skeleton or shell, save only in this general way that, by their hardness, toughness and rigidity, they give rise to a fabric more slender and delicate than we find among calcareous organisms.

A slight exception to this rule is found in the presence of true crystals, which occur within the central capsules of certain Radiolaria, for instance the genus *Collosphaera*.[1] Johannes Müller (whose knowledge and insight never fail to astonish us)[2] remarked that these were identical in form with crystals of celestine, a sulphate of strontium and barium; and Bütschli's discovery of sulphates of strontium and of barium in kindred forms renders it all but certain that they are actually true crystals of celestine.[3]

In its typical form, the radiolarian body consists of a spherical mass of protoplasm, around which, and separated from it by some sort of porous 'capsule', lies a frothy protoplasm, bubbled up into a multitude of alveoli or vacuoles, filled with a fluid which can scarcely differ much from sea-water.[4] According to their surface-tension conditions, these vacuoles may appear more or less isolated and spherical, or joined together in a 'froth' of polyhedral cells; and in the latter, which is the commoner condition, the cells tend to be of equal size, and the resulting polygonal meshwork beautifully regular. In some cases a large number of such simple individual organisms are associated together, forming a floating colony; and it is probable that many others, with whose scattered skeletons we are alone acquainted, had likewise formed part of a colonial organism.

In other words a great part of the body of the Radiolarian, and especially that outer portion to which Haeckel has given the name of the 'calymma', is built up of a mass of 'vesicles', forming a sort of stiff froth, and equivalent in the physical though not necessarily in the biological sense to 'cells', inasmuch as the little vesicles have their own well-defined boundaries, and their own surface phenomena. All that we have said of cell-surfaces and cell-conformations in our discussion of cells and of tissues will apply in like manner, and under

[1] For figures of these crystals see Brandt, *F. u. Fl. d. Golfes von Neapel*, **13**, *Radiolaria* (1885), pl. v. Cf. Johannes Müller, 'Ueber die Thalassicollen, etc.', *Abh. K. Akad. Wiss. Berl.* (1858).

[2] It is interesting to think of the lesser discoveries or inventions, due to men famous for greater things. Johannes Müller first used the tow-net, and Edward Forbes first borrowed the oyster-man's dredge. When we watch a living polyp under the microscope in its tiny aquarium of a glass-cell, we are doing what John Goodsir was the first to do; and the microtome itself was the invention of that best of laboratory-servants, 'Old Stirling', Goodsir's right-hand man.

[3] Celestine, or celestite, is $SrSO_4$ with some BaO replacing SrO.

[4] With the colloid chemists, we may adopt (as Rhumbler has done) the terms *spumoid* or *emulsoid* to denote an agglomeration of fluid-filled vesicles, restricting the name *froth* to such vesicles when filled with air or some other gas.

appropriate conditions, to these. In certain cases, even in so common and so simple a one as the vacuolated substance of an *Actinosphaerium*, we may see a close resemblance, or formal analogy, to a cellular or parenchymatous tissue in the close-packed arrangement and consequent configuration of these vesicles, and even at times in a slight membranous hardening of their walls. Leidy has figured some curious little bodies like small masses of consolidated froth, which seem to be nothing else than the dead and empty husks, or filmy skeletons, of *Actinosphaerium*;[1] and Carnoy has demonstrated

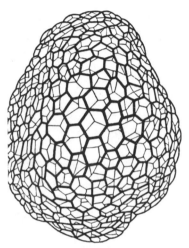

Fig. 55. '*Reticulum plasmatique.*'
After Carnoy.

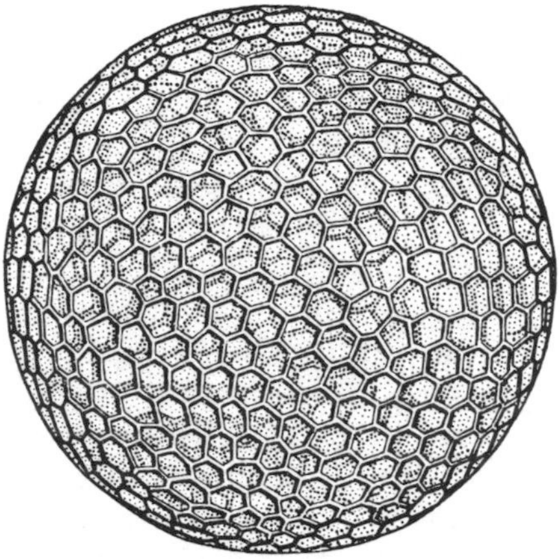

Fig. 56. *Aulonia hexagona* Hkl.

in certain cell-nuclei an all but precisely similar framework, of extreme minuteness and tenuity, formed by adsorption or partial solidification of interstitial matter in a close-packed system of alveoli (Fig. 55).[2] In short, we are again dealing or about to deal with a network or basketwork, whose meshes correspond to the boundary lines between associated cells or vesicles. It is just in those boundary walls or films, still more in their edges or at their corners, that surface-energy will be concentrated and adsorption will be hard at work; and the whole arrangement will follow, or tend to follow, the rules of *areae minimae*—the partition-walls meeting at co-equal angles, three by three in an edge, and their edges meeting four by four in a corner.

[1] J. Leidy, *Fresh-water Rhizopods of North America* (1879), p. 262, pl. xli, figs. 11, 12.
[2] Carnoy, *Biologie Cellulaire*, p. 244, fig. 108; cf. Dreyer, *op. cit.* (1892), fig. 185.

Hexagonal Skeletons

Let us suppose the outer surface of our radiolarian to be covered by a layer of froth-like vesicles, uniform in size or nearly so. We know that their mutual tensions will *tend* to conform them into the fashion of a honeycomb, or regular meshwork of hexagons, and that the free end of each hexagonal prism will be a little spherical cap. Suppose now that it be at the outer surface of the protoplasm (in contact with the surrounding sea-water) that the siliceous particles have a tendency to be secreted or adsorbed; the distribution

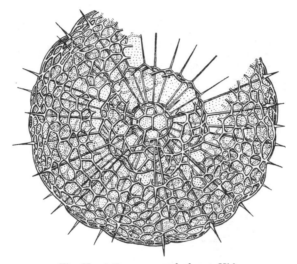

Fig. 57. *Actinomma arcadophorum* Hkl.

of surface-energy will lead them to accumulate in the grooves which separate the vesicles, and the result will be the development of a delicate sphere composed of tiny rods arranged, or apparently arranged, in a hexagonal network after the fashion of Carnoy's *reticulum plasmatique*, only more solid, and still more neat and regular. Just such a spherical basket, looking like the finest imaginable Chinese ivory ball, is found in the siliceous skeleton of *Aulonia*, another of Haeckel's Radiolaria from the 'Challenger'.

But here a strange thing comes to light. *No system of hexagons can enclose space*; whether the hexagons be equal or unequal, regular or irregular, it is still under all circumstances mathematically impossible. So we learn from Euler: the array of hexagons may be extended as far as you please, and over a surface either plane or curved, but

157

it never closes in. Neither our *reticulum plasmatique* nor what seems the very perfection of hexagonal symmetry in *Aulonia* are as we are wont to conceive them; hexagons indeed predominate in both, but a certain number of facets are and must be other than hexagonal. If we look carefully at Carnoy's careful drawing we see that both pentagons and heptagons are shown in his reticulum, and Haeckel actually states, in his brief description of his *Aulonia hexagona*, that a few square or pentagonal facets are to be found among the hexagons.

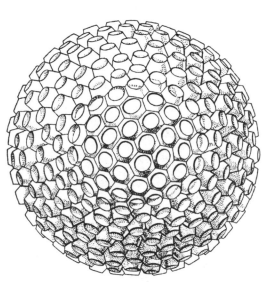

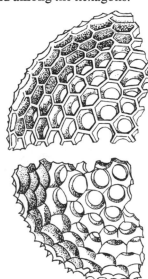

Fig. 58. *Ethmosphaera conosiphonia* Hkl.

Fig. 59. Portions of shells of two 'species' of *Ceno-sphaera*: upper figure, *C. favosa*; lower, *C. ves-paria* Hkl.

Such skeletal conformations are common: and Nature, as in all her handiwork, is quick to ring the changes on the theme. Among its many variants may be found cases (e.g. *Actinomma*) where the vesicles have been less regular in size; and others in which the mesh-work has been developed not on an outer surface only but at succes-sive levels, producing a system of concentric spheres. If the siliceous material be not limited to the linear junctions of the cells but spread over a portion of the outer spherical surfaces or caps, then we shall have the condition represented in Fig. 58 (*Ethmosphaera*), where the shell appears perforated by circular instead of hexagonal apertures and the circular pores are set on slight spheroidal eminences; and, interconnected with such types as this, we have others in which the accumulating pellicles of skeletal matter have extended from the edges into the substances of the boundary walls and have so produced a system of films, normal to the surface of the sphere, constituting a very perfect honeycomb, as in *Cenosphaera favosa* and *vesparia*.[1]

In one or two simple forms, such as the fresh-water *Clathrulina*, just such a spherical perforated shell is produced out of some organic, acanthin-like substance; and in some examples of *Clathrulina* the chitinous lattice-work of the shell is just as regular and delicate, with the meshes for the most part as beautifully hexagonal as in the siliceous shells of the oceanic Radiolaria. This is only another proof (if proof be needed) that the peculiar form and character of these little skeletons are due not to the material of which they are composed, but to the moulding of that material upon an underlying vesicular structure.

Other Skeletal Configurations

Let us next suppose that another and outer layer of cells or vesicles develops upon some such lattice-work as has just been described; and that instead of forming a second hexagonal lattice-work, the skeletal matter tends to be developed normally to the surface of the sphere, that is to say along the *radial* edges where the external vesicles (now compressed into hexagonal prisms) meet one another three by three. The result will be that, if the vesicles be removed, a series of radiating spicules will be left, directed outwards from the angles of the original polyhedron meshwork, all as is seen in Fig. 60. And it may further happen that these radiating skeletal rods branch at their outer ends into divergent rays, forming a triple fork, and corresponding (after the fashion which we have already described as occurring in certain sponge-spicules) to the superficial furrows between the three adjacent cells; this is, as it were, a half-way stage between simple rods or radial spicules and the full completion of another sphere of latticed hexagons. Another possible case, among many, is when the large, uniform vesicles of the outer protoplasm are replaced by smaller vesicles, piled on one another in concentric layers. In this case the radial rods will no longer be straight, but will be bent zig-zag, with their angles in three vertical planes corresponding to the alternate contacts of the successive layers of cells (Fig. 61).

[1] In all these latter cases we recognise a relation to, or extension of, the principle of Plateau's *bourrelet*, or van der Mensbrugghe's *masse annulaire*, or Gibbs's ring, of which we have had much to say.

The solid skeleton is confined, in all these cases, to the boundary-lines, or edges, or grooves between adjacent cells or vesicles, but adsorptive energy may extend throughout the intervening walls. This happens in not a few Radiolaria, and in a certain group called the Nassellaria it produces geometrical forms of peculiar elegance and mathematical beauty.

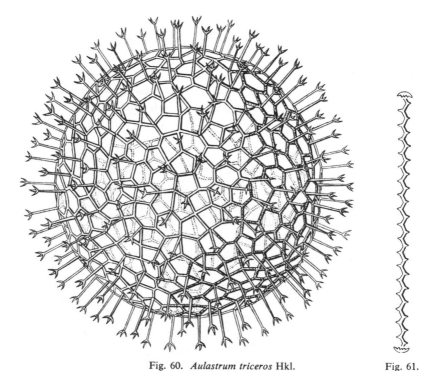

Fig. 60. *Aulastrum triceros* Hkl.   Fig. 61.

When Plateau made the wire framework of a regular tetrahedron and dipped it in soap-solution, he obtained in an instant a beautifully symmetrical system of six films, meeting three by three in four edges, and these four edges running from the corners of the figure to its centre of symmetry. Here they meet, two by two, at the Maraldi angle; and the films meet three by three, to form the re-entrant solid angle which we have called a 'Maraldi pyramid' in our account of the architecture of the honeycomb. The very same configuration is easily recognised in the minute siliceous skeleton of *Callimitra*. There are two discrepancies, neither of which need raise any difficulty. The figure is not a rectilinear but a *spherical tetrahedron*, such as might

160

be formed by the boundary-edges of a tetrahedral cluster of four co-equal bubbles; and just as Plateau extended his experiment by blowing a small bubble in the centre of his tetrahedral system, so we have a central bubble also here.

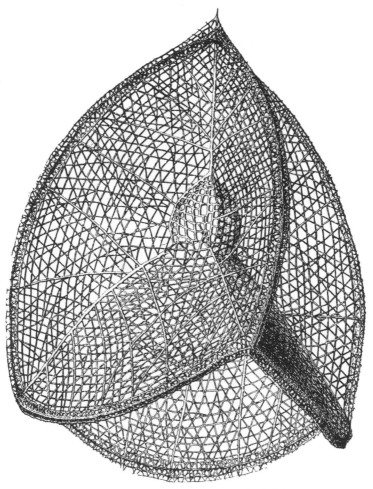

Fig. 62. A Nassellarian skeleton, *Callimitra agnesae* Hkl. (0·15 mm. diameter).

This bubble may be of any size;[1] but its situation (if it be present at all) is always the same, and its shape is always such as to give the Maraldi angles at its own four corners. The tensions of its own walls,

[1] Plateau introduced the central bubble into his cube or tetrahedron by dipping the cage a second time, and so adding an extra face-film; under these circumstances the bubble has a definite magnitude.

and those of the films by which it is supported or slung, all balance one another. Hence the bubble appears in plane projection as a curvilinear equilateral triangle; and we have only got to convert this plane diagram into the corresponding solid to obtain the spherical tetrahedron we have been seeking to explain (Fig. 63).

The geometry of the little inner tetrahedron is not less simple and elegant. Its six edges and four faces are all equal. The films attaching it to the outer skeleton are all planes. Its faces are spherical,

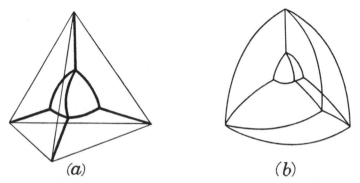

(a) (b)

Fig. 63. Diagrammatic construction of *Callimitra.* (*a*) A bubble suspended within a tetrahedral cage; (*b*) another bubble within a skeleton of the former bubble.

and each has its centre in the opposite corner. The edges are circular arcs, with cosine $\frac{1}{3}$; each is in a plane perpendicular to the chord of the arc opposite, and each has its centre in the middle of that chord. Along each edge the two intersecting spheres meet each other at an angle of $120°$.[1]

This completes the elementary geometry of the figure; but one or two points remain to be considered.

We may notice that the outer edges of the little skeleton are thickened or intensified, and these thickened edges often remain whole or strong while the rest of the surfaces show signs of imperfection or of breaking away; moreover, the four corners of the tetrahedron are not re-entrant (as in a group of bubbles) but a surplus of material forms a little point or cusp at each corner. In all this there is nothing anomalous, and nothing new. For we have already seen that it is at the margins or edges, and *a fortiori* at the corners,

---

[1] For proof, see Lamarle, *op. cit.* pp. 6–8. Lamarle showed that the sphere can be so divided in seven ways, but of these seven figures the tetrahedron alone is stable. The other six are the cube and the regular dodecahedron; prisms, triangular and pentagonal, with equilateral base and a certain ratio of base to height; and two polyhedra constructed of pentagons and quadrilaterals.

that the surface-energy reaches its maximum—with the double effect of accumulating protoplasmic material in the form of a Gibbs's ring or bourrelet, and of intensifying along the same lines the adsorptive secretion of skeletal matter. In some other tetrahedral systems analogous to *Callimitra*, the whole of the skeletal matter is concentrated along the boundary-edges, and none left to spread over the boundary-planes or interfaces: just as among our spherical Radiolaria it was at the boundary-edges of their many cells or vesicles, and often there alone, that skeletal formation occurred, and gave rise to the spherical skeleton and its meshwork of hexagons. In the beautiful form which Haeckel calls *Archiscenium* the boundary edges disappear, the four edges converging on the median point are intensified,

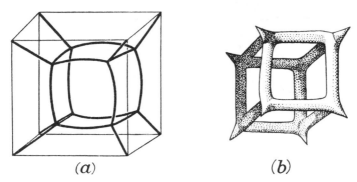

(a)   *(b)*

Fig. 64. (*a*) Bubble suspended within a cubical cage;
(*b*) *Lithocubus geometricus* Hkl.

and only three of the six convergent facets are retained; but, much as the two differ in appearance, the geometry of this and of *Callimitra* remain essentially the same.

We learned also from Plateau that, just as a tetrahedral bubble can be inserted within the tetrahedral skeleton or cage, so may a cubical bubble be introduced within a cubical cage; and the edges of the inner cube will be just so curved as to give the Maraldi angles at the corners. We find among Haeckel's Radiolaria one (he calls it *Lithocubus geometricus*) which precisely corresponds to the skeleton of this inner cubical bubble; and the little spokes or spikes which project from the corners are parts of the edges which once joined the corners of the enclosing figure to those of the bubble within (Fig. 64).

Again, if we construct a cage in the form of an equilateral triangular prism, and proceed as before, we shall probably see a vertical edge in the centre of the prism connecting two nodes near either end,

163

in each of which the Maraldi figure is displayed. But if we gradually shorten our prism there comes a point where the two nodes disappear, a plane curvilinear triangle appears horizontally in the middle of the figure, and at each of its three corners four curved edges meet at the familiar angle. Here again we may insert a central bubble, which will now take the form of a curvilinear equilateral triangular prism; and Haeckel's *Prismatium tripodium* repeats this configuration (Fig. 65).

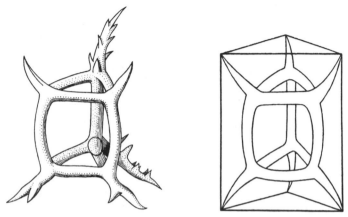

Fig. 65. *Prismatium tripodium* Hkl. From P. Grassé, *Traité de Zoologie* (Paris: Masson et Cie).

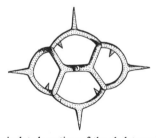

Fig. 66. An isolated portion of the skeleton of *Dictyocha*.

In a framework of two crossed rectangles, we may insert one bubble after another, producing a chain of superposed vesicles whose shapes vary as we alter the relative positions of the rectangular frames. Various species of *Triolampas, Theocyrtis*, etc. are more or less akin to these complicated figures of equilibrium. A very beautiful series of forms may be made by introducing successive bubbles within the film-system formed by a tetrahedron or a parallelepipedon. The shape and the curvature of the bubbles and of their suspensory films become extremely beautiful, and we have certain of them reproduced

164

unmistakably in various Nassellarian genera, such as *Podocyrtis* and its allies.

In Fig. 66 we see a curious little skeletal structure or complex spicule, whose conformation is easily accounted for. Isolated spicules such as this form the skeleton in the genus *Dictyocha*, and occur scattered over the spherical surface of the organism (Fig. 67). The

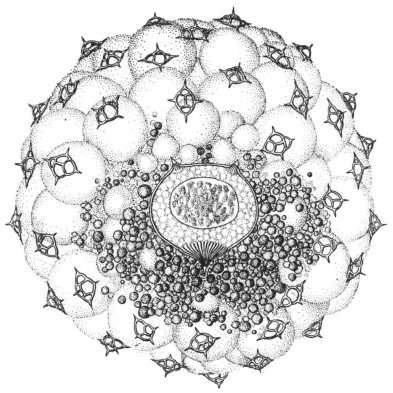

Fig. 67. *Dictyocha stapedia* Hkl. From K. G. Grell, *Protozoologie* (Berlin: Springer-Verlag).

basket-shaped spicule has evidently been developed about a cluster of four cells or vesicles, lying in or on the surface of the organism, and therefore arranged, not in the three-dimensional, tetrahedral form of *Callimitra*, but in the manner in which four contiguous cells lying side by side in one plane normally set themselves, like the four cells of a segmenting egg: that is to say with an intervening 'polar furrow', whose ends mark the meeting place, at equal angles, of the four cells in groups of three. The little projecting spokes, or

165

spikes, which are set normally to the main basketwork, seem to be uncompleted portions of a larger basket, corresponding to a more numerous aggregation of cells. Similar but more complex formations, all explicable as basket-like frameworks developed around a cluster of cells, and adsorbed or secreted in the grooves common to adjacent cells or bubbles, are found in great variety.

The *Dictyocha*-spicule, laid down as a siliceous framework in the grooves between a few clustered cells, is too simple and natural to be confined to one group of animals. We have already seen it, as a calcareous spicule, in the holothurian genus *Thyone*, and we may find it again, in many various forms, in the protozoan group known as the Silicoflagellata (Fig. 68).[1] Nothing can better illustrate the

Fig. 68. Various species of *Distephanus* (Silicoflagellata).
From Deflandre, after Ehrenberg.

physico-mathematical character of these configurations than their common occurrence in diverse groups of organisms. And the simple fact is, that we seem to know less and less of these things on the biological side, the more we come to understand their physical and mathematical characters. I have lost faith in Haeckel's four thousand 'species' of Radiolaria.

In the foregoing examples of Radiolaria, the symmetry which the organism displays seems identical with that symmetry of forces which results from the play and interplay of surface-tensions in the whole system: this symmetry being displayed, in one class of cases, in a more or less spherical mass of froth, and in another class in a simpler aggregation of a few, otherwise isolated, vesicles. In either

[1] See (*int. al.*) G. Deflandre, 'Les Silicoflagellés, etc.', *Bull. Soc. Fr. Micr.* **1** (1932), 1; the figures in which article are mostly drawn from Ehrenberg's *Mikrogeologie* (1854).

case skeletons are formed, in great variety, by one and the same kind of surface-action, namely by the adsorptive deposition of silica in walls and edges, corresponding to the manifold surfaces and interfaces of the system. But among the vast number of known Radiolaria, there are certain forms (especially among the Phaeodaria and Acantharia) which display a no less remarkable symmetry, the origin of which is by no means clear, though surface-tension may play a part in its causation. Even this is doubtful; for the fact that three-way nodes are no longer to be seen at the junctions of the cells suggests that another law than that of minimal areas has been in action here. They are cases in which (as in some of those already described) the skeleton consists (1) of radiating spicular rods, definite in number and position, and (2) of interconnecting rods or plates, tangential to the more or less spherical body of the organism, whose form becomes, accordingly, that of a geometric, polyhedral solid. The great regularity, the numerical symmetries and the apparent simplicity of these latter forms makes of them a class apart, and suggests problems which have not been solved or even investigated.

The matter is partially illustrated by the accompanying figures (Fig. 69) from Haeckel's *Monograph of the Challenger Radiolaria*.[1] In one of these we see a regular octahedron, in another a regular, or pentagonal, dodecahedron, in a third a regular icosahedron. In all cases the figure appears to be perfectly symmetrical, though neither the triangular facets of the octahedron and icosahedron, nor the pentagonal facets of the dodecahedron, are necessarily plane surfaces. In all of these cases, the radial spicules correspond to the corners of the figure; and they are, accordingly, six in number in the octahedron, twenty in the dodecahedron, and twelve in the icosahedron. If we add to these three figures the regular tetrahedron which we have just been studying, and the cube (which is represented, at least in outline, in the skeleton of the hexactinellid sponges), we have completed the series of the five regular polyhedra known to geometers, the *Platonic bodies*[2] of the older mathematicians. It is at first sight all the more remarkable that we should here meet with the whole five regular polyhedra, when we remember that, among the vast variety of crystalline forms known among minerals, the regular dodecahedron and icosahedron, simple as they are from the mathematical point of view, never occur. Not only do these latter

---

[1] Of the many thousand figures in the hundred and forty plates of this beautifully illustrated book, there is scarcely one which does not depict some subtle and elegant *geometrical* configuration.

[2] They were known long before Plato: Πλάτων δὲ καὶ ἐν τούτοις πυθαγορίζει.

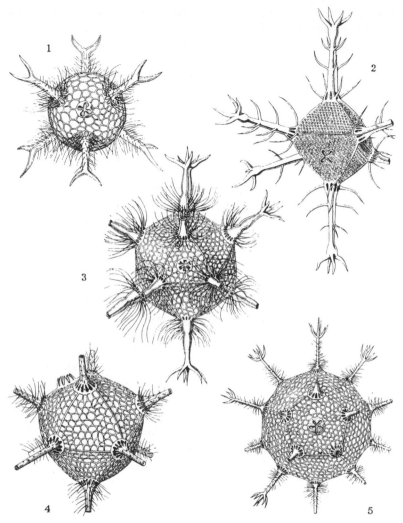

Fig. 69. Skeletons of various Radiolarians, after Haeckel. (1) *Circoporus sexfurcus*; (2) *C. octahedrus*; (3) *Circogonia icosahedra*; (4) *Circospathis novena*; (5) *Circorrhegma dodecahedra*.

never occur in crystallography, but (as is explained in textbooks of that science) it has been shown that they cannot occur, owing to the fact that their indices (or numbers expressing the relation of the faces to the three primary axes) involve an irrational quantity: whereas it is a fundamental law of crystallography, involved in the whole theory of space-partitioning, that 'the indices of any and every face of a crystal are small whole numbers.'[1] At the same time, an imperfect pentagonal dodecahedron, whose pentagonal sides are non-equilateral, is common among crystals. If we may safely judge from Haeckel's figures, the pentagonal dodecahedron of the Radiolarian (*Circorrhegma*) is perfectly regular, and we may rest assured, accordingly, that it is not brought about by principles of space-partitioning similar to those which manifest themselves in the phenomenon of crystallisation. It will be observed that in all these radiolarian poly-hedral shells, the surface of each external facet is formed of a minute hexagonal network, whose probable origin, in relation to a vesicular structure, is such as we have already discussed.

Polarity and Liquid Crystals

In many cases it is the arrangement of the axial rods—the 'polar symmetry' of the entire organism—which lies at the root of the matter; and which, if only we could account for it, would make it comparatively easy to explain the superficial configuration. But there are no obvious mechanical forces by which we can so explain this peculiar polarity. This at least is evident, that it arises in the central mass of protoplasm, which is the essential living portion of the organism as distinguished from that frothy peripheral mass whose structure has helped us to explain so many phenomena of the super-ficial or external skeleton. To say that the arrangement depends upon a specific polarisation of the cell is merely to refer the problem to other terms, and to set it aside for future solution. But it is possible that we may learn something about the lines in which *to seek for* such a solution by considering the case of Lehmann's 'fluid crystals', and the light which they throw upon the phenomena of molecular aggregation.

The phenomenon of 'fluid crystallisation' is found in a number of

[1] If the equation of any plane face of a crystal be written in the form $hx + ky + lz = 1$, then h, k, l are the indices of which we are speaking. They are the reciprocals of the parameters, or reciprocals of the distances from the origin at which the plane meets the several axes. In the case of the regular or pentagonal dodecahedron these indices are $2, 1 + \sqrt{5}, 0$. Kepler described as follows, briefly but adequately, the common characteristics of the dodecahedron and icosahedron: 'Duo sunt corpora regularia, dodeca-edron et icosaedron, quorum illud quinquangulis figuratur expresse, hoc triangulis quidem sed in quinquanguli formam coaptatis. Utriusque horum corporum ipsiusque adeo quinquanguli *structura perfici non potest sine proportione illa, quam hodierni geometrae divinam appellant*' (*De nive sexangula* (1611), Opera, ed. Fritsch, VII, 723). Here Kepler was dealing, somewhat after the manner of Sir Thomas Browne, with the mysteries of the quincunx, and also of the hexagon; and was seeking for an explanation of the mysterious or even mystical beauty of the 5-petalled or 3-petalled flower—*pulchritudinis aut proprietatis figurae, quae animam harum plantarum characterisavit.*

chemical bodies; it is exhibited at a specific temperature for each substance; and it would seem to be limited to bodies in which there is an elongated, or 'long-chain' arrangement of the atoms in the molecule. Such bodies, at the appropriate temperature, tend to aggregate themselves into masses, which are sometimes spherical drops or globules (the so-called 'spherulites'), and sometimes have the definite form of needle-like or prismatic crystals. In either case they remain liquid, and are also doubly refractive, polarising light in brilliant colours. Together with them are formed ordinary solid crystals, also with characteristic polarisation, and into such solid crystals all the fluid material ultimately turns. It seems that in these liquid crystals, though the molecules are freely mobile, just as are those of water, they are yet subject to, or endowed with, a 'directive force', a force which confers upon them a definite configuration or 'polarity', the 'Gestaltungskraft' of Lehmann.

Such an hypothesis as this has been gradually extruded from the theories of mathematical crystallography;[1] and it has come to be understood that the symmetrical conformation of a homogeneous crystalline structure is sufficiently explained by the mere mechanical fitting together of appropriate structural units along the easiest and simplest lines of 'close packing': just as a pile of oranges becomes definite, both in outward form and inward structural arrangement, without the play of any *specific* directive force. But while our conceptions of the tactical arrangement of crystalline molecules remain the same as before, and our hypotheses of 'modes of packing' or of 'space-lattices' remain as useful and as adequate as ever for the definition and explanation of the molecular arrangements, a new conception is introduced when we find something like such space-lattices maintained in what has hitherto been considered the molecular freedom of a liquid field; and Lehmann would persuade us, accordingly, to postulate a specific molecular force, or 'Gestaltungskraft' (not unlike Kepler's 'facultas formatrix'), to account for the phenomenon.[2]

Now just as some sort of specific 'Gestaltungskraft' had been of old the *deus ex machina* accounting for all crystalline phenomena (*gnara totius geometriae, et in ea exercita*, as Kepler said), and as

[1] Cf. Tutton, *Crystallography* (1911), p. 932.
[2] Kepler, if I understand him aright, saw his way to account for the shape of the bee's cell or the pomegranate-seed; and it was for want of any such mechanical explanation, and as little more than a confession of ignorance, that he fell back on a *facultas formatrix* to account for the six rays of the snow-crystal or the five petals of the flower. He was equally ready, unfortunately, to explain, by the same *facultas formatrix in aere*, the appearance of a plague of locusts or a swarm of flies.

such an hypothesis, after being dethroned and repudiated, has now fought its way back and claims a right to be heard, so it may be also in biology. We begin by an easy and general assumption of *specific properties*, by which each organism assumes its own specific form; we learn later (as it is the purpose of this book to show) that throughout the whole range of organic morphology there are innumerable phenomena of form which are not peculiar to living things, but which are more or less simple manifestations of ordinary physical law. But every now and then we come to deep-seated signs of protoplasmic symmetry or polarisation, which *seem* to lie beyond the reach of the ordinary physical forces. It by no means follows that the forces in question are not essentially physical forces, more obscure and less familiar to us than the rest; and this would seem to be a great part of the lesson for us to draw from Lehmann's beautiful discovery. For Lehmann claims to have demonstrated, in non-living, chemical bodies, the existence of just such a determinant, just such a 'Gestaltungskraft', as would be of infinite help to us if we might postulate it for the explanation (for instance) of our Radiolarian's axial symmetry. Further than this we cannot go; such analogy as we seem to see in the Lehmann phenomenon soon evades us, and refuses to be pressed home. The symmetry of crystallisation, which Haeckel tried hard to discover and to reveal in these and other organisms, resolves itself into remote analogies from which no conclusions can be drawn. Many a beautiful protozoan form has lent itself to easy physicomathematical explanation; others, no less simple and no more beautiful, prove harder to explain. That Nature keeps some of her secrets longer than others—that she tells the secret of the rainbow and hides that of the northern lights—is a lesson taught me when I was a boy.

CHAPTER VI

THE EQUIANGULAR SPIRAL

Spirals in Nature

The very numerous examples of spiral conformation which we meet with in our studies of organic form are peculiarly adapted to mathematical methods of investigation. But ere we begin to study them we must take care to define our terms, and we had better also attempt some rough preliminary classification of the objects with which we shall have to deal.

In general terms, a Spiral is a curve which, starting from a point of origin, continually diminishes in curvature as it recedes from that point; or, in other words, whose *radius of curvature* continually increases. This definition is wide enough to include a number of different curves, but on the other hand it excludes at least one which in popular speech we are apt to confuse with a true spiral. This latter curve is the simple *screw*, or cylindrical *helix*, which curve neither starts from a definite origin nor changes its curvature as it proceeds. The 'spiral' thickening of a woody plant-cell, the 'spiral' thread within an insect's tracheal tube, or the 'spiral' twist and twine of a climbing stem are not, mathematically speaking, *spirals* at all, but *screws* or *helices*. They belong to a distinct, though not very remote, family of curves.

Of true organic spirals we have no lack.[1] We think at once of horns of ruminants, and of still more exquisitely beautiful molluscan shells —in which (as Pliny says) *magna ludentis Naturae varietas*. Closely related spirals may be traced in the florets of a sunflower; a true spiral, though not, by the way, so easy of investigation, is seen in the outline of a cordiform leaf; and yet again, we can recognise typical though transitory spirals in a lock of hair, in a staple of wool,[2] in the coil of an elephant's trunk, in the 'circling spires' of a snake, in the coils of a cuttle-fish's arm, or of a monkey's or a chameleon's tail.

---

[1] A great number of spiral forms, both organic and artificial, are described and beautifully illustrated in Sir T. A. Cook's *Spirals in Nature and Art* (1903) and *Curves of Life* (1914).

[2] On this interesting case see, for example, J. E. Duerden, in *Science* (25 May 1934).

Among such forms as these, and the many others which we might easily add to them, it is obvious that we have to do with things which, though mathematically similar, are biologically speaking fundamentally different; and not only are they biologically remote, but they are also physically different, in regard to the causes to which they are severally due. For in the first place, the spiral coil of the elephant's trunk or of the chameleon's tail is, as we have said, but a transitory configuration, and is plainly the result of certain muscular forces acting upon a structure of a definite, and normally an essentially different, form. It is rather a position, or an *attitude*, than a *form*, in

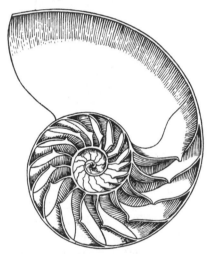

Fig. 70. The shell of *Nautilus pompilius*. From J. C. Chenu.

the sense in which we have been using this latter term; and, unlike most of the forms which we have been studying, it has little or no direct relation to the phenomenon of growth.

Again, there is a difference between such a spiral conformation as is built up by the separate and successive florets in the sunflower, and that which, in the snail or *Nautilus* shell, is apparently a single and indivisible unit. And a similar if not identical difference is apparent between the *Nautilus* shell and the minute shells of the Foraminifera which so closely simulate it: inasmuch as the spiral shells of these latter are composite structures, combined out of successive and separate chambers, while the molluscan shell, though it may (as in *Nautilus*) become secondarily subdivided, has grown as one continuous tube. It follows from all this that there cannot be a

173

physical or dynamical, though there may well be a mathematical *law of growth*, which is common to, and which defines, the spiral form in *Nautilus*, in *Globigerina*, in the ram's horn, and in the inflorescence of the sunflower. Nature at least exhibits in them all '*un reflet des formes rigoureuses qu'étudie la géometrie*'.[1]

Of the spiral forms which we have now mentioned, every one (with the single exception of the cordate outline of the leaf) is an example of the remarkable curve known as the equiangular or logarithmic spiral. But before we enter upon the mathematics of the equiangular spiral, let us carefully observe that the whole of the organic forms in which it is clearly and permanently exhibited, however different they may be from one another in outward appearance, in nature and

Fig. 71. Three views of a foraminiferal shell (*Trochamma inflata*). After Brady. From P. Grassé, *Traité de Zoologie* (Paris: Masson et Cie).

in origin, nevertheless all belong, in a certain sense, to one particular class of conformations. In the great majority of cases, when we consider an organism in part or whole, when we look (for instance) at our own hand or foot, or contemplate an insect or a worm, we have no reason (or very little) to consider one part of the existing structure as *older* than another; through and through, the newer particles have been merged and commingled among the old; the outline, such as it is, is due to forces which for the most part are still at work to shape it, and which in shaping it have shaped it as a whole. But the horn, or the snail-shell, is curiously different; for in these the presently existing structure is, so to speak, partly old and partly new. It has been conformed by successive and continuous increments; and each successive stage of growth, starting from the origin, remains as an integral and unchanging portion of the growing structure.

We may go further, and see that horn and shell, though they belong to the living, are in no sense alive.[2] They are by-products of the

[1] Haton de la Goupillière, in the introduction to his important study of the *Surfaces Nautiloides, Annaes sci. da Acad. Polytechnica do Porto*, Coimbra, III, 1908.
[2] For Oken and Goodsir the logarithmic spiral had a profound significance, for they saw in it a manifestation of life itself. For a like reason Sir Theodore Cook spoke of the

animal; they consist of 'formed material', as it is sometimes called; their growth is not of their own doing, but comes of living cells beneath them or around. The many structures which display the logarithmic spiral increase, or accumulate, rather than grow. The shell of nautilus or snail, the chambered shell of a foraminifer, the elephant's tusk, the beaver's tooth, the cat's claws or the canary-bird's—all these show the same simple and very beautiful spiral curve. And all alike consist of stuff secreted or deposited by living cells; all grow, as an edifice grows, by accretion of accumulated material; and in all alike the parts once formed remain in being, and are thenceforward incapable of change.

In a slightly different, but closely cognate way, the same is true of the spirally arranged florets of the sunflower. For here again we are regarding serially arranged portions of a composite structure, which portions, similar to one another in form, *differ in age*; and differ also in magnitude in the strict ratio of their age. Somehow or other, in the equiangular spiral the *time-element* always enters in; and to this important fact, full of curious biological as well as mathematical significance, we shall afterwards return.

The Spiral of Archimedes

In the elementary mathematics of a spiral, we speak of the point of origin as the pole (O); a straight line having its extremity in the pole, and revolving about it, is called the radius vector; and a point (P), travelling along the radius vector under definite conditions of velocity, will then describe our spiral curve.

Of several mathematical curves whose form and development may be so conceived, the two most important (and the only two with which we need deal) are those which are known as (1) the equable spiral, or spiral of Archimedes, and (2) the equiangular or logarithmic spiral.

The former may be roughly illustrated by the way a sailor coils a rope upon the deck; as the rope is of uniform thickness, so in the whole spiral coil is each whorl of the same breadth as that which precedes and as that which follows it. Using its ancient definition, we may define it by saying, that 'If a straight line revolve uniformly about its extremity, a point which likewise travels uniformly along it will

Curves of Life; and Alfred Lartigues says (in his *Biodynamique générale*, 1930, p. 60): 'Nous verrons la Conchyliologie apporter une magnifique contribution à la Stéréo-dynamique du tourbillon vital.' The fact that the spiral is always formed of non-living matter helps to contradict these mystical conceptions.

175

describe the equable spiral'.[1] Or, putting the same thing into our more modern words, 'If, while the radius vector revolve uniformly about the pole, a point (P) travel with uniform velocity along it, the curve described will be that called the equable spiral, or spiral of Archimedes'. It is plain that the spiral of Archimedes may be compared, but again roughly, to a *cylinder* coiled up. It is plain also that a radius ($r = OP$), made up of the successive and equal whorls, will increase in *arithmetical* progression: and will equal a certain constant quantity (*a*) multiplied by the whole number of whorls or (more strictly speaking) multiplied by the whole angle (*θ*) through which it has revolved: so that $r = a\theta$. And it is also plain that the radius meets the curve (or its tangent) at an angle which changes slowly but continuously, and which tends towards a right angle as the whorls increase in number and become more and more nearly circular.

The equiangular Spiral

But, in contrast to this, in the equiangular spiral of the *Nautilus* or the snail-shell or *Globigerina*, the whorls continually increase in breadth, and do so in a steady and unchanging ratio. Our definition is as follows: 'If, instead of travelling with a *uniform* velocity, our point move along the radius vector with a velocity *increasing as its distance from the pole*, then the path described is called an equiangular spiral.' Each whorl which the radius vector intersects will be broader than its predecessor in a definite ratio; the radius vector will increase in length in *geometrical* progression, as it sweeps through successive equal angles; and the equation to the spiral will be $r = a^{\theta}$. As the spiral of Archimedes, in our example of the coiled rope, might be looked upon as a coiled cylinder, so (but equally roughly) may the equiangular spiral, in the case of the shell, be pictured as a *cone* coiled upon itself; and it is the conical shape of the elephant's trunk or the chameleon's tail which makes them coil into a rough simulacrum of an equiangular spiral.

While the one spiral was known in ancient times, and was investigated if not discovered by Archimedes, the other was first recognised by Descartes, and discussed in the year 1638 in his letters to Mersenne.[2] Starting with the conception of a growing curve which should cut each radius vector at a constant angle—just as a circle does— Descartes showed how it would necessarily follow that radii at equal

---

[1] Leslie's *Geometry of Curved Lines* (1821), p. 417. This is practically identical with Archimedes' own definition (ed. Torelli, p. 219); cf. Cantor, *Geschichte der Mathematik* (1880), ɪ, 262.   [2] *Œuvres*, ed. Adam et Tannery (Paris, 1898), p. 360.

angles to one another at the pole would be in continued proportion; that the same is therefore true of the parts cut off from a common radius vector by successive whorls or convolutions of the spire; and furthermore, that distances measured along the curve from its origin, and intercepted by any radii, as at *B, C*, are proportional to the lengths

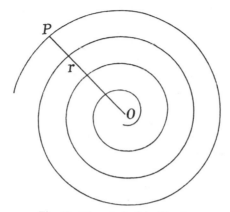

Fig. 72. The spiral of Archimedes.

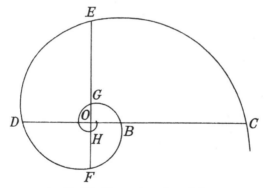

Fig 73. The equiangular spiral.

of these radii, *OB, OC*. It follows that the sectors cut off by successive radii, at equal vectorial angles, are similar to one another in every respect; and it further follows that the figure may be conceived as growing continuously without ever changing its shape the while.

The many specific properties of the equiangular spiral are so inter-related to one another that we may choose pretty well any one of them as the basis of our definition, and deduce the others from it either by analytical methods or by elementary geometry. In algebra,

177

when $m^x = n$, x is called the logarithm of n to the base m. Hence, in this instance, the equation $r = a$ may be written in the form $\log r = \theta \log a$, or $\theta = \log r / \log a$, or (since a is a constant) $\theta = k \log r$.[1] Which is as much as to say that (as Descartes discovered) the vector angles about the pole are proportional to the logarithms of the successive radii; from which circumstance the alternative name of the 'logarithmic spiral' is derived.

Moreover, for as many properties as the curve exhibits, so many names may it more or less appropriately receive. James Bernoulli called it the logarithmic spiral, as we still often do; P. Nicolas called it the geometrical spiral, because radii at equal polar angles are in geometrical progression; Halley, the proportional spiral, because the parts of a radius cut off by successive whorls are in continued proportion; and lastly, Roger Cotes, going back to Descartes' first description or first definition of all, called it the equiangular spiral.[2] We may also recall Newton's remarkable demonstration that, had the force of gravity varied inversely as the *cube* instead of the *square* of the distance, the planets, instead of being bound to their ellipses, would have been shot off in spiral orbits from the sun, the equiangular spiral being one case thereof.[3]

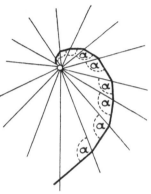

Fig. 74. Spiral path of an insect, as it draws towards a light. From Wigglesworth (after van Buddenbroek).

A singular instance of the same spiral is given by the route which certain insects follow towards a candle. Owing to the structure of their compound eyes, these insects do not look straight ahead but make for a light which they see abeam, at a certain angle. As they continually adjust their path to this constant angle, a spiral pathway brings them to their destination at last.[4]

[1] Instead of $r = a^\theta$, we might write $r = r_0 a^\theta$; in which case r_0 is the value of r for zero value of θ.

[2] James Bernoulli, in *Acta Eruditorum* (1691), p. 282; P. Nicolas, *De novis spiralibus* (Tolosae, 1693), p. 27; E. Halley, *Phil. Trans.* **19** (1696), 58; Roger Cotes, *ibid.* (1714), and *Harmonia Mensurarum* (1722), p. 19. For the further history of the curve see (e.g.) Gomes de Teixeira, *Traité des courbes remarquables* (Coimbra, 1909), pp. 76–86; Gino Loria, *Spezielle algebräische Kurven* (1911), II, 60 seq.; R. C. Archibald (to whom I am much indebted) in *Amer. Math. Mon.* **25** (1918), 189–93, and in Jay Hambidge's *Dynamic Symmetry* (1920), pp. 146–57.

[3] *Principia*, I, 9; II, 15. On these 'Cotes's spirals' see Tait and Steele, p. 147.

[4] Cf. W. Buddenbroek, *Sitzungsber. Heidelb. Akad.* (1917); V. B. Wigglesworth, *The Principles of Insect Physiology* (1939), p. 167.

In mechanical structures, *curvature* is essentially a mechanical phenomenon. It is found in flexible structures as the result of *bending*, or it may be introduced into the construction for the purpose of resisting such a bending-moment. But neither shell nor tooth nor claw are flexible structures; they have not been *bent* into their peculiar curvature, they have *grown* into it.

In the growth of a shell, we can conceive no simpler law than this, namely, that it shall widen and lengthen in the same unvarying proportions: and this simplest of laws is that which Nature tends to follow. The shell, like the creature within it, grows in size *but does not change its shape*; and the existence of this constant relativity of growth, or constant similarity of form, is of the essence, and may be made the basis of a definition, of the equiangular spiral.[1]

Such a definition, though not commonly used by mathematicians, has been occasionally employed; and it is one from which the other properties of the curve can be deduced with great ease and simplicity. In mathematical language it would run as follows: 'Any [plane] curve proceeding from a fixed point (which is called the pole), and such that the arc intercepted between any two radii at a given angle to one another is always similar to itself, is called an equiangular, or logarithmic, spiral.'

In this definition, we have the most fundamental and 'intrinsic' property of the curve, namely the property of continual similarity, and the very property by reason of which it is associated with organic growth in such structures as the horn or the shell. For it is peculiarly characteristic of the spiral shell, for instance, that it does not alter as it grows; each increment is similar to its predecessor, and the whole, after every spurt of growth, is just like what it was before. We feel no surprise when the animal which secretes the shell, or any other animal whatsoever, grows by such symmetrical expansion as to preserve its form unchanged; though even there, as we have already seen, the unchanging form denotes a nice balance between the rates of growth in various directions, which is but seldom accurately maintained for long. But the shell retains its unchanging form in spite of its *asymmetrical* growth; it grows at one end only, and so does the horn. And this remarkable property of increasing by *terminal* growth, but nevertheless retaining unchanged the form of the entire figure, is characteristic of the equiangular spiral, and of no

[1] See an interesting paper by W. A. Whitworth, 'The Equiangular Spiral, its Chief Properties Proved Geometrically', *Messenger of Mathematics* (1), **1** (1862), 5. The celebrated Christian Wiener gave an explanation on these lines of the logarithmic spiral of the shell, in his highly original *Grundzüge der Weltordnung* (1863)

other mathematical curve. It well deserves the name, by which James Bernoulli was wont to call it, of *spira mirabilis*.

We may at once illustrate this curious phenomenon by drawing the outline of a little *Nautilus* shell within a big one. We know, or we may see at once, that they are of precisely the same shape; so that, if we look at the little shell through a magnifying glass, it becomes identical with the big one. But we know, on the other hand, that the little *Nautilus* shell grows into the big one, not by growth or magnification in all parts and directions, as when the boy grows into the man, but by growing *at one end only*.[1]

Though of all plane curves, this property of continued similarity is found only in the equiangular spiral, there are many rectilinear figures in which it may be shown. For instance, it holds good of any cone; for evidently, in Fig. 75, the little inner cone (represented in its triangular section) may become identical with the larger one either by magnification all round (as in *a*), or by an increment at one end (as in *b*); or for that matter on the rest of its surface, represented by the other two sides (as in *c*). All this is associated with the fact, which we have already noted, that the *Nautilus* shell is but a cone rolled up; that, in other words, the cone is but a particular variety, or 'limiting case', of the spiral shell.

[1] If we should want further proof or illustration of the fact that the spiral shell remains of the same shape while increasing in magnitude by its terminal growth, we may find it by help of our ratio $W:L^3$, which remains constant so long as the shape remains unchanged. Here are weights and measurements of a series of small land-shells (*Clausilia*):

W (mg)	L (mm)	$\sqrt[3]{W}/L$
50	14·4	2·56
53	15·1	2·49
56	15·2	2·52
56	15·2	2·52
56	15·4	2·44
58	15·5	2·50
61	16·4	2·40
63	16·0	2·49
67	16·0	2·54
69	16·1	2·56
	Mean	2·50

In 100 specimens of *Clausilia* the mean value of $\sqrt[3]{W}/L$ was found to be 2·517, the coefficient of variation 0·092, and the standard deviation 3·6. That is to say, over 90 per cent grouped themselves about a mean value of 2·5 with a deviation of less than 4 per cent. Cf. C. Petersen, *Das Quotientengesetz* (1921), p. 55.

Gnomons

This singular property of continued similarity, which we see in the cone, and recognise as characteristic of the logarithmic spiral, would seem, under a more general aspect, to have engaged the particular attention of ancient mathematicians even from the days of Pythagoras, and so, with little doubt, from the still more ancient days of that Egyptian school whence he derived the foundations of his learning;[1] and its bearing on our biological problem of the shell, however indirect, is close enough to deserve our very careful consideration.

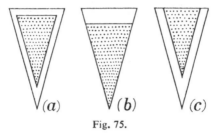

Fig. 75.

There are certain things, says Aristotle, which suffer no alteration (save of magnitude) when they grow.[2] Thus if we add to a square an L-shaped portion, shaped like a carpenter's square, the resulting figure is still a square; and the portion which we have so added, with this singular result, is called in Greek a 'gnomon'.

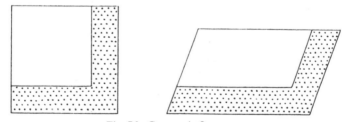

Fig. 76. Gnomonic figures.

Euclid extends the term to include the case of any parallelogram,[3] whether rectangular or not (Fig. 76); and Hero of Alexandria specifically defines a gnomon (as indeed Aristotle had implicitly

---

[1] I am well aware that the debt of Greek science to Egypt and the east is vigorously denied by many scholars, some of whom go so far as to believe that the Egyptians never had any science, save only some 'rough rules of thumb for measuring fields and pyramids' (Burnet's *Greek Philosophy*, 1914, p. 5).

[2] *Categ.* 14, 15 *a*, 30: ἔστι τινὰ αὐξανόμενα ἃ οὐκ ἀλλοιοῦται, οἷον τὸ τετράγωνον, γνώμονος περιτεθέντος, ηὔξηται μὲν ἀλλοιότερον δὲ οὐδὲν γεγένηται.

[3] Euclid (II, def. 2).

defined it), as any figure which, being added to any figure whatsoever, leaves the resultant figure similar to the original. Included in this important definition is the case of numbers, considered geometrically; that is to say, the εἰδητικοὶ ἀριθμοί, which can be translated into *form*, by means of rows of dots or other signs (cf. Arist. *Metaph.* 1092 b 12), or in the pattern of a tiled floor: all according to 'the mystical way of Pythagoras, and the secret magick of numbers'. For instance, the triangular numbers, 1, 3, 6, 10, etc., have the natural numbers for their 'differences'; and so the natural numbers may be called their gnomons, because they keep the triangular numbers still triangular. In like manner the square numbers have the successive odd numbers for their gnomons, as follows:

$$0+1 = 1^2$$
$$1^2 +3 = 2^2$$
$$2^2 +5 = 3^2$$
$$3^2 +7 = 4^2 \text{ etc.}$$

And this gnomonic relation we may illustrate graphically (σχηματο-γραφεῖν) by the dots whose addition keeps the annexed figures perfect squares:[1]

There are other gnomonic figures more curious still. For example, if we make a rectangle (Fig. 77) such that the two sides are in the ratio of $1:\sqrt{2}$, it is obvious that, on doubling it, we obtain a similar figure; for $1:\sqrt{2}::\sqrt{2}:2$; and each half of the figure, accordingly, is now a gnomon to the other. Were we to make our paper of such a shape (say, roughly, 10 in. × 7 in.), we might fold and fold it, and the shape of folio, quarto and octavo pages would be all the same. For another elegant example, let us start with a rectangle (*A*) whose sides are in the proportion of the 'divine' or 'golden section'[2] that is to say as $1:\frac{1}{2} (\sqrt{5}-1)$, or, approximately, as $1:0.618....$ The gnomon to this rectangle is the square (*B*) erected on its longer side, and so on successively (Fig. 78).

In any triangle, as Hero of Alexandria tells us, one part is always a gnomon to the other part. For instance, in the triangle *ABC* (Fig. 79), let us draw *BD*, so as to make the angle *CBD* equal to the angle *A*. Then the part *BCD* is a triangle similar to the whole triangle *ABC*, and *ABD* is a gnomon to *BCD*. A very elegant case is when

[1] Cf. Treutlein, *Z. Math. Phys.* **28** (1883), 209.   [2] Euclid, II, 11.

the original triangle *ABC* is an isosceles triangle having one angle of 36°, and the other two angles, therefore, each equal to 72° (Fig. 80). Then, by bisecting one of the angles of the base, we subdivide the large isosceles triangle into two isosceles triangles, of which one is similar to the whole figure and the other is its gnomon.[1] There is good reason to believe that this triangle was especially studied by the

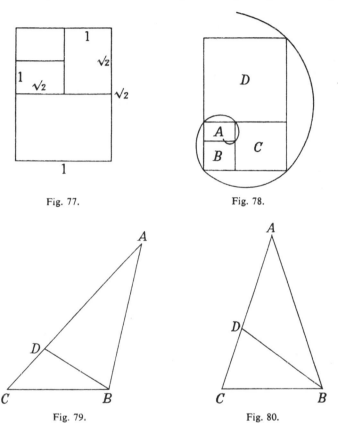

Fig. 77.   Fig. 78.

Fig. 79.   Fig. 80.

Pythagoreans; for it lies at the root of many interesting geometrical constructions, such as the regular pentagon, and its mystical 'pent-alpha', and a whole range of other curious figures beloved of the ancient mathematicians:[2] culminating in the regular, or pentagonal,

[1] This is the so-called *Dreifachgleichschenkelige Dreieck*; cf. Naber, *op. cit. infra.* The ratio 1:0·618 is again not hard to find in this construction.

[2] See, on the mathematical history of the gnomon, Heath's *Euclid* (1908), I, *passim*; Zeuthen, *Théorème de Pythagore* (Genève, 1904); also a curious and interesting book, *Das Theorem des Pythagoras*, by Dr H. A. Naber (Haarlem, 1908).

dodecahedron, which symbolised the universe itself, and with which Euclidean geometry ends.

If we take any one of these figures, for instance the isosceles triangle which we have just described, and add to it (or subtract from it) in succession a series of gnomons, so converting it into larger and larger (or smaller and smaller) triangles all similar to the first, we find that the apices (or other corresponding points) of all these triangles have their *locus* upon an equiangular spiral: a result which follows directly from that alternative definition of the equiangular spiral which I have quoted from Whitworth (p. 179).

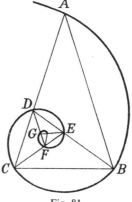

Fig. 81.

If in this, or any other isosceles triangle, we take corresponding median lines of the successive triangles, by joining C to the mid-point (M) of AB, and D to the mid-point (N) of BC, then the pole of the spiral, or centre of similitude of ABC and BCD, is the point of intersection of CM and DN.[1]

Again, we may build up a series of right-angled triangles, each of which is a gnomon to the preceding figure; and here again, an equiangular spiral is the locus of corresponding points in these successive triangles. And lastly, whensoever we fill up space with a collection of equal and similar figures, as in Figs. 82, 83, there we can always discover a series of equiangular spirals in their successive multiples.

Once more, then, we may modify our definition, and say that: 'Any plane curve proceeding from a fixed point (or pole), and such that the vectorial area of any sector is always a gnomon to the whole preceding figure, is called an equiangular, or logarithmic, spiral.' And we may now introduce this new concept and nomenclature into our description of the *Nautilus* shell and other related organic forms, by saying that: (1) if a growing structure be built up of successive parts, similar in form, magnified in geometrical progression, and similarly situated with respect to a centre of similitude, we can always trace through corresponding points a series of equiangular spirals; and (2) it is characteristic of the growth of the horn, of the shell, and of all other organic forms in which an equiangular spiral can be

[1] I owe this simple but novel construction, like so much else, to Dr G. T. Bennett.

recognised, that *each successive increment of growth is similar, and similarly magnified, and similarly situated to its predecessor, and is in consequence a gnomon to the entire pre-existing structure.* Conversely

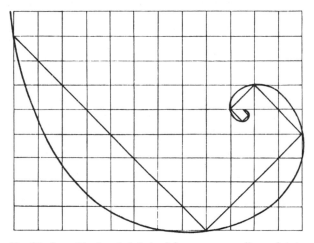

Fig. 82. Logarithmic spiral derived from corresponding points in a system of squares.

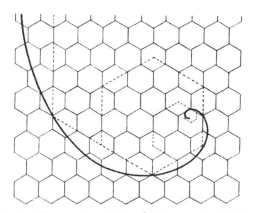

Fig. 83. The same in a system of hexagons. From Naber.

(3) it follows that in the spiral outline of the shell or of the horn we can always inscribe an endless variety of other gnomonic figures, having no necessary relation, save as a mathematical accident, to the nature or mode of development of the actual structure.[1] But

[1] For many beautiful geometrical constructions based on the molluscan shell, see S. Colman and C. A. Coan, *Nature's Harmonic Unity*, ch. ix, Conchology (New York, 1912).

185

observe that the gnomons to a square may form increments of any size, and the same is true of the gnomons to a *Haliotis*-shell; but in the higher symmetry of a chambered *Nautilus*, or of the successive triangles in Fig. 81, growth goes on by a progressive series of gnomons, each one of which is the gnomon to another.

Of these three propositions, the second is of great use and advantage for our easy understanding and simple description of the molluscan shell, and of a great variety of other structures whose mode of growth is analogous, and whose mathematical properties are therefore identical. We see that the successive chambers of a spiral *Nautilus* or of a straight *Orthoceras*, each whorl or part of a whorl of a periwinkle or other gastropod, each new increment of the operculum

Fig. 84. A shell of *Haliotis*, showing the many lines of growth, or generating curves: the areas bounded by these lines of growth being in all cases gnomons to the pre-existing shell. From J. C. Chenu.

of a gastropod, each additional increment of an elephant's tusk, or each new chamber of a spiral foraminifer, has its leading characteristic at once described and its form so far explained by the simple statement that it constitutes a *gnomon* to the whole previously existing structure. And herein lies the explanation of that 'time-element' in the development of organic spirals of which we have spoken already; for it follows as a simple corollary to this theory of gnomons that we must never expect to find the logarithmic spiral manifested in a structure whose parts are simultaneously produced, as for instance in the margin of a leaf, or among the many curves that make the contour of a fish. But we must look for it wherever the organism retains, and still presents at a single view, the successive phases of preceding growth: the successive magnitudes attained, the successive outlines occupied, as growth pursued the even tenor of its way. And it follows from this that it is in the hard parts of organisms, and not the soft, fleshy, actively growing parts, that this spiral is commonly and characteristically found: not in the fresh mobile tissue whose form is constrained merely by the active forces of the moment; but in things like shell and tusk, and horn and claw, visibly composed of parts successively and permanently laid down. The shell-less molluscs are never spiral; the snail is spiral but not the slug.[1] In short, it is the shell which curves the snail, and not the snail which curves the shell. The logarithmic spiral is characteristic, not of the living tissues, but of the dead. And for the same reason it will always or nearly always be accompanied, and adorned, by a pattern formed of 'lines of growth', the lasting record of successive stages of form and magnitude.

Spirals in Plants

The cymose inflorescences of the botanists are analogous in a curious and instructive way to the equiangular spiral. In Fig. 85 (*b*) (which represents the *Cicinnus* of Schimper, or *cyme unipare scorpioide* of Bravais, as seen in the Borage), we begin with a primary shoot from which is given off, at a certain definite angle, a secondary shoot: and from that in turn, on the same side and at the same angle, another shoot, and so on. The deflection, or curvature, is continuous and progressive, for it is caused by no external force but only by causes intrinsic in the system. And the whole system is symmetrical: the angles at which the successive shoots are given off being all equal, and the lengths of the shoots diminishing *in constant ratio*. The result is that the successive shoots, or successive increments of growth, are tangents to a curve, and this curve is a true logarithmic spiral. Or

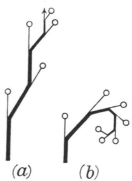

(*a*)   (*b*)

Fig. 85. (*a*) a scorpoid; (*b*) a helicoid cyme.

in other words, we may regard each successive shoot as forming, or defining, a gnomon to the preceding structure. While in this simple case the successive shoots are depicted as lying in *a plane*, it may also happen that, in addition to their successive angular divergence from one another within that plane, they also tend to diverge by successive equal angles *from* that plane of reference; and by this means, there will be superposed upon the equiangular spiral a twist or screw. And, in the particular case where this latter angle of divergence is just equal to 180°, or two right angles, the successive shoots will once

[1] Note also that *Chiton*, where the pieces of the shell are disconnected, shows no sign of spirality.

more come to lie in a plane, but they will appear to come off from one another on *alternate* sides, as in Fig. 85, *a*. This is the *Schraube* or *Bostryx* of Schimper, the *cyme unipare hélicoïde* of Bravais. The equiangular spiral is still latent in it, as in the other; but is concealed from view by the deformation resulting from the helicoid. Many botanists did not recognise (as the brothers Bravais did) the mathematical significance of the latter case, but were led by the snail-like spiral of the scorpoid cyme to transfer the name 'helicoid' to it.[1]

The Molluscan Shell

The paper in which, more than a hundred years ago, Canon Moseley[2] gave a simple mathematical account of the spiral forms of univalve shells, is one of the classics of Natural History. But other students before, and sometimes long before, him had begun to recognise the same simplicity of form and structure. About the year 1818 Reinecke had declared *Nautilus* to be a well-defined geometrical figure, whose chambers followed one another in a constant ratio or continued proportion; and Leopold von Buch and others accepted and even developed the idea.

Long before, Swammerdam[3] had grasped with a deeper insight the root of the whole matter; for, taking a few diverse examples, such as *Helix* and *Spirula*, he showed that they and all other spiral shells whatsoever were referable to one common type, namely to that of a simple tube, variously curved according to definite mathematical laws; that all manner of ornamentation, in the way of spines, tuberosities, colour-bands and so forth, might be superposed upon them, but the type was one throughout and specific differences were of a geometrical kind.

Nay more, we may go back yet another hundred years and find Sir Christopher Wren contemplating the architecture of a snail-shell, and finding in it the logarithmic spiral. For Wallis,[4] after defining and describing this curve with great care and simplicity, tells us that Wren not only conceived the spiral shell to be a sort of cone

[1] The names of these structures have been often confused and misunderstood; cf. S. H. Vines, 'The History of the Scorpoid Cyme', *J. Bot.* (n.s.), **10** (1881), 3–9.

[2] The Rev. Henry Moseley (1801–72), of St John's College, Cambridge, Canon of Bristol, Professor of Natural Philosophy in King's College, London, was a man of great and versatile ability. He was father of H. N. Moseley, naturalist on board the *Challenger* and Professor of Zoology in Oxford; and he was grandfather of H. G. J. Moseley (1887–1915)—Moseley of the Moseley numbers—whose death at Gallipoli, long ere his prime, was one of the major tragedies of the Four Years War.

[3] *Biblia Naturae sive Historia Insectorum* (Leyden, 1737), p. 152.

[4] Joh. Wallis, *Tractatus duo, de Cycloide*, etc. (Oxon., 1659), pp. 107, 108.

or pyramid coiled round a vertical axis, but also saw that on the magnitude of *the angle of the spire* depended the specific form of the shell.

The surface of any shell, whether discoid or turbinate, may be imagined to be generated by the revolution about a fixed axis of a closed curve, which, remaining always geometrically similar to itself, increases its dimensions continually: and, since the scale of the figure increases in geometrical progression while the angle of rotation increases in arithmetical, and the centre of similitude remains fixed, the curve traced in space by corresponding points in the generating curve is, in all such cases, an equiangular spiral. In discoid shells, the generating figure revolves in a plane perpendicular to the axis, as in the Nautilus, the Argonaut and the Ammonite. In turbinate shells, it follows a skew path with respect to the axis of revolution, and the curve in space generated by any given point makes a constant angle to the axis of the enveloping cone, and partakes, therefore, of the character of a helix, as well as of a logarithmic spiral; it may be strictly entitled a helico-spiral. Such turbinate or helico-spiral shells include the snail, the periwinkle and all the common typical gastropods.

The generating figure may be taken as any section of the shell, whether parallel, normal, or otherwise inclined to the axis. It is very commonly assumed to be identical with the mouth of the shell; in which case it is sometimes a plane curve of simple form; in other and more numerous cases, it becomes complicated in form and its boundaries do not lie in one plane: but in such cases as these we may replace it by its 'trace', on a plane at some definite angle to the direction of growth, for instance by its form as it appears in a section through the axis of the helicoid shell. The generating curve is of very various shapes. It

Fig. 86. Section of a spiral univalve, *Triton corrugatus* Lam. From Woodward.

is circular in *Scalaria* or *Cyclostoma*, and in *Spirula*; it may be considered as a segment of a circle in *Natica* or in *Planorbis*. It is triangular in *Conus* or *Thatcheria*, and rhomboidal in *Solarium* or *Potamides*. It is very commonly more or less elliptical: the long axis of the ellipse being parallel to the axis of the shell in *Oliva* and *Cypraea*;

189

all but perpendicular to it in many Trochi; and oblique to it in many well-marked cases, such as *Stomatella, Lamellaria, Sigaretus haliotoides* (Fig. 87) and *Haliotis*. In *Nautilus pompilius* it is approximately a semi-ellipse, and in *N. umbilicatus* rather more than a semi-ellipse, the long axis lying in both cases perpendicular to the axis of the shell.[1] Its form is seldom open to easy mathematical expression, save when it is an actual circle or ellipse; but an exception to this rule may be found in certain Ammonites, forming the group 'Cordati', where (as Blake points out) the curve is very nearly represented by a cardioid, whose equation is $r = a(1 + \cos \theta)$.

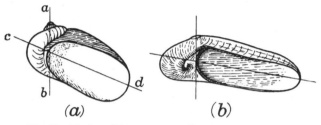

Fig. 87. (*a*) *Lamellaria perspicua*; (*b*) *Sigaretus haliotoides*.
After Woodward.

When the generating curves of successive whorls cut one another, the line of intersection forms the conspicuous helico-spiral or loxo-dromic curve called the *suture* by conchologists.

The generating curve may grow slowly or quickly; its growth-factor is very slow in *Dentalium* or *Turritella*, very rapid in *Nerita*, or *Pileopsis*, or *Haliotis* or the Limpet. It may contain the axis in its plane, as in *Nautilus*; it may be parallel to the axis, as in the majority of gastropods; or it may be inclined to the axis, as it is in a very marked degree in *Haliotis*. In fact, in *Haliotis* the generating curve is so oblique to the axis of the shell that the latter appears to grow by additions to one margin only (cf. Fig. 84).

The general appearance of the entire shell is determined (apart from the form of its generating curve) by the magnitude of three angles; and these in turn are determined by the ratios of certain velocities of growth. These angles are (1) the constant angle of the

[1] In *Nautilus*, the 'hood' has somewhat different dimensions in the two sexes, and these differences are impressed upon the shell, that is to say upon its 'generating curve'. The latter constitutes a somewhat broader ellipse in the male than in the female. But this difference is not to be detected in the young; in other words, the form of the generating curve perceptibly alters with advancing age. Somewhat similar differences in the shells of Ammonites were long ago suspected, by d'Orbigny, to be due to sexual differences. (Cf. Willey, *Natural Science*, 1895, vi, 411; *Zoological Results*, 1902, p. 742.)

equiangular spiral (α); (2) in turbinate shells, the enveloping angle of the cone, or (taking half that angle) the angle (β) which a tangent to the whorls makes with the axis of the shell; and (3) an angle called the 'angle of retardation' (γ), which expresses the retardation in growth of the inner as compared with the outer part of each whorl, and therefore measures the extent to which the one whorl overlaps, or the extent to which it is separated from, another.

The spiral angle (α) is very small in a limpet, where it is usually taken as $=0°$; but it is evidently of a significant amount, though obscured by the shortness of the tubular shell. In *Dentalium* it is still small, but sufficient to give the appearance of a regular curve; it amounts here probably to about 30° to 40°. In *Haliotis* it is from about 70° to 75°; in *Nautilus* about 80°; and it lies between 80° and 85° or even more, in the majority of gastropods.[1]

The case of *Fissurella* is curious. Here we have, apparently, a conical shell with no trace of spiral curvature, or (in other words) with a spiral angle which approximates to 0°; but in the minute embryonic shell (as in that of the limpet) a spiral convolution is distinctly to be seen. It would seem, then, that what we have to do with here is an unusually large growth-factor in the generating curve, causing the shell to dilate into a cone of very wide angle, the apical portion of which has become lost or absorbed, and the remaining part of which is too short to show clearly its intrinsic curvature. In the closely allied *Emarginula*, there is likewise a well-marked spiral in the embryo, which however is still manifested in the curvature of the adult, nearly conical, shell. In both cases we have to do with a very wide-angled cone, and with a high retardation-factor for its inner, or posterior, border. The series is continued, from the apparently simple cone to the complete spiral, through such forms as *Calyptraea*.

The angle α is not always, nor rigorously, a constant angle. In some Ammonites it may increase with age, the whorls becoming closer and closer; in others it may decrease rapidly and even fall to zero, the coiled shell then straightening out, as in *Lituites* and similar forms. It diminishes somewhat, also, in many Orthocerata, which are slightly curved in youth but straight in age. It tends to increase notably in some common land-shells, the *Pupae* and *Bulimi*; and it decreases in *Succinea*.

A variation with advancing age of β is common, but (as Blake

[1] What is sometimes called, as by Leslie, the *angle of deflection* is the complement of what we have called the *spiral angle* (α), or obliquity of the spiral. When the angle of deflection is 6° 17′ 41″, or the spiral angle 83° 42′ 19″, the radiants, or breadths of successive whorls, are doubled at each entire circuit.

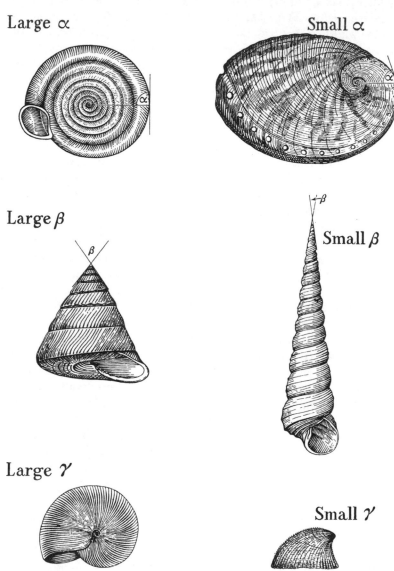

Large α

Small α

Large β

Small β

Large γ

Small γ

Fig. 88. Various gastropods showing the effect of the alteration of different angles. In the top row the shells have large and small spiral angles (α); in the middle row they have large and small enveloping angles of the conical ends (β); in the bottom row there are large and small angles of retardation (γ) which govern the extent to which the whorls overlap. From J. C. Chenu.

points out) it is often not to be distinguished or disentangled from an alteration of α. Whether alone, or combined with a change in α, we find it in all those many gastropods whose whorls cannot all be touched by the same enveloping cone, and whose spire is accordingly described as *concave* or *convex*. The former condition, as we have it in *Cerithium*, and in the cusp-like spire of *Cassis, Dolium* and some Cones, is much the commoner of the two.

The angle of retardation (γ) is very small in *Dentalium* and *Patella*; it is very large in *Haliotis*; it becomes infinite in *Argonauta* and in *Cypraea*. Connected with the angle of retardation are the various possibilities of contact or separation, in various degrees, between adjacent whorls in the discoid shell, and between both adjacent and opposite whorls in the turbinate.

Foraminiferan Spirals

It is characteristic and even diagnostic of the Foraminifera (1) that development proceeds by a well-marked alternation of rest and of activity—of activity during which the protoplasm increases, and of rest during which the shell is formed; (2) that the shell is formed at the outer surface of the protoplasmic organism, and tends to constitute a continuous or all but continuous covering; and it follows (3) from these two factors taken together that each successive increment is added on outside of and distinct from its predecessors, that the successive parts or chambers of the shell are of different and successive ages, so that one part of the shell is always relatively new, and the rest old in various grades of seniority.

That the shell in the Foraminifera should tend towards a spiral form need not surprise us; for we have learned that one of the fundamental conditions of the production of a concrete spiral is just precisely what we have here, namely the development of a structure by means of successive graded increments superadded to its exterior, which then form part, successively, of a permanent and rigid structure. This condition is obviously forthcoming in the foraminiferal, but not at all in the radiolarian, shell. Our second fundamental condition of the production of a logarithmic spiral is that each successive increment shall be so posited and so conformed that its addition to the system leaves the form of the whole system unchanged. We have now to enquire into this latter condition; and to determine whether the successive increments, or successive chambers, of the foraminiferal shell actually constitute *gnomons* to the entire structure.

193

This may be best examined by investigating the various configurations to be found among the Foraminifera.

Firstly we have the typically spiral shells, which occur in great variety, and which (for our present purpose) we need hardly describe further. We may merely notice how in certain cases, for instance *Globigerina*, the individual chambers are little removed from spheres;

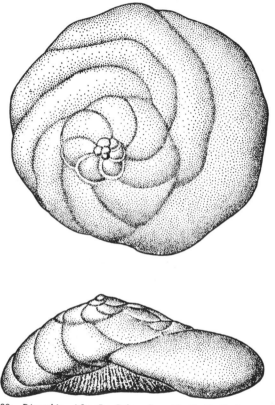

Fig. 89. *Discorbis*. After Le Calvez from K. G. Grell, *Protozoologie* (Berlin: Springer-Verlag).

in other words, the area of contact between the adjacent chambers is small. In such forms as *Cyclammina* and *Pulvinulina*, on the other hand, each chamber is greatly overlapped by its successor, and the spherical form of each is lost in a marked asymmetry. Furthermore, in *Globigerina* and some others we have a tendency to the development of a gauche spiral in space, as in so many of our univalve molluscan shells. The mathematical problem of how a shell should grow, under the assumptions which we have made, would probably find its most general statement in such a case as that of *Globigerina*, where the whole organism lives and grows freely poised in a medium whose density is little different from its own.

The majority of spiral forms, on the other hand, are plane or discoid spirals, and we may take it that in these cases some force has exercised a controlling influence, so as to keep all the chambers in a plane. This is especially the case in forms like *Rotalia* or *Discorbis* (Fig. 89), where the organism lives attached to a rock or a frond of sea-weed; for here (just as in the case of the coiled tubes which little worms such as *Serpula* and *Spirorbis* make, under similar conditions) the spiral disc is itself asymmetrical, its whorls being markedly flattened on their attached surfaces.

We may also conceive, among other conditions, the very curious case in which the protoplasm may entirely overspread the surface of the shell without reaching a position of equilibrium; in which case a new shell will be formed *enclosing* the old one, whether the old one be in the form of a single, solitary chamber, or have already attained to the form of a chambered or spiral shell. This is precisely what often happens in the case of *Orbulina*, when within the spherical shell we find a small, but perfectly formed, spiral '*Globigerina*'.[1]

The various Miliolidae only differ from the typical spiral, or rotaline forms, in the large angle subtended by each chamber, and the consequent abruptness of their inclination to each other. In these cases the *outward* appearance of a spiral tends to be lost; and it behoves us to recollect, all the more, that our spiral curve is not necessarily identical with the *outline* of the shell, but is always a line drawn through corresponding *points* in the successive chambers of the latter.

We reach a limiting case of the logarithmic spiral when the chambers are arranged in a straight line; and the eye will tend to associate with this limiting case the much more numerous forms in which the spiral angle is small, and the shell only exhibits a gentle curve with no succession of enveloping whorls. This constitutes the Nodosarian type (Fig. 23); and here again, we must postulate some force which has tended to keep the chambers in a rectilinear series: such for instance as gravity, acting on a system of 'hanging drops'.

In *Textularia* and its allies (Fig. 90) we have a precise parallel to the helicoid cyme of the botanists (cf. p. 187): that is to say we have a screw translation, perpendicular to the plane of the underlying

---

[1] Cf. G. Schacko, 'Ueber *Globigerina*-Einschluss bei *Orbulina*', *Wiegmann's Archiv*, **49** (1883), 428; Brady, *Chall. Rep.* (1884), p. 607.

logarithmic spiral. In other words, in tracing a genetic spiral through the whole succession of chambers, we do so by a continuous vector rotation through successive angles of 180° (or 120° in some cases), while the pole moves along an axis perpendicular to the original plane of the spiral.

Another type is furnished by the 'cyclic' shells of the Orbitolitidae, where small and numerous chambers tend to be added on round and round the system, so building up a circular flattened disc. This again we perceive to be, mathematically, a limiting case of the logarithmic spiral; the spiral has become wellnigh a circle and the constant angle is wellnigh 90°.

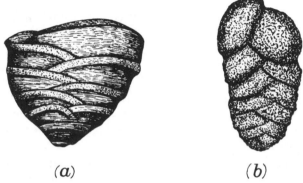

(a)   *(b)*

Fig. 90. (*a*) *Textularia trochus* d'Orb. (*b*) *T. concava* Karrer.

Variation among Foraminifera

Lastly there are a certain number of Foraminifera in which, without more ado, we may simply say that the arrangement of the chambers is irregular, neither the law of constant ratio of magnitude nor that of constant form being obeyed. The chambers are heaped pell-mell upon one another, and such forms are known to naturalists as the Acervularidae.

While in these last we have an extreme lack of regularity, we must not exaggerate the regularity or constancy which the more ordinary forms display. We may think it hard to believe that the simple causes, or simple laws, which we have described should operate, and operate again and again, in millions of individuals to produce the same delicate and complex conformations. But we are taking a good deal for granted if we assert that they do so, and in particular we are assuming, with very little proof, the 'constancy of species' in this group of animals. Just as Verworn has shown that the typical *Amoeba proteus*, when a trace of alkali is added to the water in which it lives, tends, by alteration of surface-tensions, to protrude the more delicate pseudopodia characteristic of *A. radiosa*—and again when the water is rendered a little more alkaline, to turn apparently into the so-called *A. limax*—so it is evident that a very slight modification in the surface-energies concerned might tend to turn one so-called species into another among the Foraminifera. To what extent this process actually occurs, we do not know.

But that this, or something of the kind, does actually occur we can scarcely doubt. For example, in the genus *Peneroplis*, the first portion of the shell consists of a series of chambers arranged in a spiral or nautiloid series; but as age advances the spiral is apt to be modified in various ways.[1] Sometimes the successive chambers grow rapidly broader, the whole shell becoming fan-shaped. Sometimes the chambers become narrower, till they no longer enfold the earlier chambers but only come in contact each with its immediate predecessor: the result being that the shell straightens out, and (taking into account the earlier spiral portion) may be described as crozier-shaped. Between these extremes of shape, and in regard to other variations of thickness or thinness, roughness or smoothness, and so on, there are innumerable gradations passing one into another and intermixed without regard to geographical distribution: 'wherever Peneroplides abound this wide variation exists, and nothing can be more easy than to pick out a number of striking specimens and give to each a distinctive name, but *in no other way can they be divided into "species"*'.[2] Some writers have wondered at the peculiar variability of this particular shell;[3] but for all we know of the life-history of the Foraminifera, it may well be that a great number of the other forms which we distinguish as separate species and even genera are no more than temporary manifestations of the same variability.

If we can comprehend and interpret on some such lines as these the form and mode of growth of the foraminiferal shell we may also begin to understand two striking features of the group, on the one hand the large number of diverse types or families which exist and

---

[1] Cf. H. B. Brady, *Chall. Rep.*, *Foraminifera* (1884), p. 203, pl. xiii.

[2] Brady, *op. cit.* p. 206; Batsch, one of the earliest writers on Foraminifera, had already noticed that this whole series of ear-shaped and crozier-shaped shells was filled in by gradational forms; *Conchylien des Seesandes* (1791), p. 4, pl. vi, fig. 15 *a–f*. See also, in particular, Dreyer, *Peneroplis; eine Studie zur biologischen Morphologie und zur Speciesfrage* (Leipzig, 1898); also Eimer und Fickert, 'Artbildung und Verwandschaft bei den Foraminiferen', *Tübing. zool. Arbeiten*, 3 (1899), 35.

[3] Doflein, *Protozoenkunde* (1911), p. 263: 'Was diese Art veranlässt in dieser Weise gelegentlich zu variiren, ist vorläufig noch ganz räthselhaft.'

the large number of species and varieties within each, and on the other the persistence of forms which in many cases seem to have undergone little change or none at all from the Cretaceous or even from earlier periods to the present day. In few other groups, perhaps only among the Radiolaria, do we seem to possess so nearly complete a picture of all possible transitions between form and form, and of the whole branching system of the evolutionary tree: as though little or nothing of it had ever perished, and the whole web of life, past and present, were as complete as ever. It leads one to imagine that these shells have grown according to laws so simple, so much in harmony with their material, with their environment, and with all the forces internal and external to which they are exposed, that none is better than another and none fitter or less fit to survive. It invites one also to contemplate the possibility of the lines of possible variation being here so narrow and determinate that identical forms may have come independently into being again and again.

While we can trace in the most complete and beautiful manner the passage of one form into another among these little shells, and ascribe them all at last (if we please) to a series which starts with the simple sphere of *Orbulina* or with the amoeboid body of *Astrorhiza*, the question stares us in the face whether this be an 'evolution' which we have any right to correlate with historic *time*. The mathematician can trace one conic section into another, and 'evolve' for example, through innumerable graded ellipses, the circle from the straight line: which tracing of continuous steps is a true 'evolution', though time has no part therein. It was after this fashion that Hegel, and for that matter Aristotle himself, was an evolutionist—to whom evolution was a mental concept, involving order and continuity in thought but not an actual sequence of events in time. Such a conception of evolution is not easy for the modern biologist to grasp, and is harder still to appreciate. And so it is that even those who, like Dreyer[1] and like Rhumbler, study the foraminiferal shell as a physical system, who recognise that its whole plan and mode of growth is closely akin to the phenomena exhibited by fluid drops under particular conditions, and who explain the conformation of the shell by help of the same physical principles and mathematical laws—yet all the while abate no jot or tittle of the ordinary postulates of modern biology, nor doubt the validity and universal applicability of the concepts of Darwinian evolution. For these writers the *biogenetisches*

[1] F. Dreyer, 'Prinzipien der Gerüstbildung bei Rhizopoden, etc.', *Jena Z. Naturw.* **26** (1892), 204–468.

Grundgesetz remains impregnable. The Foraminifera remain for them a great family tree, whose actual pedigree is traceable to the remotest ages; in which historical evolution has coincided with progressive change; and in which structural fitness for a particular function (or functions) has exercised its selective action and ensured 'the survival of the fittest.' By successive stages of historic evolution we are supposed to pass from the irregular *Astrorhiza* to a *Rhabdammina* with its more concentrated disc; to the forms of the same genus which consist of but a single tube with central chamber; to those where this chamber is more and more distinctly segmented; so to the typical many-chambered Nodosariae; and from these, by another definite advance and later evolution, to the spiral Trochamminae. After this fashion, throughout the whole varied series of the Foraminifera, Dreyer and Rhumbler (following Neumayr) recognise so many successions of related forms, one passing into another and standing towards it in a definite relationship of ancestry or descent. Each evolution of form, from simpler to more complex, is deemed to have been attended by an advantage to the organism, an enhancement of its chances of survival or perpetuation; hence the historically older forms are on the whole structurally the simpler; or conversely, the simpler forms, such as the simple sphere, were the first to come into being in primeval seas; and finally, the gradual development and increasing complication of the individual within its own lifetime is held to be at least a partial recapitulation of the unknown history of its race and dynasty.[1]

We encounter many difficulties when we try to extend such concepts as these to the Foraminifera. We are led for instance to assert, as Rhumbler does, that the increasing complexity of the shell, and of the manner in which one chamber is fitted on another, makes for advantage; and the particular advantage on which Rhumbler rests his argument is *strength*. Increase of strength, *die Festigkeitssteigerung*, is according to him the guiding principle in foraminiferal evolution, and marks the historic stages of their development in geologic time. But in days gone by I used to see the beach of a little Connemara bay bestrewn with millions upon millions of foraminiferal shells, simple Lagenae, less simple Nodosariae, more complex Rotaliae: all drifted by wave and gentle current from their sea-cradle to their sandy grave: all lying bleached and dead: one more delicate

---

[1] A difficulty arises in the case of forms (like *Peneroplis*) where the young shell appears to be more complex than the old, the first-formed portion being closely coiled while the later additions become straight and simple: 'die biformen Arten verhalten sich, kurz gesagt, gerade umgekehrt als man nach dem biogenetischen Grundgesetz erwarten sollte', Rhumbler, *Foraminiferen der Plankton-Expedition*, 1911, p. 33, etc.

than another, but all (or vast multitudes of them) perfect and un-broken. And so I am not inclined to believe that niceties of form affect the case very much: nor in general that foraminiferal life involves a struggle for existence wherein breakage is a danger to be averted, and strength an advantage to be ensured.

In the course of the same argument Rhumbler remarks that Foraminifera are absent from the coarse sands and gravels, as Williamson indeed had observed many years ago: so averting, or at least escaping, the dangers of concussion. But this is after all a very simple matter of mechanical analysis. The coarseness or fineness of the sediment on the sea-bottom is a measure of the current: where the current is strong the larger stones are washed clean, where there is perfect stillness the finest mud settles down; and the light, fragile shells of the Foraminifera find their appropriate place, like every other graded sediment in this spontaneous order of levigation.

The theorem of Organic Evolution is one thing; the problem of deciphering the lines of evolution, the order of phylogeny, the degrees of relationship and consanguinity, is quite another. Among the higher organisms we arrive at conclusions regarding these things by weighing much circumstantial evidence, by dealing with the resultant of many variations, and by considering the probability or improbability of many coincidences of cause and effect; but even then our conclusions are at best uncertain, our judgments are continually open to revision and subject to appeal, and all the proof and confirmation we can ever have is that which comes from the direct, but fragmentary evidence of palaeontology.

But in so far as forms can be shown to depend on the play of physical forces, and the variations of form to be directly due to simple quantitative variations in these, just so far are we thrown back on our guard before the biological conception of consanguinity, and compelled to revise the vague canons which connect classification with phylogeny.

The physicist explains in terms of the properties of matter, and classifies according to a mathematical analysis, all the drops and forms of drops and associations of drops, all the kinds of froth and foam, which he may discover among inanimate things; and his task ends there. But when such forms, such conformations and configurations, occur among *living* things, then at once the biologist introduces his concepts of heredity, of historical evolution, of succession in time, of recapitulation of remote ancestry in individual growth, of common origin (unless contradicted by direct evidence) of similar forms remotely separated by geographic space or geologic time, of fitness for a function, of adaptation to an environment, of higher and lower, of 'better' and 'worse'. This is the fundamental difference between the 'explanations' of the physicist and those of the biologist.

In the order of physical and mathematical complexity there is no question of the sequence of historic time. The forces that bring about the sphere, the cylinder or the ellipsoid are the same yesterday and to-morrow. A snow-crystal is the same to-day as when the first snows fell. The physical forces which mould the forms of *Orbulina*, of *Astrorhiza*, of *Lagena* or of *Nodosaria* to-day were still the same, and for aught we have reason to believe the physical conditions under which they worked were not appreciably different, in that yesterday which we call the Cretaceous epoch; or, for aught we know, throughout all that duration of time which is marked, but not measured, by the geological record.

In a word, the minuteness of our organism brings its conformation as a whole within the range of the molecular forces; the laws of its growth and form appear to lie on simple lines; what Bergson calls[1] the 'ideal kinship' is plain and certain, but the 'material affiliation' is problematic and obscure; and, in the end and upshot, it seems to me by no means certain that the biologist's usual mode of reasoning is appropriate to the case, or that the concept of continuous historical evolution must necessarily, or may safely and legitimately, be employed.

That things not only alter but improve is an article of faith, and the boldest of evolutionary conceptions. How far it be true were very hard to say; but I for one imagine that a pterodactyl flew no less well than does an albatross, and that Old Red Sandstone fishes swam as well and easily as the fishes of our own seas.

---

[1] The evolutionist theory, as Bergson puts it, 'consists above all in establishing relations of ideal kinship, and in maintaining that wherever there is this relation of, so to speak, *logical* affiliation between forms, *there is also a relation of chronological succession between the species in which these forms are materialised*' (*Creative Evolution*, 1911, p. 26).

CHAPTER VII

THE SHAPES OF HORNS AND OF
TEETH OR TUSKS

Horns

We have had so much to say on the subject of shell-spirals that we must deal briefly with the analogous problems which are presented by the horns of sheep, goats, antelopes and other horned quadrupeds; and all the more, because these horn-spirals are on the whole less symmetrical, less easy of measurement than those of the shell, and in other ways also are less easy of investigation. Let us dispense altogether in this case with mathematics; and be content with a very simple account of the configuration of a horn.

There are three types of horn which deserve separate consideration: firstly, the horn of the rhinoceros; secondly, the horns of the sheep, the goat, the ox or the antelope, that is to say, of the so-called hollow-horned ruminants; and thirdly, the solid bony horns, or 'antlers', which are characteristic of the deer.

The horn of the rhinoceros presents no difficulty. It is physiologically equivalent to a mass of consolidated hairs, and, like ordinary hair, it consists of non-living or 'formed' material, continually added to by the living tissues at its base. In section the horn is elliptical, with the long axis fore-and-aft, or in some species nearly circular. Its longitudinal growth proceeds with a maximum velocity anteriorly, and a minimum posteriorly; and the ratio of these velocities being constant, the horn curves into the form of a logarithmic spiral in the manner that we have already studied. The spiral is of small angle, but in the longer-horned species, such as the great white rhinoceros (*Ceratorhinus*), the spiral curvature is distinctly recognised. As the horn occupies a median position on the head—a position, that is to say, of symmetry in respect to the field of force on either side—there is no tendency towards a lateral twist, and the horn accordingly develops as a *plane* logarithmic spiral. When two median horns coexist, the hinder one is much the smaller of the two: which is as much as to say that the force, or rate, of growth diminishes as we pass backwards, just as it does within the limits of the single horn. And

202

accordingly, while both horns have *essentially* the same shape, the spiral curvature is less manifest in the second one, by the mere reason of its shortness.

The paired horns of the ordinary hollow-horned ruminants, such as the sheep or the goat, grow under conditions which are in some respects similar, but which differ in other and important respects from the conditions under which the horn grows in the rhinoceros. As regards its structure, the entire horn now consists of a bony core with a covering of skin; the inner, or dermal, layer of the latter is richly supplied with nutrient blood-vessels, while the outer layer, or epidermis, develops the fibrous or chitinous material, chemically and morphologically akin to a mass of cemented or consolidated hairs, which constitutes the 'sheath' of the horn. A zone of active growth at the base of the horn keeps adding to this sheath, ring by ring, and the specific form of this annular zone may be taken as the 'generating curve' of the horn.[1] Each horn no longer lies, as it does in the rhinoceros, in the plane of symmetry of the animal of which it forms a part; and the limited field of force concerned in the genesis and growth of the horn is bound, accordingly, to be more or less laterally asymmetrical. But the two horns are in symmetry one with another; they form 'conjugate' spirals, one being the 'mirror-image' of the other. Just as in the hairy coat of the animal each hair, on either side of the median 'parting', tends to have a certain definite direction of its own, inclined away from the median axial plane of the whole system, so is it both with the bony core of the horn and with the consolidated mass of hairs or hair-like substance which constitutes its sheath; the primary axis of the horn is more or less inclined to, and may even be nearly perpendicular to, the axial plane of the animal.

The growth of the horny sheath is not continuous, but more or less definitely periodic: sometimes, as in the sheep, this periodicity is particularly well-marked, and causes the horny sheath to be composed of a series of all but separate rings, which are supposed to be formed year by year, and so to record the age of the animal.[2]

[1] In this chapter we keep to Moseley's way of regarding the equiangular spiral in space, of shell or horn, as generated by a certain figure which (*a*) grows, (*b*) revolves about an axis, and (*c*) is translated along or parallel to the said axis, all at certain appropriate and specific velocities. This method is simple, and even adequate, from the naturalist's point of view; but not so, or much less so, from the mathematician's.

[2] Cf. R. S. Hindekoper, *On the Age of the Domestic Animals* (Philadelphia and London, 1891), p. 173. In the case of the ram's horn, the assumption that the rings are annual is probably justified. In cattle they are much less conspicuous, but are sometimes well-marked in the cow; and in Sweden they are then called 'calf-rings', from a belief that they record the number of offspring. That is to say, the growth of the horn

203

Just as Moseley sought for the true generating curve in the orifice, or 'lip', of the molluscan shell, so we begin by assuming that in the spiral horn the generating curve corresponds to the lip or margin of one of the horny rings or annuli. This annular margin, or boundary of the ring, is usually a sinuous curve, not lying in a plane, but such as would form the boundary of an anticlastic surface of great complexity; to the meaning and origin of which phenomenon we shall return presently. But in the case of the molluscan shell, the complexities of the lip itself, or of the corresponding lines of growth upon the shell, need not concern us in our study of the development

Fig. 91. The Argali sheep: *Ovis ammon.* From Cook's
Spirals in Nature and Art.

of the spiral: inasmuch as we may substitute for these actual boundary lines, their 'trace', or projection on a plane perpendicular to the axis—in other words the simple outline of a transverse section of the whorl. In the horn, this transverse section is often circular or nearly so, as in the oxen and many antelopes: it now and then becomes of somewhat complicated polygonal outline, as in a Highland ram; but in many antelopes, and in most of the sheep, the outline is that of an isosceles or sometimes nearly equilateral triangle, a form which is typically displayed, for instance, in *Ovis ammon.* The horn in this latter case is a trihedral prism, whose three faces are (1) an upper, or frontal face, in continuation of the plane of the frontal bone; (2) an outer, or orbital, starting from the upper margin of the orbit; and (3) an inner, or nuchal, abutting on the parietal bone.[1] Along these

is supposed to be retarded during gestation, and to be accelerated after parturition, when superfluous nourishment seeks a new outlet (cf. Lönnberg, *Proc. Zool. Soc. Lond.* (1900), p. 689).

[1] Cf. Sir V. Brooke, 'On the Large Sheep of the Thian Shan', *Proc. Zool. Soc. Lond.* (1875), p. 511.

three faces, and their corresponding angles or edges, we can trace in the fibrous substance of the horn a series of homologous spirals, such as we have called in a preceding chapter the '*ensemble* of generating spirals' which define or constitute the surface.

The case of the horn differs in ways of its own from that of the molluscan shell. For one thing, the horn is always tubular—its generating curve is actually, as well as theoretically, a closed curve; there is no such thing as 'involution', or the wrapping of one whorl within another, or successive intersection of the generating curve. Again, while the calcareous substance of the shell is laid down once for all, fixed and immovable, there is reason to believe that the young horn has, to begin with, a certain measure of flexibility, a certain freedom, even though it be slight, to bend or fold or wrinkle. And this being so, while it is no harder in the horn than in the shell to recognise the general field of force or general direction of growth, the actual conditions are somewhat more complex.

In some few cases, of which the male musk ox is one of the most notable, the horn is not developed in a continuous spiral curve. It changes its shape as growth proceeds; and this, as we have seen, is enough to show that it does not constitute a logarithmic spiral. The reason is that the bony exostoses, or horn-cores, about which the horny sheath is shaped and moulded, neither grow continuously nor even remain of constant size after attaining their full growth. But as the horns grow heavy the bony core is bent downwards by their weight, and so guides the growth of the horn in a new direction. Moreover as age advances, the core is further weakened and to a great extent absorbed: and the horny sheath or horn proper, deprived of its support, continues to grow, but in a flattened curve very different from its original spiral.[1] The chamois is a somewhat analogous case. Here the terminal, or oldest, part of the horn is curved; it tends to assume a spiral form, though from its comparative shortness it seems merely to be bent into a hook. But later on the bony core within, as it grows and strengthens, stiffens the horn and guides it into a straighter course or form. The same phenomenon of change of curvature, manifesting itself at the time when, or the place where, the horn is freed from the support of the internal core, is seen in a good many other antelopes (such as the hartebeest) and in many buffaloes; and the cases where it is most manifest appear to be those where the bony core is relatively short, or relatively weak. All these

[1] Cf. E. Lönnberg, 'On the Structure of the Musk Ox', *Proc. Zool. Soc. Lond.* (1900), pp. 686–718.

illustrate the cardinal difference between the growth of the horn and that of the bone below: the one dead, the other alive; the one adding and retaining its successive increments, the other mobile, plastic, and in continual flux throughout.

But in the great majority of horns we have no difficulty in recognising a continuous logarithmic spiral, nor in correlating it with an unequal rate of growth (parallel to the axis) on two opposite sides of the horn, the inequality maintaining a constant ratio as long as growth proceeds. In certain antelopes, such as the gemsbok, the spiral angle is very small, or in other words the horn is very nearly

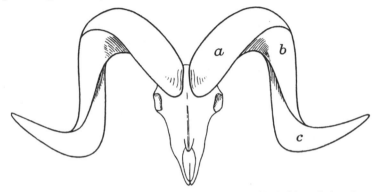

Fig. 92. Diagram of ram's horns. (*a*) Frontal; (*b*) orbital; (*c*) nuchal surface. After Sir Vincent Brooke, from the *Proc. Zool. Soc. Lond.*

straight; in other species of the same genus *Oryx*, such as the Beisa antelope and the Leucoryx, a gentle curve (not unlike though generally less than that of a *Dentalium* shell) is evident; and the spiral angle, according to the few measurements I have made, is found to measure from about 20° to nearly 40°. In some of the large wild goats, such as the Scinde wild goat, we have a beautiful logarithmic spiral, with a constant angle of rather less than 70°; and we may easily arrange a series of forms, such for example as the Siberian ibex, the moufflon, *Ovis ammon*, etc., and ending with the long-horned Highland ram: in which, as we pass from one to another, we recognise precisely homologous spirals with an increasing angular constant, the spiral angle being, for instance, about 75° or rather less in *Ovis ammon*, and in the Highland ram a very little more. We have already seen that in the neighbourhood of 70° or 80° a small change of angle makes a marked difference in the appearance of the spire; and we know also that the actual length of the horn makes a very striking difference, for the spiral becomes especially conspicuous to the eye when horn or shell is long enough to show several whorls, or at least a considerable part of one entire convolution.

Even in the simplest cases, such as the wild goats, the spiral is never a plane but always a *gauche* spiral: in greater or less degree there is always superposed upon the plane logarithmic spiral a helical spiral in space. Sometimes the latter is scarcely apparent, for the horn (though long, as in the said wild goats) is not nearly long enough to show a complete convolution: at other times, as in the

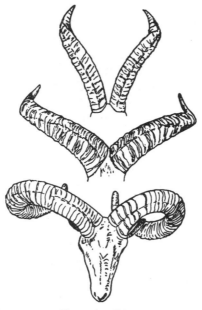

Fig. 93. Different types of horns in wild goats. Top, *Capra pyrenaica*; middle, *C. caucasica caucasica*; bottom, *C. caucasica cylindricornis*. From P. Grassé, *Traité de Zoologie* (Paris: Masson et Cie).

ram, and still better in many antelopes such as the koodoo, the corkscrew curve of the horn becomes its most characteristic feature. So we may study, as in the molluscan shell, the helicoid component of the spire—in other words the variation in what we have called the angle β. This factor it is which, more than the constant angle of the logarithmic spiral, imparts a characteristic appearance to the various species of sheep, for instance to the various closely allied species of Asiatic wild sheep, or Argali. In all of these the constant angle of the logarithmic spiral is very much the same, but the enveloping angle of the cone differs greatly. Thus the long drawn out horns of *Ovis poli*, 4 ft or more from tip to tip, differ conspicuously from those of *Ovis ammon* or *O. hodgsoni*, in which a very similar logarithmic spiral is wound (as it were) round a much blunter cone.

Let us continue to dispense with mathematics, for the mathematical treatment of a gauche spiral is never very simple, and let us deal with the matter by experiment. We have seen that the generating curve, or transverse section, of a typical ram's horn is triangular in form. Measuring (along the curve of the horn) the length of the three edges of the trihedral structure in a specimen of *Ovis ammon*, and calling them respectively the outer, inner, and hinder edges (from their position at the base of the horn, relatively to the skull), I find the outer edge to measure 80 cm, the inner 74 cm, and the posterior 45 cm; let us say that, roughly, they are in the ratio of 9:8:5. Then, if we make a number of little cardboard triangles, equip each with three little legs (I make them of cork), whose relative lengths are as 9:8:5, and pile them up and stick them all together, we straightway build up a curve of double curvature precisely analogous to the ram's horn: except only that, in this first approximation, we have not allowed for the gradual increment (or decrement) of the triangular surfaces, that is to say, for the *tapering* of the horn due to the magnification of the generating curve.

In this case then, and in most other trihedral or three-sided horns, one of the three components, or three unequal velocities of growth, is of relatively small magnitude, but the other two are nearly equal one to the other; it would involve but little change for these latter to become precisely equal; and again but little to turn the balance of inequality the other way. But the immediate consequence of this altered ratio of growth would be that the horn would appear to wind the other way, as it does in the antelopes, and also in certain goats, e.g. the markhor, *Capra falconeri.*

Summarised then in a very few words, the argument by which we account for the spiral conformation of the horn is as follows: The horn elongates by dint of continual growth within a narrow zone, or annulus, at its base. If the rate of growth be identical on all sides of this zone, the horn will grow straight; if it be greater on one side than on the other, the horn will become curved; and it probably *will* be greater on one side than on the other, because each single horn occupies an unsymmetrical field with reference to the plane of symmetry of the animal. If the maximal and minimal velocities of growth be precisely at opposite sides of the zone of growth, the resultant spiral will be a plane spiral; but if they be not precisely or diametrically opposite, then the spiral will be a gauche spiral in space;

and it is by no means likely that the maximum and minimum *will* occur at precisely opposite ends of a diameter, for no such plane of symmetry is manifested in the field of force to which the growing annulus corresponds or appertains.

Now we must carefully remember that the rates of growth of which we are here speaking are the net rates of longitudinal increment, in which increment the activity of the living cells in the zone of growth at the base of the horn is only one (though it is the fundamental) factor. In other words, if the horny sheath were continually being added to with equal rapidity all round its zone of active growth,

Fig. 94. Marco Polo's sheep: *Ovis poli*. From Cook.

but at the same time had its elongation more retarded on one side than the other (prior to its complete solidification) by varying degrees of adhesion or membranous attachment to the bony core within, then the net result would be a spiral curve precisely such as would have arisen from initial inequalities in the rate of growth itself. It seems probable that this is an important factor, and sometimes even the chief factor in the case. The same phenomenon of attachment to the bony core, and the consequent friction or retardation with which the sheath slides over its surface, will lead to various subsidiary phenomena: among others to the presence of transverse folds or corrugations upon the horn, and to their unequal distribution upon its several faces or edges. And while it is perfectly true that nearly all the characters of the horn can be accounted for by unequal velocities of longitudinal growth upon its different sides, it is also plain that the actual field of force is a very complicated one indeed. For example, we can easily see (at least in the great majority of cases) that the direction of growth of the horny fibres of the sheath is by no means parallel to the axis of the core within; accordingly these fibres will tend to wind in a system of helicoid curves around the core, and not

209

only this helicoid twist but any other tendency to spiral curvature on the part of the sheath will tend to be opposed or modified by the resistance of the core within. On the other hand living bone is a very plastic structure, and yields easily though slowly to any forces tending to its deformation; and so, to a considerable extent, the bony core itself will tend to be modelled by the curvature which the growing sheath assumes, and the final result will be determined by an equilibrium between these two systems.

While it is not very safe, perhaps, to lay down any general rule as to what horns are more and what are less spirally curved, I think

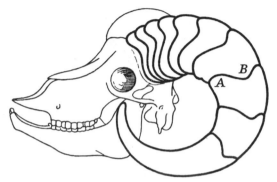

Fig. 95. Head of *Ovis ammon*, showing St Venant's curves.

it may be said that, on the whole, the thicker the horn the greater is its spiral curvature. It is the slender horns, of such forms as the Beisa antelope, which are gently curved, and it is the robust horns of goats or of sheep in which the curvature is more pronounced. Other things being the same, this is what we should expect to find; for it is where the transverse section of the horn is large that we may expect to find the more marked differences in the intensity of the field of force, whether of active growth or of retardation, on opposite sides or in different sectors thereof.

But there is yet another and a very remarkable phenomenon which we may discern in the growth of a horn when it takes the form of a curve of double curvature, namely, an effect of torsional strain; and this it is which gives rise to the sinuous 'lines of growth', or sinuous boundaries of the separate horny rings, of which we have already spoken. It is not at first sight obvious that a mechanical strain of torsion is necessarily involved in the growth of the horn. In our experimental illustration we built up a twisted coil of separate elements, and no torsional strain attended the development of the

210

system. So would it be if the horny sheath grew by successive annular increments, free save for their relation to one another and having no attachment to the solid core within. But as a matter of fact there is such an attachment, by subcutaneous connective tissue, to the bony core; and accordingly a torsional strain will be set up in the growing horny sheath, again provided that the forces of growth therein be directed more or less obliquely to the axis of the core; for a 'couple' is thus introduced, giving rise to a strain which the sheath would not experience were it free (so to speak) to slip along, impelled only by the pressure of its own growth from below. And furthermore, the successive small increments of the growing horn (that is to say, of the horny sheath) are not instantaneously converted from living to solid and rigid substance; but there is an intermediate stage, probably long-continued, during which the new-formed horny substance in the neighbourhood of the zone of active growth is still plastic and capable of deformation.

Now we know, from the celebrated experiments of St Venant,[1] that in the torsion of an elastic body, other than a cylinder of circular section, a very remarkable state of strain is introduced. If the body be thus cylindrical (whether solid or hollow), then a twist leaves each circular section unchanged, in dimensions and in figure. But in all other cases, such as an elliptic rod or a prism of any particular sectional form, forces are introduced which act parallel to the axis of the structure, and which warp each section into a complex 'anticlastic' surface. Thus in the case of a triangular and equilateral prism, such as is shown in section in Fig. 96a, if the part of the rod represented in the section be twisted by a force acting in the direction of the arrow, then the originally plane section will be warped as indicated in the diagram—where the full contour-lines represent elevation above, and the dotted lines represent depression below, the original level. On the external surface of the prism, then, contour-lines which were originally parallel and horizontal will be found warped into sinuous curves, such that, on each of the three faces, the curve will be convex upwards on one half, and concave upwards on the other half of the face. The ram's horn, and still better that of *Ovis ammon*, is comparable to such a prism, save that in section it is not quite equilateral, and that its three faces are not plane. The warping is therefore not precisely identical on the three faces of the horn; but,

---

[1] St Venant, 'De la torsion des prismes, avec des considérations sur leur flexion, etc.', *Mém. Savants Étrangers, Paris*, **14** (1856), 233–560. Karl Pearson dedicated part of his *History of the Theory of Elasticity* to the memory of this ingenious and original man. For a modern account of the subject see Love's *Elasticity* (2nd ed.), ch. xiv.

211

in the general distribution of the curves, it is in complete accordance with theory.[1] Similar anticlastic curves are well seen in many antelopes; but they are conspicuous by their absence in the *cylindrical* horns of oxen.

The better to illustrate this phenomenon, the nature of which is indeed obvious enough from a superficial examination of the horn, I made a plaster cast of one of the horny rings in a horn of *Ovis ammon*, so as to get an accurate pattern of its sinuous edge: and then, filling the mould up with wet clay, I modelled an anticlastic surface, such as to correspond as nearly as possible with the sinuous

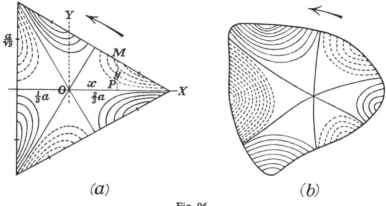

(*a*)   (*b*)

Fig. 96.

outline.[2] Finally, after making a plaster cast of this sectional surface, I drew its contour-lines (as shown in Fig. 96*b*) with the help of a simple form of spherometer. It will be seen that in great part this diagram is precisely similar to St Venant's diagram of the cross-section of a twisted triangular prism; and this is especially the case in the neighbourhood of the sharp angle of our prismatic section. That in parts the diagram is somewhat asymmetrical is not to be wondered at: and (apart from inaccuracies due to the somewhat rough means by which it was made) this asymmetry can be sufficiently accounted for by anisotropy of the material, by inequalities in thickness of different parts of the horny sheath, and especially (I think) by unequal distributions of rigidity due to the presence of the smaller corrugations of

[1] The case of a thin conical shell under torsion is more complicated than either that of the cylinder or of a prismatic rod; and the more tapering horns doubtless deserve further study from this point of view. Cf. R. V. Southwell, 'On the Torsion of Conical Shells', *Proc. Roy. Soc.* A, **163** (1937), 337–55.

[2] This is not difficult to do, with considerable accuracy, if the clay be kept well wetted or semi-fluid, and the smoothing be done with a large wet brush.

the horn. It is on account of these minor corrugations that in such horns as the Highland ram's, where they are strongly marked, the main St Venant effect is not nearly so well shown as in smoother horns, such as those of *Ovis ammon* and its congeners.[1]

The distribution of forces which manifest themselves in the growth and configuration of a horn is no simple nor merely superficial matter. One thing is co-ordinated with another; the direction of the axis of the horn, the form of its sectional boundary, the specific rates of growth in the mean spiral and at various parts of its periphery—all these play their parts, controlled in turn by the supply of nutriment which the character of the adjacent tissues and the distribution of the blood-vessels combine to determine. To suppose that this or that size or shape of horn has been produced or altered, acquired or lost, *by Natural Selection*, whensoever one type rather than another proved serviceable for defence or attack or any other purpose, is an hypothesis harder to define and to substantiate than some imagine it to be.

There are still one or two small matters to speak of before we leave these spiral horns. It is the way of sportsmen to keep record of big game by measuring the length along the curve of the horn and the span from tip to tip. Now if we study such measurements (as they may be found in Mr Rowland Ward's book),[2] we shall soon see that the two measurements do not tally one with the other: but that a pair of horns, the longer when measured along the curve, may be the shorter from tip to tip, and vice versa. We might set this down to mere variability of form, but the true reason is simpler still. If the axes of the two horns stood straight out, at right angles to the median plane, then growth in length and in width of span would go on together. But if the two horns diverge at any lesser angle, then as the horns grow their spiral curvature will tend to bring their tips nearer and farther apart alternately.

There is one last, but not least curious property to be seen in a ram's horns. However large and heavy the horns may be—and in *Ovis poli* 50 or 60 lb is no unusual weight for the pair to grow to—the ram carries them with grace and ease, and they neither endanger his poise nor encumber his movements. The reason is that head and horns are very perfectly *balanced*, in such a way that no bending-moment tends to turn the head up or down about its fulcrum in the atlas vertebra; if one puts two fingers into the foramen magnum one may lift up the heavy skull, and find it hang in perfect equilibrium.

[1] The curves are well shown in most of Sir V. Brooke's figures of the various species of Argali, in the paper quoted on p. 204. [2] *Records of Big Game* (9th ed. 1928).

213

Moreover, the horns go on growing, but this equipoise is never lost nor changed; for the centre of gravity of the logarithmic spiral remains constant. There are other cases where heavy horns, well balanced as they doubtless are, yet visibly affect the set and balance of the head. The stag carries his head higher than a horse, and an Indian buffalo tilts his muzzle higher than a cow.

Teeth, Beak and Claw

In a fashion similar to that manifested in the shell or the horn, we find the equiangular spiral to be implicit in a great many other organic structures where the phenomena of growth proceed in a similar way: that is to say, where about an axis there is some asymmetry leading to unequal rates of longitudinal growth, and where the structure is of such a kind that each new increment is added on as a permanent and unchanging part of the entire conformation. Nail and claw, beak and tooth, all come under this category. The logarithmic spiral *always* tends to manifest itself in such structures as these, though it usually only attracts our attention in elongated structures, where (that is to say) the radius vector has described a considerable angle. When the canary-bird's claws grow long from lack of use, or when the incisor tooth of a rabbit or a rat grows long by reason of disease or of injury of the opponent tooth against which it was wont to bite,[1] we know that the tooth or claw tends to grow into a spiral curve, and we speak of it as a malformation.[2] But there has been no fundamental change of form, only an abnormal increase in length; the elongated tooth or claw has the selfsame curvature which it had when it was short, but the spiral becomes more and more manifest the longer it grows. It is only natural, but nevertheless it is curious to see, how closely a rabbit's abnormally overgrown teeth come to resemble the tusks of swine or elephants, of which the normal state is one of hypertrophy. A curiously analogous case is that of the New Zealand Huia bird, in which the beak of the male is comparatively short and straight, while that of the female is long and curved; it is easy to see

[1] Cf. John Hunter, *Natural History of the Human Teeth* (3rd ed., 1808), p. 110: 'Where a tooth has lost its opposite, it will in time become really so much longer than the rest as the others grow shorter by abrasion.' Cf. James Murie, 'Notes on some Diseased Dental Conditions in Animals', *Trans. Odont. Soc., Lond.* (1867–8), pp. 37–69, 257–98. We now know that a *Coenurus* cyst in a rabbit's masseter muscle may twist the jaw sideways, so that the incisors fail to meet, and grow accordingly: H. A. Baylis, *Trans. R. Soc. Trop. Med. Hyg.* 33 (1939), 4.

[2] See Professor W. C. McIntosh's paper on 'Abnormal Teeth in Certain Mammals, especially in the Rabbit', *Trans. Roy. Soc. Edinb.* 56, 333–407, for a large collection o instances admirably illustrated.

that there is a slight but identical curve also in the beak of the male, and that the beak of the female shows nothing but an extension or prolongation of the same. In the case of the more curved beaks, such as those of an eagle or a parrot, we may, if we please, determine the constant angle of the logarithmic spiral, just as we have done in the case of the *Nautilus* shell; and here again, as the bird grows older or the beak longer, the spiral nature of the curve becomes more and more apparent, as in the hooked beak of an old eagle, or in the great beak of a hyacinthine macaw.

Let us glance at one or two instances to illustrate the spiral curvature of teeth.

A dentist knows that every tooth has a curvature of its own, and that in pulling the tooth he must follow the direction of the curve; but in an ordinary tooth this curvature is scarcely visible, and is least so when the diameter of the tooth is large compared with its length. In simple, more or less conical teeth, such as those of the dolphin, and in the more or less similarly shaped canines and incisors of mammals in general, the curvature of the tooth is particularly well seen. We see it in the little teeth of a hedgehog, and in the canines of a dog or a cat it is very obvious indeed. When the great canine of the carnivore becomes still further enlarged or elongated, as in *Machairodus*, it grows into the strongly curved sabre-tooth of that extinct tiger; and the boar's canine grows into the spiral tusk of wart-hog or babirussa. In rodents, it is the incisors which undergo elongation; their rate of growth differs, though but slightly, on the two sides of the axis, and by summation of these slight differences in the rapid growth of the tooth an unmistakable logarithmic spiral is gradually built up; we see it admirably in the beaver, or in the great ground-rat *Geomys*. The elephant is a similar case, save that the tooth or tusk remains, owing to comparative lack of wear, in a more perfect condition. In the rodent (save only in those abnormal cases mentioned on the last page) the tip, or first-formed part of the tooth, wears away as fast as it is added to from behind; and in the grown animal, all those portions of the tooth near to the pole of the logarithmic spiral have long disappeared. In the elephant, on the other hand, we see, practically speaking, the whole unworn tooth, from point to root; and its actual tip nearly coincides with the pole of the spiral. If we assume (as with no great inaccuracy we may do) that the tip actually coincides with the pole, then we may very easily construct the continuous spiral of which the existing tusk constitutes a part; and by so doing, we see the short, gently curved tusk of our ordinary elephant growing gradually into the spiral tusk of the mammoth. No doubt, just as in the case of our molluscan shells, we have a tendency to variation, both individual and specific, in the constant angle of the spiral; some elephants, and some species of elephant, undoubtedly have a higher spiral angle than others. But in most cases, the angle would seem to be such that a spiral configuration would become very manifest indeed if only the tusk pursued its steady growth, unchanged otherwise in form, till it attained the dimensions which we meet with in the mammoth. In a species such as *Mastodon angustidens*, or *M. arvernensis*, the specific angle is low and the tusk comparatively straight; but the American mastodons and the existing species of elephant have tusks which do not differ appreciably, except in size, from the great spiral tusks of the mammoth, though from their comparative shortness the spiral is little developed and only appears to the eye as a gentle curve. Wherever the tooth is very long indeed, as in the mammoth or the beaver, the effect of some slight and all but inevitable lateral asymmetry in the rate of growth begins to show itself: in other words, the spiral is seen to lie not absolutely in a plane, but to be a gauche curve, like a twisted horn. We see this condition very well in the huge canine tusks of the babirussa; it is a conspicuous feature in the mammoth, and it is more or less perceptible in any large tusk of the ordinary elephants.

The Narwhal's 'Horn'

The 'horn' or tusk of the narwhal is a very remarkable and a very anomalous thing. It is the only tooth in the creature's head to come to maturity; it grows to an immense and apparently unwieldy size, say to 8 or even 9 ft long; it never curves nor bends, but grows as straight as straight can be—a very singular and exceptional thing; it looks as though it were twisted, but really carries on its straight axis a *screw* of several contiguous low-pitched threads; and (last and most anomalous thing of all) when, as happens now and then, two tusks are developed instead of one, one on either side, these two do not form a *conjugate* or symmetrical pair, they are not mirror-images of one another, but are *identical screws, with both threads running the same way.*[1]*

[1] The male narwhal carries the horn, the female being tuskless; but the whalers say that the rare two-horned specimens are all females. A famous two-horned skull in the

* There are a number of other interesting oddities concerning narwhals and their tusks, see M. P. Porsild, *J. Mammol.* 3 (1922), 8.

All ordinary teeth, as we have seen, have their own natural curvature, less or more, which becomes more manifest and conspicuous the longer they grow. We cannot suppose that the field of force (internal and external) in which the narwhal's tusk develops is so simple and uniform as to allow it to grow in perfect symmetry, year after year, without the least bias or intention toward either side; we must rather suppose that the resistances which the growing tusk encounters average out and cancel one another, and leave no one-sided resultant. The long, straight, tapering tooth is commonly said to have a 'spiral twist', but there is no twist at all; the ivory is straight-grained and uniform, through and through. The tusk, in short, is a straight, left-handed, low-pitched screw or helix, with several threads; which threads, in the form of alternate grooves and ridges, wind evenly and continuously from one end of the tusk to the other, *even extending to its root*, deep-set in the socket or alveolus of the upper jaw.

How this composite spiral thread is formed is quite unknown. We have just seen that it is not due to any twisting of the dentinal axis of the tooth. That it is uniform and unbroken from end to end shows that the tooth somehow fashions it as a whole; and that it extends deep down within the alveolus is enough to show that it is not impressed or graven on the tooth by any external agency. We note, as a minor feature, that the several grooves or ridges which constitute the composite thread have their individual or accidental differences; a broader or a narrower groove continues unchanged and recognisable from one end of the tooth to the other; in other words, whatever makes each ridge or groove goes on acting in the selfsame way, as long as growth goes on. A *screw* is made, in general, by compounding a translatory with a rotatory motion, and by bringing the latter into relation with the mould or matrix by which the thread is fashioned or imposed; and I cannot see how to avoid believing that the narwhal's tooth must revolve in like manner, very slowly on its longitudinal axis, all the while it grows—however strange, anomalous and hard to imagine such a mode of growth may be. We know that the tooth grows throughout life in its longitudinal direction, the open root and 'permanent pulp' accounting for this;

Hamburg Museum is known to have belonged to a pregnant female. It was brought home in 1684, and is one of the oldest museum specimens in the world; the tusks measure 242 and 236 cm. During my thirty years' close acquaintance with the Dundee whalers, only four two-horned narwhals passed through my hands. Bateson (*Problems of Genetics*, 1913, p. 44) makes the curious remark that 'the Narwhal's tusks, in being both twisted in the same direction, are highly anomalous, and are *comparable with pairs of twins*'.

and only by a simultaneous and equally continuous rotation (so far as I can see) can we account for the perfect straightness of the tusk, for the grooving or 'rifling' of the surface accompanied by no internal twist, for the extension of that rifling to the alveolar portion of the tusk within the jaw, and for the fact that the several associate grooves and ridges preserve their individual character as they pass along and wind their way around. A very slow rotation is all we need demand—say four or five complete revolutions of the tusk in the whole course of a lifetime.

The progress of a whale or dolphin through the water may be explained as the reaction to a wave which is caused to run from head to tail, the creature moving through the water somewhat slower than the wave travels. The same is true, so far, of a fish; but the wave tends to be in one plane in the fish, the dorsal and ventral fins helping to keep it so; while in the dolphin it may be said to be 'circularly polarised', or resoluble into two oscillations in planes normal to one another, and caused by tail and tail-end swishing around in circular orbits which alter in phase from one transverse section to another. Just as in the case of a screw-propeller, or as in a torpedo (where it is specially corrected or compensated), this mode of action entails a certain waste of energy; it comes of the development of a 'harmful moment', which tends to rotate the body about its axis, and to *screw* the animal along its course. A slight left-handed curvature of the dolphin's tail goes some little way towards correcting this tendency. M. Shuleikin's study of the kinematics of the dolphin[1]—a fine piece of work both on its experimental and its theoretical side—shows the dolphin to be a better swimmer than the fish, inasmuch as its speed of progression comes nearer to the velocity of the wave which is propagated along its body; the so-called 'step', or fraction of the body-length travelled in a single period, is found to be about 0·7 in the dolphin, against 0·57 in a fast-swimming fish (tunny or mackerel).

Shuleikin makes the curious remark that the asymmetry of the skull (discernible in some Cetacea), which in the dolphin shows a screw-twist with a pitch about equal to the length of the body, acts as a compensatory check to the screw-component in the creature's movement of progression, and that 'the till now obscure purpose of the skull's asymmetry' is accordingly explained. I should put this differently, and suggest that this counter-spirality of the skull

[1] Wassilev Shuleikin, 'Kinematics of a Dolphin (Russian)', *Bull. Acad. Sci. U.R.S.S.* (*Cl. sci. math. phys.*) (1935), pp. 651–71; also, 'Dynamics, External and Internal, of a Fish', *ibid.* (1934), pp. 1151–86. On the latter subject see James Gray, *Croonian Lecture,* (1940), and other papers.

is the direct *result* of the spiral component in locomotion. It implies, I take it, a lagging and incomplete response in the fore-part of the body to the rotatory impulse of the parts behind: or, in the plain words of the engineer, a *torque of inertia*.

This tendency, dimly seen in the dolphin's skull, is clearly demonstrated in the narwhal's 'horn', and gives a complete explanation of its many singularities. The narwhal and its horn are joined together, and move together as one piece—nearly, but not quite! Stiff, straight and heavy, the great tusk has its centre of inertia well ahead of the animal, and far from the driving impulse of its tail. At each powerful stroke of the tail the creature not only darts forward, but twists or slews all of a sudden to one side; and the heavy horn, held only by its root, responds (so to speak) with difficulty. For at its slender base the 'couple', by which it has to follow the twisting of the body, works at no small disadvantage. A 'torque of inertia' is bound to manifest itself. The horn does not twist round in perfect synchronism with the animal; but the animal (so to speak) goes slowly, slowly, little by little, round its own horn! The play of motion, the lag, between head and horn is slight indeed; but it is repeated with every stroke of the tail. It is felt just at the growing root, the permanent pulp, of the tooth; and it puts a strain, or exercises a torque, at the very seat, and during the very process, of calcification.

Suppose that at every sweep of the tail there be a lag of no more than a fifth part of a second of arc[1] between the rotation of the tusk and of the body, that small amount would amply suffice to account, on a rough estimate of the age and of the activity of the animal, for as many turns of the screw as a fair-sized tusk is found to exhibit.

According to this explanation, or hypothesis, the slow rotation of the tusk corrects all tendency to flexure or curvature in one direction or another; the grooves and ridges which constitute the 'thread' of the screw are the result of irregularities or inequalities within the alveolus, which 'rifle' the tusk as it grows; and the identity of direction in the two horns of a pair is at once accounted for.

Beautiful as the spiral pattern of the tusk is, it obviously falls short, in regularity and elegance, of what we find, for instance, in a long tapering *Terebra* or *Turritella*, or any other spiral gastropod shell. In the narwhal we have, as we suppose, only a *general* and never a precise agreement between rate of torsion and rate of growth; for these two velocities—of translation and rotation—are separate and

---

[1] Or say a hundred-thousandth part of the angle subtended by a minute on the clock.

independent, and their resultant keeps fairly steady but no more. In the snail-shell, on the other hand, actual tissue-growth is the common cause of both longitudinal and torsional displacements, and the resultant spiral is very perfect and regular.

The ingenious argument of D'Arcy Thompson to account for the twist of the narwhal's tusk has numerous possible difficulties although none is really damaging to his hypothesis. In particular he has been criticised* for his supporting argument that the skull of cetaceans is asymmetrical because of the 'spiral component in locomotion', an idea, according to Mayer, that originated with Kükenthal (1908) who attributed the asymmetry to 'unequal water pressure'. The major stumbling-block is that the asymmetry is found in the embryo as well as the adult and this is presumed to rule out the possibility that mechanical forces arising during swimming could directly cause an alteration of skull shape.

First, I should like to point out that even were this true it need not have any bearing on the problem of the twist of the narwhal's tusk. Secondly, it is likely not a valid argument for there are other cases where a particular structure can be induced by physical forces and inherited as well. Since the phenomenon has even greater pertinence to the next chapter, its significance will be discussed there.

* E. Mayer, *Anat. Rec.* **85** (1943), 115; G. E. Hutchinson, *Amer. Sci.* **36** (1948), 581.

CHAPTER VIII

ON FORM AND MECHANICAL EFFICIENCY

This chapter is a discussion of 'direct adaptations', instances where mechanical forces operate upon a living structure in such a way as to modify it and make it mechanically efficient. There immediately arises the question, rather carefully avoided by D'Arcy Thompson, of the relations of these adaptations to the problem of inheritance. That there is a relation follows not only from the requisites of the Darwinian conception of evolution, but also from direct evidence.

In some instances a particular structure may not be inherited (such as the configuration of bone trabeculae in a poorly set broken leg), but the sensitivity of the cells to physical forces surely is a factor capable of inheritance and obviously of adaptive value.

In other cases, a structure is both stimulated into existence by mechanical factors and present in the embryo before the mechanical factors could possibly have operated. The soles of the feet are already thickened in the human foetus, although clearly the abrasion of walking barefoot vastly exaggerates the embryonic beginning. We have already suggested an explanation of how the organism might inherit a reactivity to the environment that would possess adaptive value, and now comes the question of how the structure itself could be directly inherited without use.

The simplest possibility (sometimes referred to as the Baldwin effect)* is that if certain gene combinations appeared that produced a structure that was identical to the one produced by mechanical factors, obviously it would be advantageous and be retained by the population. While this possibility is logical and may even be the whole solution, it suffers from the difficulty that some find it hard to imagine that a fortuitous combination of genes producing a complex structure would ever have arisen by mere chance, the only other alternative being a Lamarckian interpretation, which is obviously objectionable.

Another possibility has been suggested by various authors, but stated most explicitly by Waddington† who bases his conclusion on a series of particularly interesting experiments. For instance, he subjected a certain strain of fruit flies to a temperature shock during their development and found that a low percentage of the flies produced wings that lacked a cross vein. If he now selected these cross-veinless flies and repeated the experiment, the percentage of cross-veinless individuals increased, and they continued to do so over a series of successive generations, each with a shock treatment. The surprising thing is that after a while some of the

* G. G. Simpson, *Evolution*, 7 (1953), 110.
† *The Strategy of the Genes* (George Allen and Unwin, London, 1957).

flies appeared cross-veinless without the temperature shock; the environmental prodding was no longer needed.

The analogy to the cases involving mechanical factors is striking, for here, in the fruit flies, by selecting those individuals that react, he has produced flies that respond without the stimulus, just as the soles of unborn infants are thickened. From this Waddington suggests that it is the developmental processes themselves that are the objects of selection and that these cases should be interpreted in terms of a progressive selection of reactivity to a particular stimulus, until ultimately the reaction fires off (that is, a particular course of development ensues) without any provocation. It is like filing down a hair trigger until finally the gun fires off of its own accord.

This is no place for a critical discussion of these interesting arguments; the important point here is not so much whether we have the final explanation of direct adaptations and their genetic assimilation, but simply that the phenomenon with which D'Arcy Thompson is concerned in this chapter can be shown to have a significant relation to problems of development, heredity, and evolution. They are all part of one story, a story that as yet has not been completely told.

There is a certain large class of morphological problems of which we have not yet spoken, and of which we shall be able to say but little. Nevertheless they are so important, so full of deep theoretical significance, and so bound up with the general question of form and its determination as a result of growth, that an essay on growth and form is bound to take account of them, however imperfectly and briefly. The phenomena which I have in mind are just those many cases where *adaptation* in the strictest sense is obviously present, in the clearly demonstrable form of mechanical fitness for the exercise of some particular function or action which has become inseparable from the life and well-being of the organism.

The problems associated with these phenomena are difficult at every stage, even long before we approach to the unsolved secrets of causation; and for my part I confess I lack the requisite knowledge for even an elementary discussion of the form of a fish, or of an insect, or of a bird. But in the form of a bone we have a problem of the same kind and order, so far simplified and particularised that we may to some extent deal with it, and may possibly even find, in our partial comprehension of it, a partial clue to the principles of causation underlying this whole class of phenomena.

Tension and Compression

Before we speak of the form of bone, let us say a word about the mechanical properties of the material of which it is built,[1] in relation to the strength it has to manifest or the forces it has to resist: understanding always that we mean thereby the properties of fresh or living bone, with all its organic as well as inorganic constituents, for dead, dry bone is a very different thing. In all the structures raised by the engineer, in beams, pillars and girders of every kind, provision

has to be made, somehow or other, for strength of two kinds, strength to resist compression or crushing, and strength to resist tension or pulling asunder. The evenly loaded column is designed with a view to supporting a downward pressure, the wire-rope, like the tendon of a muscle, is adapted only to resist a tensile stress; but in many or most

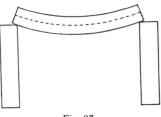

Fig. 97.

cases the two functions are very closely inter-related and combined. The case of a loaded beam is a familiar one; though, by the way, we are now told that it is by no means so simple as it looks, and indeed that 'the stresses and strains in this log of timber are so complex that the problem has not yet been solved in a manner that reasonably accords with the known strength of the beam as found by actual experiment'.[2] However, be that as it may, we know, roughly, that when the beam is loaded in the middle and supported at both ends, it tends to be bent into an arc, in which condition its lower fibres are being stretched, or are undergoing a tensile stress, while its upper fibres are undergoing compression. It follows that in some intermediate layer there is a 'neutral zone', where the fibres of the wood are subject to no stress of either kind.

The phenomenon of a compression-member side by side with a tension-member may be illustrated in many simple ways. Ruskin (in *Deucalion*) describes it in a glacier. He then bids us warm a stick of

---

[1] Cf. Sir Donald MacAlister, 'How a Bone is Built', *Engl. Ill. Mag.* (1884).

[2] Professor Claxton Fidler, *On Bridge Construction* (4th ed. 1909), p. 22; cf. (*int. al.*) Love's *Elasticity*, p. 20 (*Historical Introduction*) (2nd ed. 1906), where the bending of the beam, and the distortion or warping of its cross-section, are studied after the manner of St Venant, in his *Memoir on Torsion* (1855). How complex the question has become may be judged from such papers as Price, 'On the Structure of Wood in Relation to its Elastic Properties', *Phil. Trans.* A, **208** (1928); or D. B. Smith and R. V. Southwell, 'On the Stresses Induced by Flexure in a Deep Rectangular Beam', *Proc. Roy. Soc.* A, **143** (1934), 271–85.

223

sealing-wax and bend it in a horseshoe: 'you will then see, through a lens of moderate power, the most exquisite facsimile of glacier fissures produced by extension on its convex surface, and as faithful an image of glacier surge produced by compression on its concave side.' A still more beautiful way of exhibiting the distribution of strain is to use gelatine, into which bubbles of gas have been introduced with the help of sodium bicarbonate. A bar of such gelatine, when bent into a hoop, shows on the one side the bubbles elongated by tension and on the other those shortened by compression.[1]

In like manner a vertical pillar, if unevenly loaded (as for instance the shaft of our thigh-bone normally is), will tend to bend, and so to endure compression on its concave, and tensile stress upon its convex side. In many cases it is the business of the engineer to separate out, as far as possible, the pressure-lines from the tension-lines, in order to use separate modes of construction, or even different materials for each. In a suspension-bridge, for instance, a great part of the fabric is subject to tensile strain only, and is built throughout of ropes or wires; but the massive piers at either end of the bridge carry the weight of the whole structure and of its load, and endure all the 'compression-strains' which are inherent in the system. Very much the same is the case in that wonderful arrangement of struts and ties which constitute, or complete, the skeleton of an animal. The 'skeleton', as we see it in a Museum, is a poor and even a misleading picture of mechanical efficiency.[2] From the engineer's point of view, it is a diagram showing all the compression-lines, but by no means all the tension-lines of the construction; it shows all the struts, but few of the ties, and perhaps we might even say *none* of the principal ones; it falls all to pieces unless we clamp it together, as best we can, in a more or less clumsy and immobilised way. But in life, that fabric of struts is surrounded and interwoven with a complicated system of ties—'its living mantles jointed strong, With glistering band and silvery thong':[3] ligament and membrane, muscle and tendon, run between bone and bone; and the beauty and strength of the mechanical construction lie not in one part or in another, but in the harmonious concatenation which all the parts, soft and hard,

[1] Cf. Emil Hatschek, 'Gestalt und Orientierung von Gasblasen in Gelen', *Kolloid-zschr.* 15 (1914), 226–34.

[2] In preparing or 'macerating' a skeleton, the naturalist nowadays carries on the process till nothing is left but the whitened bones. But the old anatomists, whose object was not the study of 'comparative morphology' but the wider theme of comparative physiology, were wont to macerate by easy stages; and in many of their most instructive preparations the ligaments were intentionally left in connection with the bones, and as part of the 'skeleton'.

[3] See Oliver Wendell Holmes's *Anatomist's Hymn.*

rigid and flexible, tension-bearing and pressure-bearing, make up together.[1]

However much we may find a tendency, whether in Nature or art, to separate these two constituent factors of tension and compression, we cannot do so completely; and accordingly the engineer seeks for a material which shall, as nearly as possible, offer equal resistance to both kinds of strain.[2]

From the engineer's point of view, bone may seem weak indeed; but it has the great advantage that it is very nearly as good for a tie as for a strut, nearly as strong to withstand rupture, or tearing asunder, as to resist crushing. The strength of timber varies with the kind, but it always stands up better to tension than to compression, and wrought iron, with its greater strength, does much the same; but in cast-iron there is a still greater discrepancy the other way, for it makes a good strut but a very bad tie indeed. Mild steel, which has displaced the old-fashioned wrought iron in all engineering constructions, is not only a much stronger material, but it also possesses, like bone, the two kinds of strength in no very great relative disproportions.[3]

Form and Strength

When the engineer constructs an iron or steel girder, to take the place of the primitive wooden beam, we know that he takes advantage of the elementary principle we have spoken of, and saves weight and economises material by leaving out as far as possible all the middle portion, all the parts in the neighbourhood of the 'neutral zone'; and in so doing he reduces his girder to an upper and lower 'flange', connected together by a 'web', the whole resembling, in cross-section, an I or an ⊥.

But it is obvious that, if the strains in the two flanges are to be equal as well as opposite, and if the material be such as cast-iron or wrought iron, one or other flange must be made much thicker than the other

[1] In a few anatomical diagrams, for instance in some of the drawings in Schmaltz's *Atlas der Anatomie des Pferdes*, we may see the system of 'ties' diagrammatically inserted in the figure of the skeleton. Cf. W. K. Gregory, 'On the Principles of Quadrupedal Locomotion', *Ann. N.Y. Acad. Sci.* **22** (1912), 289.

[2] The strength of materials is not easy to discuss, and is still harder to tabulate. The wide range of qualities in each material, in timber the wide differences according to the direction in which the block is cut, and in all cases the wide difference between yield-point and fracture-point, are some of the difficulties in the way of a succinct statement.

[3] In the modern device of 'reinforced concrete', blocks of cement and rods of steel are so combined together as to resist both compression and tension in due or equal measure.

in order that they may be equally strong;[1] and if at times the two flanges have, as it were, to change places, or play each other's parts, then there must be introduced a margin of safety by making both flanges thick enough to meet that kind of stress in regard to which the material happens to be weakest. There is great economy, then, in any material which is, as nearly as possible, equally strong in both ways; and so we see that, from the engineer's or contractor's point of view, bone is a good and suitable material for purposes of construction.

The I or the H-girder or rail is designed to resist bending in one particular direction, but if, as in a tall pillar, it be necessary to resist bending in all directions alike, it is obvious that the tubular or cylindrical construction best meets the case; for it is plain that this hollow tubular pillar is but the I-girder turned round every way, in a 'solid of revolution', so that on any two opposite sides compression and tension are equally met and resisted, and there is now no need for any substance at all in the way of web or 'filling' within the hollow core of the tube. And it is not only in the supporting pillar that such a construction is useful; it is appropriate in every case where *stiffness* is required, where bending has to be resisted. A sheet of paper becomes a stiff rod when you roll it up, and hollow tubes of thin bent wood withstand powerful thrusts in aeroplane construction. The long bone of a bird's wing has little or no weight to carry, but it has to withstand powerful bending-moments; and in the arm-bone of a long-winged bird, such as an albatross, we see the tubular construction manifested in its perfection, the bony substance being reduced to a thin, perfectly cylindrical, and almost empty shell.[2] The quill of the bird's feather, the hollow shaft of a reed, the thin tube of the wheat-straw bearing its heavy burden in the ear, are all illustrations which Galileo used in his account of this mechanical principle;[3] and the working of his practical mind is

[1] This principle was recognised as soon as iron came into common use as a structural material. The great suspension bridges only became possible, in Telford's hands, when wrought iron became available.

[2] Marsigli (*op. cit.*) was acquainted with the hollow wing-bones of the pelican; and Buffon deals with the whole subject in his *Discours sur la nature des oiseaux*.

[3] Galileo, *Dialogues concerning Two New Sciences* (1638), Crew and Salvio's translation (New York, 1914), p. 150; *Opere*, ed. Favaro, VIII, 186. (According to R. A. Millikan, 'we owe our present-day civilisation to Galileo'.) Cf. Borelli, *De Motu Animalium* (1685), I, prop. CLXXX. Cf. also P. Camper, 'La structure des os dans les oiseaux', *Œuvres* (ed. 1803), 3, 459; A. Rauber, 'Galileo über Knochenformen', *Morph. J.* 7 (1881), 327, 328; Paolo Enriques, 'Della economia di sostanza nelle osse cave', *Arch. Entw. Mech.* 20 (1906), 427–65. Galileo's views on the mechanism of the human body are also discussed by O. Fischer, in his article on *Physiologische Mechanik*, in the *Encycl. mathem. Wissenschaften* (1904).

exemplified by this catalogue of varied instances which one demonstration suffices to explain.

The same principle is beautifully shown in the hollow body and tubular limbs of an insect or a crustacean; and these complicated and elaborately jointed structures have doubtless many constructional lessons to teach us. We know, for instance, that a thin cylindrical tube, under bending stress, tends to flatten before it buckles, and also to become 'lobed' on the compression side of the bend; and we often recognise both of these phenomena in the joints of a crab's leg.[1]

Two points, both of considerable importance, present themselves here, and we may deal with them before we go further on. In the first place, it is not difficult to see that in our bending beam the stress is greatest at its middle; if we press our walking-stick hard against the ground, it will tend to snap midway. Hence, if our cylindrical column be exposed to strong bending stresses, it will be prudent and economical to make its walls thickest in the middle and thinning off gradually towards the ends; and if we look at a longitudinal section of a thigh-bone, we shall see that this is just what Nature has done. The presence of a 'danger-point' has been avoided, and the thickness of the walls becomes nothing less than a diagram, or 'graph', of the bending-moments from one point to another along the length of the bone.

The second point requires a little more explanation. If we imagine our loaded beam to be supported at one end only (for instance, by being built into a wall), so as to form what is called a 'bracket' or 'cantilever', then we can see, without much difficulty, that the lines of stress in the beam run somewhat as in the accompanying diagram. Immediately under the load, the 'compression-lines' tend to run vertically downward, but where the bracket is fastened to the wall there is pressure directed horizontally against the wall in the lower part of the surface of attachment; and the vertical beginning and the horizontal end of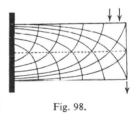

Fig. 98.

these pressure-lines must be continued into one another in the form of some even mathematical curve—which, as it happens, is part of a parabola. The tension-lines are identical in form with the compression-lines, of which they constitute the 'mirror-image'; and where the two systems intercross they do so at right angles, or 'orthogonally'

[1] Cf. L. G. Brazier, 'On the Flexure of Thin Cylindrical Shells, etc.', *Proc. Roy. Soc.* A, **116** (1927), 104.

to one another. Such systems of stress-lines as these we shall deal with again; but let us take note here of the important though well-nigh obvious fact, that while in the beam they both unite to carry the load, yet it is often possible to weaken one set of lines at the expense of the other, and in some cases to do away altogether with one set or the other. For example, when we replace our end-supported beam by a curved bracket, bent upwards or downwards as the case may be, we have evidently cut away in the one case the greater part of the tension-lines, and in the other the greater part of the compression-lines. And if instead of bridging a stream with our beam of wood we bridge it with a rope, it is evident that this new construction contains all the tension-lines, but none of the compression-lines of the old. The biological interest connected with this principle lies chiefly in the mechanical construction of the rush or the straw, or any other typically cylindrical stem. The material of which the stalk is constructed is very weak to withstand compression, but parts of it have a very great tensile strength. Schwendener, who was both botanist and engineer, has elaborately investigated the factor of strength in the cylindrical stem, which Galileo was the first to call attention to. Schwendener[1] showed that its strength was concentrated in the little bundles of 'bast-tissue', but that these bast-fibres had a tensile strength per square mm. of section not less, up to the limit of elasticity, than that of steel wire of such quality as was in use in his day.

For instance, we see in the following table the load which various fibres, and various wires, were found capable of sustaining, not up to the breaking-point but up to the 'elastic limit', or point beyond

	Stress, or load in g per sq. mm., at limit of elasticity	Stress, or load in tons per sq. inch	Strain, or amount of stretching, per mm.
Secale cereale	15–20	9·4–12·5	4·4
Lilium auratum	19	11·8	7·6
Phormium tenax	20	12·5	13·0
Papyrus antiquorum	20	12·5	15·2
Molinia coerulea	22	13·8	11·0
Pincenectia recurvata	25	15·6	14·5
Copper wire	12·1	7·6	1·0
Brass wire	13·3	8·5	1·35
Iron wire	21·9	13·7	1·0
Steel wire	24·6[1]	15·4	1·2

[1] This figure should be considerably higher for the best modern steel.

[1] S. Schwendener, *Das mechanische Princip im anatomischen Bau der Monocotyleen* (Leipzig, 1874); 'Zur Lehre von der Festigkeit der Gewächse', *Sb. Berlin Akad.* (1884), pp. 1045–70.

which complete recovery to the original length took place no longer after release of the load.

In other respects, it is true, the plant-fibres were inferior to the wires; for the former broke asunder very soon after the limit of elasticity was passed, while the iron wire could stand, before snapping, about twice the load which was measured by its limit of elasticity: in the language of a modern engineer, the bast-fibres had a low 'yield-point', little above the elastic limit. Nature seems content, as Schwendener puts it, if the strength of the fibre be ensured up to the elastic limit; for the equilibrium of the structure is lost as soon as that limit is passed, and it then matters little how far off the actual breaking-point may be.[1] But nevertheless, within certain limits, plant-fibre and wire were just as good and strong one as the other. And when Schwendener proceeds to show, in many beautiful diagrams, the various ways in which these strands of strong tensile tissue are arranged in various stems: sometimes, in the simpler cases, forming numerous small bundles arranged in a peripheral ring, not quite at the periphery, for a certain amount of space has to be left for living and active tissue; sometimes in a sparser ring of larger and stronger bundles; sometimes with these bundles further strengthened by radial balks or ridges; sometimes with all the fibres set close together in a continuous hollow cylinder. In the case figured in Fig. 99, Schwendener calculated that the resistance to bending was at least twenty-five times as great as it would have been had the six main bundles been brought close together in a solid core. In many cases the centre of the stem is altogether empty; in all other cases it is filled with soft tissue, suitable for various functions, but never such as to confer mechanical rigidity. In a tall conical stem, such as that of a palm-tree, we can see not only these principles in the construction of the cylindrical trunk, but we can observe, towards the apex, the bundles of fibre curving over and intercrossing orthogonally with one another, exactly after the fashion of our stress-lines in Fig. 98; but of course, in this case, we are still dealing with tensile members, the opposite

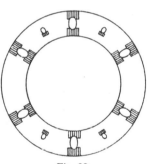

Fig. 99.

[1] The great extensibility of the plant-fibre is due to the spiral arrangement of the ultramicroscopic micellae of which the bast-fibre is built up: the spiral untwisting as the fibre stretches, in a right or left-hand spiral according to the species. Cf. C. Steinbruck, 'Die Micellartheorie auf botanischem Gebiete', *Biol. Zbl.* (1925), p. 1.

bundles taking on in turn, as the tree sways, the alternate function of resisting tensile strain.[1]

The Forth Bridge, from which the anatomist may learn many a lesson, is built of tubes, which correspond even in detail to the structure of a cylindrical branch or stem. The main diagonal struts are tubes 12 ft in diameter, and within the wall of each of these lie six T-shaped 'stiffeners', corresponding precisely to the fibro-vascular bundles of Fig. 99; in the same great tubular struts the tendency to 'buckle' is resisted, just as in the jointed stem of a bamboo, by 'stiffening rings', or perforated diaphragms set 20 ft apart within the tube. We may draw one more curious, albeit parenthetic, comparison. An engineering construction, no less than the skeleton of plant or animal, has *to grow*; but the living thing is in a sense complete during every phase of its existence, while the engineer is often hard put to it to ensure sufficient strength in his unfinished and imperfect structure. The young twig stands more upright than the old, and between winter and summer the weight of leafage affects all the curving outlines of the tree. A slight upward curvature, a matter of a few inches, was deliberately given to the great diagonal tubes of the bridge during their piecemeal construction; and it was a triumph of engineering foresight to see how, like the twig, as length and weight increased, they at last came straight and true.

The Structure of Bone

Let us now come, at last, to the mechanical structure of bone, of which we find a well-known and classical illustration in the various bones of the human leg. In the case of the tibia, the bone is somewhat widened out above, and its hollow shaft is capped by an almost flattened roof, on which the weight of the body directly rests. It is obvious that, under these circumstances, the engineer would find it necessary to devise means for supporting this flat roof, and for

[1] For further botanical illustrations, see (*int. al.*) R. Hegler, 'Einfluss der Zugkräfte auf die Festigkeit und die Ausbildung mechanischer Gewebe in Pflanzen', *SB. sächs. Ges. Wiss.* (1891), p. 638; 'Einfluss des mechanischen Zuges auf das Wachstum der Pflanze', *Cohn's Beiträge*, 6 (1893), 383–432; O. M. Ball, 'Einfluss von Zug auf die Ausbildung der Festigkeitsgewebe', *Jb. wiss. Bot.* 39 (1903), 305–41; L. Kny, 'Einfluss von Zug und Druck auf die Richtung der Scheidewände in sich teilenden Pflanzenzellen', *Ber. bot. Gesellsch.* 14 (1896), 378–91; Sachs, 'Mechanomorphose und Phylogenie', *Flora*, 78 (1894); cf. also Pflüger, 'Einwirkung der Schwerkraft, etc., über die Richtung der Zelltheilung', *Archiv*, 34 (1884); G. Haberlandt's *Physiological Plant Anatomy*, tr. by Montagu Drummond (1914), pp. 150–213. On the engineering side of the case, see Angus R. Fulton, 'Experiments to Show how Failure under Stress Occurs in Timber, etc.', *Trans. Roy. Soc. Edinb.* 48 (1912), 417–40; Fulton shows (*int. al.*) that 'the initial cause of fracture in timbers lies in the medullary rays'.

distributing the vertical pressures which impinge upon it to the cylindrical walls of the shaft.

In the long wing-bones of a bird the hollow of the bone is empty, save for a thin layer of living tissue lining the cylinder of bone; but in our own bones, and all weight-carrying bones in general, the hollow space is filled with marrow, blood-vessels and other tissues; and amidst these living tissues lies a fine lattice-work of little interlaced 'trabeculae' of bone, forming the so-called 'cancellous tissue'. The older anatomists were content to describe this cancellous tissue as a sort of spongy network or irregular honeycomb;[1] but at length its orderly construction began to be perceived, and attempts were made to find a meaning or 'purpose' in the arrangement. Sir Charles Bell had a glimpse of the truth when he asserted[2] that 'this minute lattice-work, or the cancelli which constitute the interior structure of bone, have still reference to the forces acting on the bone'; but he did not succeed in showing what these forces are, nor how the arrangement of the cancelli is related to them.

Jeffries Wyman, of Boston, came much nearer to the truth in a paper long neglected and forgotten.[3] He gives the gist of the whole matter in two short paragraphs: '1. The cancelli of such bones as assist in supporting the weight of the body are arranged either in the direction of that weight, or in such a manner as to support and brace those cancelli which are in that direction. In a mechanical point of view they may be regarded in nearly all these bones as a series of "studs" and "braces". 2. The direction of these fibres in some of the bones of the human skeleton is characteristic and, it is believed, has a definite relation to the erect position which is naturally assumed by man alone.' A few years afterwards the story was told again, and this time with convincing accuracy. It was shown by Hermann Meyer (and afterwards in greater detail by Julius Wolff and others) that the trabeculae, as seen in a longitudinal section of the femur, spread in beautiful curving lines from the head to the hollow shaft of the bone; and that these linear bundles are crossed by others, with so nice a regularity of arrangement that each intercrossing is as nearly as possible an orthogonal one: that is to say, the one set of fibres or

---

[1] Sir John Herschel described a bone as a 'framework of the most curious carpentry: in which occurs not a single straight line nor any known geometrical curve, yet all evidently systematic, and constructed by rules which defy our research' (*On the Study of Natural Philosophy*, 1830, p. 203).

[2] In *Animal Mechanics, or Proofs of Design in the Animal Frame* (1827).

[3] 'Animal Mechanics: on the Cancellated Structure of some of the Bones of the Human Body', *Boston Soc. of Nat. Hist.* (1849). Reprinted, together with Sir C. Bell's work, by Morrill Wyman (Cambridge, Mass., 1902).

cancelli cross the other everywhere at right angles. A great engineer, Professor Culmann of Zürich, to whom by the way we owe the whole modern method of 'graphic statics', happened (in the year 1866) to come into his colleague Meyer's dissecting-room, where the anatomist was contemplating the section of a bone.[1] The engineer, who had been busy designing a new and powerful crane, saw in a moment that

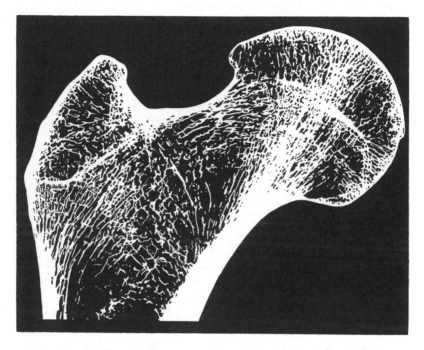

Fig. 100. Head of the human femur in section. After Schäfer, from a photo by Professor A. Robinson.

the arrangement of the bony trabeculae was nothing more nor less than a diagram of the lines of stress, or directions of tension and compression, in the loaded structure: in short, that Nature was strengthening the bone in precisely the manner and direction in which strength was required; and he is said to have cried out, 'That's my crane!' In the accompanying diagram of Culmann's crane-head, we recognise a simple modification, due entirely to the

[1] The first metatarsal, rather than the femur, is said to have been the bone which Meyer was demonstrating when Culmann first recognised the orthogonal intercrossing of the cancelli in tension and compression; cf. A. Kirchner, 'Architektur der Metatarsalien des Menschen', *Arch. Entw. Mech.* **24** (1907), 539–616.

curved shape of the structure, of the still simpler lines of tension and compression which we have already seen in our end-supported beam, as represented in Fig. 98. In the shaft of the crane the concave or inner side, overhung by the loaded head, is the 'compression-member'; the outer side is the 'tension-member'; the pressure-lines, starting from the loaded surface, gather themselves together, always in the direction of the resultant pressure, till they form a close bundle running down the compressed side of the shaft: while the tension-lines, running upwards along the opposite side of the shaft, spread

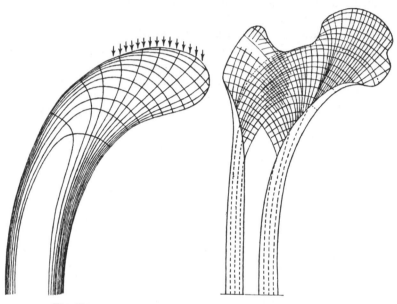

Fig. 101. Crane-head and femur. After Culmann and J. Wolff.

out through the head, orthogonally to, and linking together, the system of compression-lines. The head of the femur (Fig. 100) is a little more complicated in form and a little less symmetrical than Culmann's diagrammatic crane, from which it chiefly differs in the fact that its load is divided into two parts, that namely which is borne by the head of the bone, and that smaller portion which rests upon the great trochanter; but this merely amounts to saying that a *notch* has been cut out of the curved upper surface of the structure, and we have no difficulty in seeing that the anatomical arrangement of the trabeculae follows precisely the mechanical distribution of compressive and tensile stress or, in other words, accords perfectly

233

with the theoretical stress-diagram of the crane. The lines of stress are bundled close together along the sides of the shaft, and lost or concealed there in the substance of the solid wall of bone; but in and near the head of the bone, a peripheral shell of bone does not suffice to contain them, and they spread out through the central mass in the actual concrete form of bony trabeculae.[1]

Mutatis mutandis, the same phenomenon may be traced in any other bone which carries weight and is liable to flexure; and in the *os calcis* and the tibia, and more or less in all the bones of the lower limb, the arrangement is found to be very simple and clear.

Thus, in the *os calcis*, the weight resting on the head of the bone has to be transmitted partly through the backward-projecting heel to the ground, and partly forwards through its articulation with the cuboid bone, to the arch of the foot. We thus have, very much as in a triangular roof-tree, two compression-members sloping apart from one another; and these have to be bound together by a 'tie' or tension-member, corresponding to the third, horizontal member of the truss.

It is a simple corollary, confirmed by observation, that the trabeculae have a very different distribution in animals whose actions and

---

[1] Among other works on the mechanical construction of bone see: Bourgery, *Traité de l'anatomie* (*I. Ostéologie*), 1832 (with admirable illustrations of trabecular structure); L. Fick, *Die Ursachen der Knochenformen* (Göttingen, 1857); H. Meyer, 'Die Architektur der Spongiosa', *Arch. Anat. Phys.* 47 (1867), 615–28; *Statik u. Mechanik des menschlichen Knochengerüstes* (Leipzig, 1873); H. Wolfermann, 'Beitrag zur K. der Architektur der Knochen', *Arch. Anat. Phys.* (1872), p. 312; J. Wolff, 'Die innere Architektur der Knochen', *Arch. Anat. Phys.* 50 (1870); *Das Gesetz der Transformation bei Knochen* (1892); Y. Dwight, 'The Significance of Bone-architecture', *Mem. Boston Soc. Nat. Hist.* 4 (1886), 1; V. von Ebner, 'Der feinere Bau der Knochensubstanz', *Wien. Bericht*, 72 (1875); Anton Rauber, *Elastizität , nd Festigkeit der Knochen* (Leipzig, 1876); O. Meserer, *Elast. u. Festigk. d. menschlichen Knochen* (Stuttgart, 1880); Sir Donald MacAlister, 'How a Bone is Built', *English Illustr. Mag.* (1884), pp. 640–9; Rasumowsky, 'Architektonik des Fußskelets', *Int. Monatsschr. Anat.* (1889), p. 197; Zschokke, *Weitere Unters. über das Verhältnis der Knochenbildung zur Statik und Mechanik des Vertebratenskelets* (Zürich, 1892); W. Roux, *Ges. Abhandlungen über Entwicklungsmechanik der Organismen, Bd. I, Funktionelle Anpassung* (Leipzig, 1895); J. Wolff, 'Die Lehre von der funktionellen Knochengestalt', *Virchows Arch.* 155 (1899); R. Schmidt, 'Vergl. anat. Studien über den mechanischen Bau der Knochen und seine Vererbung', *Z. wiss. Zool.* 65 (1899), 65; B. Solger, 'Der gegenwärtige Stand der Lehre von der Knochenarchitektur', in Moleschott's *Unters. Naturlehre Menschen*, 16 (1899), 187; H. Triepel, 'Die Stossfestigkeit der Knochen', *Arch. Anat. Phys.* (1900); Gebhardt, 'Funktionell wichtige Anordnungsweisen der feineren und gröberen Bauelemente des Wirbelthierknochens, etc.', *Arch. Entw. Mech.* (1900–10); Revenstorf, 'Ueber die Transformation der Calcaneus-architektur', *Arch. Entw. Mech.* 23 (1907), 379; H. Bernhardt, *Vererbung der inneren Knochenarchitektur beim Menschen, und die Teleologie bei J. Wolff* (Inaug. Diss., München, 1907); Herm. Triepel, 'Die trajectoriellen Structuren' in *Einf. in die Physikalische Anatomie* (1908); A. F. Dixon, 'Architecture of the Cancellous Tissue forming the Upper End of the Femur', *J. Anat., Lond.* (3), 44 (1910), 223–30; A. Benninghoff, 'Ueber Leitsystem der Knochencompacta; Studien zur Architektur der Knochen', *Beitr. Anat. funktioneller Systeme*, 1 (1930).

attitudes are materially different, as in the aquatic mammals, such as the beaver and the seal.[1] And in much less extreme cases there are lessons to be learned from a study of the same bone in different animals, as the loads alter in direction and magnitude. The gorilla's heelbone resembles man's, but the load on the heel is much less, for the erect posture is imperfectly achieved: in a common monkey the heel is carried high, and consequently the direction of the trabeculae is still more changed. The bear walks on the sole of his foot, though less perfectly than does man, and the lie of the trabeculae is plainly analogous in the two; but in the bear more powerful strands than in

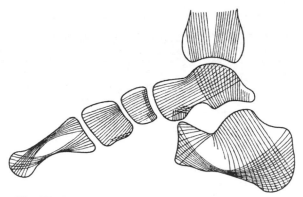

Fig. 102. Diagram of stress-lines in the human foot. From Sir D. McAlister, after H. Meyer.

the *os calcis* of man transmit the load forward to the toes, and less of it through the heel to the ground. In the leopard we see the full effect of tip-toe, or digitigrade, progression. The long hind part (or tuberosity) of the heel is now more a mere lever than a pillar of support; it is little more than a stiffened rod, with compression-members and tension-members in opposite bundles, inosculating orthogonally at the two ends.[2]

In the bird the small bones of the hand, dwarfed as they are in size, have still a deal to do in carrying the long primary flight-feathers, and in forming a rigid axis for the terminal part of the wing. The simple tubular construction, which answers well for the long, slender arm-bones, does not suffice where a still more efficient stiffening is required. In all the mechanical side of anatomy nothing can be more

[1] Cf. G. de M. Rudolf, 'Habit and the architecture of the mammalian femur', *J. Anat., Lond.* **56** (1922), 139–46.
[2] Cf. Fr. Weidenreich, 'Ueber formbestimmende Ursachen am Skelett, und die Erblichkeit der Knochenform', *Arch. Entw. Mech.* **51** (1922), 438–81.

beautiful than the construction of a vulture's metacarpal bone, as figured here (Fig. 103). The engineer sees in it a perfect Warren's truss, just such a one as is often used for a main rib in an aeroplane. Not only so, but the bone is better than the truss; for the engineer has to be content to set his V-shaped struts all in one plane, while in the bone they are put, with obvious but inimitable advantage, in a three-dimensional configuration.

So far, dealing wholly with the stresses and strains due to tension and compression, we have omitted to speak of a third very important factor in the engineer's calculations, namely what is known as

Fig. 103. Metacarpal bone from a vulture's wing; stiffened after the manner of a Warren's truss. From O. Prochnow, *Formenkunst der Natur*.

'shearing stress'. A shearing force is one which produces 'angular distortion' in a figure, or (what comes to the same thing) which tends to cause its particles to slide over one another. A shearing stress is a somewhat complicated thing, and we must try to illustrate it (however imperfectly) in the simplest possible way. If we build up a pillar, for instance, of flat horizontal slates, or of a pack of cards, a vertical load placed upon it will produce compression, but will have no tendency to cause one card to slide, or shear, upon another; and in like manner, if we make up a cable of parallel wires and, letting it hang vertically, load it evenly with a weight, again the tensile stress produced has no tendency to cause one wire to slip or shear upon another. But the case would have been very different if we had built up our pillar of cards or slates lying obliquely to the lines of pressure, for then at once there would have been a tendency for the elements of the pile to slip and slide asunder, and to produce what the geologists call 'a fault' in the structure.

This is as much as to say, that under this form of loading there is no shearing stress along or perpendicular to the lines of principal stress, or along the lines of maximum compression or tension; but shear has a definite value on all other planes, and a maximum value when it is inclined at 45° to the cross-section. This may be further

236

illustrated in various simple ways. When we submit a cubical block of iron to compression in the testing machine, it does not tend to give way by crumbling all to pieces, but always disrupts by shearing, and along some plane approximately at 45° to the axis of compression; this is known as Coulomb's Theory of Fracture, and, while subject to many qualifications, it is still an important first approximation to the truth. Again, in the beam which we have already considered under a bending-moment, we know that if we substitute for it a pack of cards, they will be strongly sheared on one another; and the shearing stress is greatest in the 'neutral zone', where neither tension nor compression is manifested: that is to say in the line which cuts at equal angles of 45° the orthogonally intersecting lines of pressure and tension.

In short we see that, while shearing *stresses* can by no means be got rid of, the danger of rupture or breaking-down under shearing stress is lessened the more we arrange the materials of our construction along the pressure-lines and tension-lines of the system; for *along these lines* there is no shear.[1]

To apply these principles to the growth and development of our bone, we have only to imagine a little trabecula (or group of trabeculae) being secreted and laid down fortuitously in any direction within the substance of the bone. If it lie in the direction of one of the pressure-lines, for instance, it will be in a position of comparative equilibrium, or minimal disturbance; but if it be inclined obliquely to the pressure-lines, the shearing force will at once tend to act upon it and move it away. This is neither more nor less than what happens when we comb our hair, or card a lock of wool: filaments lying in the direction of the comb's path remain where they were; but the others, under the influence of an oblique component of pressure, are sheared out of their places till they too come into coincidence with the lines of force. So straws show how the wind blows—or rather how it has been blowing. For every straw that lies askew to the wind's path tends to be sheared into it; but as soon as it has come to lie the way of the wind it tends to be disturbed no more, save (of course) by a violence such as to hurl it bodily away.

In the biological aspect of the case, we must always remember that our bone is not only a living, but a highly plastic structure; the little trabeculae are constantly being formed and deformed, demolished and formed anew. Here, for once, it is safe to say that 'heredity' need not and cannot be invoked to account for the configuration and

[1] It is also obvious that a free surface is always a region of zero-shear.

237

arrangement of the trabeculae: for we can see them at any time of life in the making, under the direct action and control of the forces to which the system is exposed. If a bone be broken and so repaired that its parts lie somewhat out of their former place, so that the pressure- and tension-lines have now a new distribution, before many weeks are over the trabecular system will be found to have been entirely remodelled, so as to fall into line with the new system of forces. And as Wolff pointed out, this process of reconstruction extends a long way off from the seat of injury, and so cannot be looked upon as a mere accident of the physiological process of healing and repair; for instance, it may happen that, after a fracture of the *shaft* of a long bone, the trabecular meshwork is wholly altered and reconstructed within the distant *extremities* of the bone. Moreover, in cases of transplantation of bone, for example when a diseased metacarpal is repaired by means of a portion taken from the lower end of the ulna, with astonishing quickness the plastic capabilities of the bony tissue are so manifested that neither in outward form nor inward structure can the old portion be distinguished from the new.

Herein then lies, so far as we can discern it, a great part at least of the physical causation of what at first sight strikes us as a purely functional adaptation: as a phenomenon, in other words, whose physical cause is as obscure as its final cause or end is apparently manifest.

Stress and Strain

Partly associated with the same phenomenon, and partly to be looked upon (meanwhile at least) as a fact apart, is the very important physiological truth that a condition of *strain*, the result of a *stress*, is a direct stimulus to growth itself. This indeed is no less than one of the cardinal facts of theoretical biology. The soles of our boots wear thin, but the soles of our feet grow thick, the more we walk upon them: for it would seem that the living cells are 'stimulated' by pressure, or by what we call 'exercise', to increase and multiply. The surgeon knows, when he bandages a broken limb, that his bandage is doing something more than merely keeping the parts together: and that the even, constant pressure which he skilfully applies is a direct encouragement of growth and an active agent in the process of repair. In the classical experiments of Sédillot,[1] the greater part of the shaft of the tibia was excised in some young puppies, leaving the whole weight

---

[1] Sédillot, 'De l'influence des fonctions sur la structure et la forme des organes', *C.R.* **59** (1864), 539; cf. **60** (1865), 97; **68** (1869), 1444.

of the body to rest upon the fibula. The latter bone is normally about one-fifth or -sixth of the diameter of the tibia; but under the new conditions, and under the 'stimulus' of the increased load, it grew till it was as thick or even thicker than the normal bulk of the larger bone. Among plant tissues this phenomenon is very apparent, and in a somewhat remarkable way; for a strain caused by a constant or increasing weight (such as that in the stalk of a pear while the pear is growing and ripening) produces a very marked increase of *strength* without any necessary increase of bulk, but rather by some histological, or molecular, alteration of the tissues. Hegler, Pfeffer, and others have investigated this subject, by loading the young shoot of a plant nearly to its breaking point, and then redetermining the breaking-strength after a few days. Some young shoots of the sunflower were found to break with a strain of 160 g; but when loaded with 150 g, and retested after 2 days, they were able to support 250 g; and being again loaded with something short of this, by next day they sustained 300 g, and a few days later even 400 g.[1]

The kneading of dough is an analogous phenomenon. The viscosity and perhaps other properties of the stuff are affected by the strains to which we have submitted it, and may thus be said to depend not only on the nature of the substance but on its history.[2] It is a long way from this simple instance, but we stretch across it easily in imagination, to the experimental growth of a nerve-fibre within a mass of clotted lymph: where, when we draw out the clot in one direction or another we lay down traction-lines, or tension-lines, and make of them a path for growth to follow.[3]

Such experiments have been amply confirmed, but so far as I am aware we do not know much more about the matter: we do not know, for instance, how far the change is accompanied by increase in number of the bast-fibres, through transformation of other tissues; or how far it is due to increase in size of these fibres; or whether it be not simply due to strengthening of the original fibres by some molecular change. But I should be much inclined to suspect that this last had a good deal to do with the phenomenon. We know nowadays that a railway axle, or any other piece of steel, is weakened

---

[1] *Op. cit.* Hegler's results are criticised by O. M. Ball, 'Einfluss von Zug auf die Ausbildung der Festigungsgewebe', *Jb. wiss. Bot.* 39 (1903), 305–41, and by H. Keller, 'Einfluss von Belastung und Lage auf die Ausbildung des Gewebes in Fruchtstielen' (Inaug. Diss. Kiel, 1904).

[2] Cf. R. K. Schofield and G. W. S. Blair, 'On Dough', *Proc. Roy. Soc.* A, **138**, 707; **139** (1932–3), 557; also Nadai and Wahl's *Plasticity* (1931). For analogous properties of hairs and fibres, see Shorter, *J. Textile Inst.* **15** (1924), etc.

[3] Cf. Ross Harrison's *Croonian Lecture* (1933).

by a constant succession of frequently interrupted strains; it is said to be 'fatigued', and its strength is restored by a period of rest. The converse effect of continued strain in a uniform direction may be illustrated by a homely example. The confectioner takes a mass of boiled sugar or treacle (in a particular molecular condition determined by the temperature to which it has been raised), and draws the soft sticky mass out into a rope; and then, folding it up lengthways, he repeats the process again and again. At first the rope is pulled out of the ductile mass without difficulty; but as the work goes on it gets harder to do, until all the man's force is used to stretch the rope. Here we have the phenomenon of increasing strength, following mechanically on a rearrangement of molecules, as the original isotropic condition is transmuted more and more into molecular asymmetry or anisotropy; and the rope apparently 'adapts itself' to the increased strain which it is called on to bear, all after a fashion which at least suggests a parallel to the increasing strength of the stretched and weighted fibre in the plant. For increase of strength by rearrangement of the particles we have already a rough illustration in our lock of wool or hank of tow. The tow will carry but little weight while its fibres are tangled and awry: but as soon as we have carded or 'hatchelled' it out, and brought all its long fibres parallel and side by side, we make of it a strong and useful cord.[1]

But the lessons which we learn from dough and treacle are nowadays plain enough in steel and iron, and become immensely more important in these. For here again plasticity is associated with a certain capacity for structural rearrangement, and increased strength again results therefrom. Elaborate processes of rolling, drawing, bending, hammering, and so on, are regularly employed to toughen and strengthen the material. The 'mechanical structure' of solids has become an important subject. And when the engineer talks of repeated loading, of elastic fatigue, of hysteresis, and other phenomena associated with plasticity and strain, the physiological analogues of these physical phenomena are perhaps not far away.

In some such ways as these, then, it would seem that we may co-ordinate, or hope to co-ordinate, the phenomenon of growth with certain of the beautiful structural phenomena which present themselves to our eyes as 'provisions', or mechanical adaptations,[2] for

[1] Cf. Sir Charles Bell's *Animal Mechanics*, ch. v, 'Of the tendons compared with cordage'.

[2] So P. Enriques (*Arch. Entw. Mech.* **20**, 1906), writing on the economy of material in the construction of a bone, admits that 'una certa impronta di teleologismo qua e la è rimasta, mio malgrado, in questo scritto'.

the display of strength where strength is most required. That is to say the origin, or causation, of the phenomenon would seem to lie partly in the tendency of growth to be accelerated under strain: and partly in the automatic effect of shearing strain, by which it tends to displace parts which grow obliquely to the direct lines of tension and of pressure, while leaving those in place which happen to lie parallel or perpendicular to those lines: an automatic effect which we can probably trace as working on all scales of magnitude, and as accounting therefore for the rearrangement of minute particles in the metal or the fibre, as well as for the bringing into line of the fibres within the plant, or of the trabeculae within the bone.

The Skeleton as a whole

But we may now attempt to pass from the study of the individual bone to the much wider and not less beautiful problems of mechanical construction which are presented to us by the skeleton as a whole.* Certain problems of this class are by no means neglected by writers on anatomy, and many have been handed down from Borelli, and even from older writers. For instance, it is an old tradition of anatomical teaching to point out in the human body examples of the three orders of levers;[1] again, the principle that the limb-bones tend to be shortened in order to support the weight of a very heavy animal is well understood by comparative anatomists, in accordance with Euler's law, that the weight which a column liable to flexure is capable of supporting varies inversely as the square of its length; and again, the statical equilibrium of the body, in relation for instance to the erect posture of man, has long been a favourite theme of the philosophical anatomist. But the general method, based upon that of graphic statics, to which we have been introduced in our study of a bone, has not, so far as I know, been applied to the general fabric of the skeleton. Yet it is plain that each bone plays a part in relation to the whole body, analogous to that which a little trabecula, or a little

[1] E.g. (1) the head, nodding backwards and forwards on a fulcrum, represented by the atlas vertebra, lying between the weight and the power; (2) the foot, raising on tiptoe the weight of the body against the fulcrum of the ground, where the weight is between the fulcrum and the power, the latter being represented by the *tendo Achillis*; (3) the arm, lifting a weight in the hand, with the power (i.e. the biceps muscle) between the fulcrum and the weight. (The second case, by the way, has been much disputed; cf. Haycraft in Schäfer's *Textbook of Physiology* (1900), p. 251.) Cf. (*int. al.*) G. H. Meyer, *Statik u. Mechanik der menschlichen Knochengerüste* (1873), pp. 13–25.

* For a recent discussion of the problem of the mechanics of the tetrapod skeleton see J. Gray, *J. Exp. Biol.* **20** (1944), 88.

group of trabeculae, plays within the bone itself: that is to say, in the normal distribution of forces in the body the bones tend to follow the lines of stress, and especially the pressure-lines. To demonstrate this in a comprehensive way would doubtless be difficult; for we should be dealing with a framework of very great complexity, and should have to take account of a great variety of conditions.[1] This framework is complicated as we see it in the skeleton, where (as we have said) it is only, or chiefly, the *struts* of the whole fabric which are represented; but to understand the mechanical structure in detail, we should have to follow out the still more complex arrangement of the *ties*, as represented by the muscles and ligaments, and we should also require much detailed information as to the weights of the various parts and as to the other forces concerned. Without these latter data we can only treat the question in a preliminary and imperfect way. But, to take once again a small and simplified part of a big problem, let us think of a quadruped (for instance, a horse) in a standing posture, and see whether the methods and terminology of the engineer may not help us, as they did in regard to the minute structure of the single bone. And let us note in passing that the 'standing posture', whether on two legs or on four, is no very common thing; but is (so to speak), with all its correlated anatomy, a privilege of the few.

Standing four-square upon its fore-legs and hind-legs, with the weight of the body suspended between, the quadruped at once suggests to us the analogy of a bridge, carried by its two piers. And if it occurs to us, as naturalists, that we never look at a standing quadruped without contemplating a bridge, so, conversely, a similar idea has occurred to the engineer; for Professor Fidler, in his *Treatise on Bridge-Construction*, deals with the chief descriptive part of his subject under the heading of 'The Comparative Anatomy of Bridges'.[2] The designation is most just, for in studying the various types of bridges we are studying a series of well-planned *skeletons*;[3] and (at

---

[1] Our problem is analogous to Thomas Young's problem of the best disposition of the timbers in a wooden ship (*Phil. Trans.* 1814, p. 303). He was not long in finding that the forces which act upon the fabric are very numerous and very variable, and that the best mode of resisting them, or best structural arrangement for ultimate strength, becomes an immensely complicated problem.

[2] By a bolder metaphor Fontenelle said of Newton that he had 'fait l'anatomie de la lumière'.

[3] In like manner, Clerk Maxwell could not help employing the term 'skeleton' in defining the mathematical conception of a 'frame', constituted by points and their interconnecting lines: in studying the equilibrium of which, we consider its different points as mutually acting on each other with forces whose directions are those of the lines joining each pair of points. Hence (says Maxwell), 'in order to exhibit the mechanical action of the frame in the most elementary manner, we may draw it as a *skeleton*, in which the different points are joined by straight lines, and we may indicate by numbers

242

the cost of a little pedantry) we might go even further, and study (after the fashion of the anatomist) the 'osteology' and 'desmology' of the structure, that is to say the bones which are represented by 'struts', and the ligaments, etc., which are represented by 'ties'. Furthermore, after the methods of the comparative anatomist, we may classify the families, genera and species of bridges according to their distinctive mechanical features, which correspond to certain definite conditions and functions.

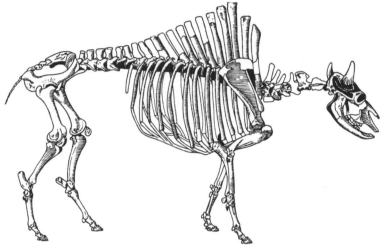

Fig. 104. Skeleton of a fossil bison. From O. P. Hay, Iowa Geological Survey Annual Report, 1912.

In more ways than one, the quadrupedal bridge is a remarkable one; and perhaps its most remarkable peculiarity is that it is a jointed and flexible bridge, remaining in equilibrium under considerable and sometimes great modifications of its curvature, such as we see, for instance, when a cat humps or flattens her back. The fact that *flexibility* is an essential feature in the quadrupedal bridge, while it is the last thing which an engineer desires and the first which he seeks to provide against, will impose certain important limiting conditions upon the design of the skeletal fabric. But let us begin by considering the quadruped at rest, when he stands upright and motionless upon his feet, and when his legs exercise no function save only to carry the weight of the whole body. So far as that function is concerned, we

attached to these lines the tensions or compressions in the corresponding pieces of the frame' (*Trans. Roy. Soc. Edinb.* **26** (1870), 1). It follows that the diagram so constructed represents a 'diagram of forces', in this limited sense that it is geometrical as regards the position and direction of the forces, but arithmetical as regards their magnitude. It is to just such a diagram that the animal's skeleton tends to approximate.

might now perhaps compare the horse's legs with the tall and slender piers of some railway bridge; but it is obvious that these jointed legs are ill-adapted to receive the *horizontal thrust* of any *arch* that may be placed atop of them. Hence it follows that the curved backbone of the horse, which appears to cross like an arch the span between his shoulders and his flanks, cannot be regarded as an *arch*, in the engineer's sense of the word. It resembles an arch in *form*, but not in *function*, for it cannot act as an arch unless it be held back at each end (as every arch is held back) by *abutments* capable of resisting the horizontal thrust; and these necessary abutments are not present in the structure. But in various ways the engineer can modify his superstructure so as to supply the place of these *external* reactions,

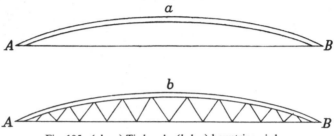

Fig. 105. (*above*) Tied arch; (*below*) bowstring girder.

which in the simple arch are obviously indispensable. Thus, for example, we may begin by inserting a straight steel tie, *AB* (Fig. 105), uniting the ends of the curved rib *AaB*; and this tie will supply the place of the external reactions, converting the structure into a 'tied arch', such as we may see in the roofs of many railway stations. Or we may go on to fill in the space between arch and tie by a 'web-system', converting it into what the engineer describes as a 'parabolic bowstring girder' (Fig. 105). In either case, the structure becomes an independent 'detached girder', supported at each end but not otherwise fixed, and consisting essentially of an upper compression-member, *AaB*, and a lower tension-member, *AB*. But again, in the skeleton of the quadruped, *the necessary tie, AB, of the simple bow-girder is not to be found*; and it follows that these comparatively simple types of bridge do not correspond to, nor do they help us to understand, the type of bridge which Nature has designed in the skeleton of the quadruped. Nevertheless, if we try to look, as an engineer would look, at the actual design of the animal skeleton and the actual distribution of its load, we find that the one is most admirably adapted to the other, according to the strict principles of engineering construction. The structure is not an arch, nor a tied arch, nor a bowstring girder: but it is strictly and beautifully comparable to the main girder of a double-armed cantilever bridge.

Obviously, in our quadrupedal bridge, the superstructure does not terminate (as it did in our former diagram) at the two points of support, but it extends beyond them, carrying the head at one end and sometimes a heavy tail at the other, upon projecting arms or 'cantilevers'.

In a typical cantilever bridge, such as the Forth Bridge (Fig. 106), a certain simplification is introduced. For each pier carries, in this case, its own double-armed cantilever, linked by a short connecting girder to the next, but so jointed to it that no weight is transmitted

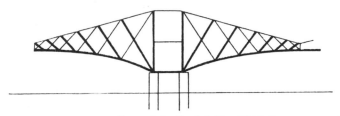

Fig. 106. A two-armed cantilever of the Forth Bridge. Thick lines, compression-members (bones); thin lines, tension-members (ligaments).

from one cantilever to another. The bridge in short is *cut* into separate sections, practically independent of one another; at the joints a certain amount of bending is not precluded, but shearing strain is evaded; and each pier carries only its own load. By this arrangement the engineer finds that design and construction are alike simplified and facilitated. In the horse or the ox, it is obvious that the two piers of the bridge, that is to say the fore-legs and the hind-legs, do not bear (as they do in the Forth Bridge) separate and independent loads, but the whole system forms a continuous structure. In this case, the calculation of the loads will be a little more difficult and the corresponding design of the structure a little more complicated. We shall accordingly simplify our problem very considerably if, to begin with, we look upon the quadrupedal skeleton as constituted of two separate systems, that is to say of two balanced cantilevers, one supported on the fore-legs and the other on the hind; and we may deal afterwards with the fact that these two cantilevers are not independent, but are bound up in one common field of force and plan of construction.

In both horse and ox it is plain that the two cantilever systems

245

into which we may thus analyse the quadrupedal bridge are unequal in magnitude and importance. The fore-part of the animal is much bulkier than its hind-quarters, and the fact that the fore-legs carry, as they so evidently do, a greater weight than the hind-legs has long been known and is easily proved; we have only to walk a horse on to a weighbridge, weigh first his fore-legs and then his hind-legs, to discover that what we may call his front half weighs a good deal more than what is carried on his hind feet, say about three-fifths of the whole weight of the animal.

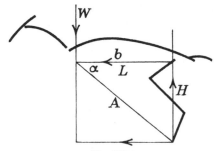

Fig. 107.

The great (or anterior) cantilever then, in the horse, is constituted by the heavy head and still heavier neck on one side of that pier which is represented by the fore-legs, and by the dorsal vertebrae carrying a large part of the weight of the trunk upon the other side; and this weight is so balanced over the fore-legs that the cantilever, while 'anchored' to the other parts of the structure, transmits but little of its weight to the hind-legs, and the amount so transmitted will vary with the attitude of the head and with the position of any artificial load.[1] Under certain conditions, as when the head is thrust well forward, it is evident that the hind-legs will be actually relieved of a portion of the comparatively small load which is their normal share.

But here we pass from the statical problem to the dynamical, from the horse at rest to the horse in motion, from the observed fact that weight lies mainly over the fore-legs to the question of what advantage is gained by such a distribution of the load. Taking the hind-legs as the main propulsive agency, as we may now safely do, the moment of propulsion is about the hind-hooves; then (as we

[1] When the jockey crouches over the neck of his racehorse, and when Tod Sloan introduced the 'American seat', the avowed object in both cases is to relieve the hind-legs of weight, and so leave them free for the work of propulsion. On the share taken by the hind-limbs in this latter duty, and other matters, cf. Stillman, *The Horse in Motion* (1882), p. 69.

see in Fig. 107) we may take the weight, $W = A \sin \alpha$, and the propulsive force,

$$f = A \cos \alpha \quad \text{and} \quad \frac{W}{f} = \frac{H}{L}, \; WL = fH$$

being the balanced condition. From the statical point of view the load must balance over the fore-legs; from the dynamical point of view it might well lie even farther forward. And when the jockey crouches over the horse's neck, and when Tod Sloan introduced the 'American seat', both show a remarkable, though perhaps unconscious, insight into the dynamical proposition.

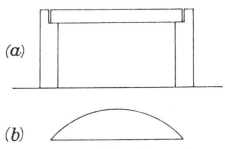

Fig. 108. (*a*) Span of proposed bridge; (*b*) stress diagram, or diagram of bending-moments.[1]

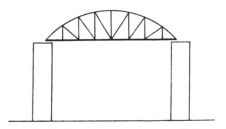

Fig. 109. The bridge constructed, as a parabolic girder.

Our next problem is to discover, in a rough and approximate way, some of the structural details which the balanced load upon the double cantilever will impress upon the fabric.

Working by the methods of graphic statics, the engineer's task is, in theory, one of great simplicity. He begins by drawing in outline

[1] This and the following diagrams are borrowed and adapted from Professor Fidler's *Bridge Construction*. We may reflect with advantage on Clerk Maxwell's saying that 'the use of diagrams is a particular instance of that method of symbols which is so powerful an aid in the advancement of science'; and on his explanation that 'a diagram differs from a picture in this respect: that in a diagram no attempt is made to represent those factors of the actual material system which are not the special objects of our study'.

the structure which he desires to erect; he calculates the stresses and bending-moments necessitated by the dimensions and load on the structure; he draws a new diagram *representing these forces*, and he designs and builds his fabric on the lines of this statical diagram. He does, in short, precisely what we have seen *Nature* doing in the case of the bone. For if we had begun, as it were, by blocking out the femur roughly, and considering its position and dimensions, its means of support and the load which it has to bear, we could have proceeded at once to draw the system of stress-lines which must occupy that field of force: and to precisely those stress-lines has Nature kept in the building of the bone, down to the minute arrangement of its trabeculae.

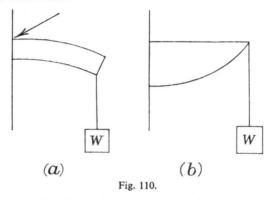

(a)   (b)

Fig. 110.

The essential function of a bridge is to stretch across a certain span, and carry a certain definite load; and this being so, the chief problem in the designing of a bridge is to provide due resistance to the 'bending-moments' which result from the load. These bending-moments will vary from point to point along the girder, and taking the simplest case of a uniform load, whether supported at one or both ends, they will be represented by points on a parabola. If the girder be of uniform depth and section, that is to say if its two flanges, respectively under tension and compression, be equal and parallel to one another, then the stress upon these flanges will vary as the bending-moments, and will accordingly be very severe in the middle and will dwindle towards the ends. But if we make the *depth* of the girder everywhere proportional to the bending-moments, that is to say if we copy in the girder the outlines of the bending-moment diagram, then our design will automatically meet the circumstances of the case, for the horizontal stress in each flange will now be uniform throughout the length of the girder. In short, in Professor Fidler's words, 'Every diagram of moments represents the outline of a framed structure which will carry the given load with a uniform horizontal stress in the principal members'.

In the above diagrams (Fig 110, *a*, *b*) (which are taken from the original ones of Culmann), we see at once that the loaded beam or bracket (*a*) has a 'danger-point' close to its fixed base, that is to say at the point remotest from its load. But in the parabolic bracket (*b*) there is no danger-point at all, for the dimensions of the structure

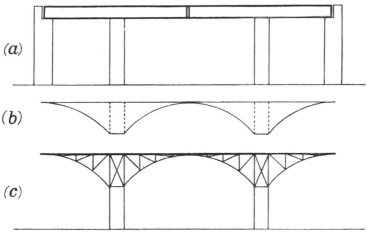

Fig. 111.

are made to increase *pari passu* with the bending-moments: stress and resistance vary together. Again in Fig. 111, we have a simple span (*a*), with its stress diagram (*b*); and in (*c*) we have the corresponding parabolic girder, whose stresses are now uniform throughout. In fact we see that, by a process of conversion, the stress diagram in each case becomes the structural diagram in the other.[1] Now all this is but the modern rendering of one of Galileo's most famous propositions. In the Dialogue which we have already noted more than once,[2] Sagredo says 'It would be a fine thing if one could discover the proper shape to give a solid in order to make it equally resistant

[1] The method of constructing *reciprocal diagrams*, of which one should represent the outlines of a frame and the other the system of forces necessary to keep it in equilibrium, was first indicated in Culmann's *Graphische Statik*; it was greatly developed soon afterwards by Macquorn Rankine (*Phil. Mag.* February 1864, and *Applied Mechanics, passim*), to whom the application of the principle to engineering practice is mainly due. See also Fleeming Jenkin, 'On the Practical Application of Reciprocal Figures to the Calculation of Strains in Framework', *Trans. Roy. Soc. Edinb.* 25 (1869), 441–8; and Clerk Maxwell, *ibid.* 26 (1870), 9, and *Phil. Mag.* (April, 1864).

[2] *Dialogues concerning Two New Sciences* (1638); Crew and Salvio's translation, pp. 140 *seq.*

at every point, in which case a load placed at the middle would not produce fracture more easily than if placed at any other point'.[1] And Galileo (in the person of Salviati) first puts the problem into its more general form; and then shows us how, by giving a parabolic outline to our beam, we have its simple and comprehensive solution. It was such teaching as this that led R. A. Millikan to say that 'we owe our present-day civilisation to Galileo'.

In the case of our cantilever bridge, we show the primitive girder in Fig. 111a, with its bending-moment diagram (b); and it is evident that, if we turn this diagram upside down, it will still be illustrative, just as before, of the bending-moments from point to point: for as yet it is merely a diagram, or graph, of relative magnitudes.

To either of these two stress diagrams, direct or inverted, we may fit the design of the construction, as in figs. 111 and 112.

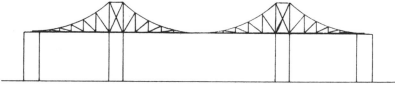

Fig. 112.

Now in different animals the amount and distribution of the load differ so greatly that we can expect no single diagram, drawn from the comparative anatomy of bridges, to apply equally well to all the cases met with in the comparative anatomy of quadrupeds; but nevertheless we have already gained an insight into the general principles of 'structural design' in the quadrupedal bridge.

In our last diagram the upper member of the cantilever is under tension; it is represented in the quadruped by the *ligamentum nuchae* on the one side of the cantilever, and by the supraspinous ligaments of the dorsal vertebrae on the other. The compression-member is similarly represented, on both sides of the cantilever, by the vertebral column, or rather by the *bodies* of the vertebrae; while the web, or 'filling', of the girders, that is to say the upright or sloping members which extend from one flange to the other, is represented on the one hand by the spines of the vertebrae, and on the other hand by the oblique interspinous ligaments and muscles—that is to say, by compression-members and tension-members inclined in opposite directions to one another. The high spines over the quadruped's withers

---

[1] As in the great case of the Eiffel Tower, *supra*, p. 21.

are no other than the high struts which rise over the supporting piers in the parabolic girder, and correspond to the position of the maximal bending-moments. The fact that these tall vertebrae of the withers usually slope backwards, sometimes steeply, in a quadruped, is easily and obviously explained.[1] For each vertebra tends to act as a 'hinged lever', and its spine, acted on by the tensions transmitted by the ligaments on either side, takes up its position as the diagonal of the parallelogram of forces to which it is exposed.

It happens that in these comparatively simple types of cantilever bridge the whole of the parabolic curvature is transferred to one or other of the principal members, either the tension-member or the compression-member as the case may be. But it is of course equally permissible to have both members curved, in opposite directions. This, though not exactly the case in the Forth Bridge, is approximately so; for here the main compression-member is curved or arched, and the main tension-member slopes downwards on either side from its maximal height above the piers. In short, the Forth Bridge (Fig. 106) is a nearer approach than either of the other bridges which we have illustrated to the plan of the quadrupedal skeleton; for the main compression-member almost exactly recalls the form of the backbone, while the main tension-member, though not so closely similar to the supraspinous and nuchal ligaments, corresponds to the plan of these in a somewhat simplified form.

We may now pass without difficulty from the two-armed cantilever supported on a single pier, as it is in each separate section of the Forth Bridge, or as we have imagined it to be in the fore-quarters of a horse, to the condition which actually exists in a quadruped, when a two-armed cantilever has its load distributed over two separate piers. This is not precisely what an engineer calls a 'continuous' girder, for that term is applied to a girder which, as a continuous structure, has three supports and crosses two or more spans, while here there is only one. But nevertheless, this girder is *effectively* continuous from the head to the tip of the tail; and at each point of support (*A* and *B*) it is subjected to the negative bending-moment due

---

[1] The form and direction of the vertebral spines have been frequently and elaborately described; cf. (e.g.) H. Gottlieb, 'Die Anticlinie der Wirbelsäule der Säugethiere', *Morph. Jb.* **49** (1915), 179–220, and many works quoted therein. According to Morita, 'Ueber die Ursachen der Richtung und Gestalt der thoracalen Dornfortsätze der Säugethierwirbelsäule' (p. 201), various changes take place in the direction or inclination of these processes in rabbits, after section of the interspinous ligaments and muscles. These changes seem to be very much what we should expect, on simple mechanical grounds. See also O. Fischer, *Theoretische Grundlagen für eine Mechanik der lebenden Körper* (Leipzig, 1906), pp. x, 372.

to the overhanging load on each of the projecting cantilever arms *AH* and *BT*. The diagram of bending-moments will (according to the ordinary conventions) lie below the base line (because the moments are negative), and must take some such form as that shown in the diagram: for the girder must suffer its greatest bending stress not at the centre, but at the two points of support *A* and *B*, where the moments are measured by the vertical ordinates. It is plain that this figure only differs from a representation of *two* independent two-armed cantilevers in the fact that there is no point midway in the span where the bending-moment vanishes, but only a region between the two piers in which it tends to diminish.

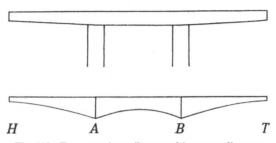

Fig. 113. Two-armed cantilever and its stress-diagram.

The diagram effects a graphic summation of the positive and negative moments, but its form may assume various modifications according to the method of graphic summation which we choose to adopt; and it is obvious also that the form of the diagram may assume many modifications of detail according to the actual distribution of the load. In all cases the essential points to be observed are these: first, that the girder which is to resist the bending-moments induced by the load must possess its two principal members—an upper tension-member or tie, represented by ligament (whose tension doubtless varies along its length), and a lower compression-member represented by bone: these members being united by a web represented by the vertebral spines with their interspinous ligaments, and being placed one above the other in the order named because the moments are negative; secondly, we observe that the depth of the web, or distance apart of the principal members—that is to say the height of the vertebral spines—must be proportional to the bending-moment at each point along the length of the girder.

In the case of an animal carrying most of his weight upon his fore-legs, as the horse or the ox do, the bending-moment diagram will be unsymmetrical, after the fashion of Fig. 114, the precise form depending on the distribution of weights and distances.

On the other hand the Dinosaur, with his light head and enormous tail, would give us a moment-diagram with the opposite kind of asymmetry, the greatest bending stress being now found over the haunches, at *B* (Fig. 115). A glance at the skeleton of Diplodocus will show us the high vertebral spines over the loins, in precise correspondence with the requirements of this diagram: just as in the horse, under the opposite conditions of load, the highest vertebral spines are those of the withers, that is to say those of the posterior cervical and anterior dorsal vertebrae.

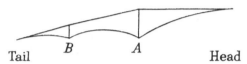

Tail   *B*   *A*   Head

'Fig. 114. Stress-diagram of horse's backbone.

We have now not only dealt with the general resemblance, both in structure and in function, of the quadrupedal backbone with its associated ligaments to a double-armed cantilever girder, but we have begun to see how the characters of the vertebral system must differ in different quadrupeds, according to the conditions imposed

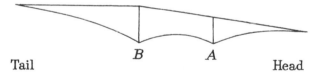

Tail   *B*   *A*   Head

Fig. 115. Stress-diagram of backbone of Dinosaur.

by the varying distribution of the load: and in particular how the height of the vertebral spines which constitute the web will be in a definite relation, as regards magnitude and position, to the bending-moments induced thereby. We should require much detailed information as to the actual weights of the several parts of the body before we could follow out quantitatively the mechanical efficiency of each type of skeleton; but in an approximate way what we have already learnt will enable us to trace many interesting correspondences between structure and function in this particular part of comparative anatomy. We must, however, be careful to note that the great cantilever system is not of necessity constituted by the vertebral column and its ligaments alone, but that the pelvis, firmly united as it is to the sacral

253

vertebrae, and stretching backwards far beyond the acetabulum, becomes an intrinsic part of the system; and helping (as it does) to carry the load of the abdominal viscera, it constitutes a great portion of the posterior cantilever arm, or even its chief portion in cases where the size and weight of the tail are insignificant, as is the case in the majority of terrestrial mammals.

We may also note here, that just as a bridge is often a 'combined' or composite structure, exhibiting a combination of principles in its construction, so in the quadruped we have, as it were, another girder supported by the same piers to carry the viscera; and consisting of an inverted parabolic girder, whose compression-member is again constituted by the backbone, its tension-member by the line of the sternum and the abdominal muscles, while the ribs and intercostal muscles play the part of the web or filling.

A very few instances must suffice to illustrate the chief variations in the load, and therefore in the bending-moment diagram, and therefore also in the plan of construction, of various quadrupeds. But let us begin by setting forth, in a few cases, the actual weights which are borne by the fore-limbs and the hind-limbs, in our quadrupedal bridge.[1]

	Gross weight (cwt)	On fore-feet (cwt)	On hind-feet (cwt)	% on fore-feet	% on hind-feet
Camel (Bactrian)	14·25	9·25	4·5	67·3	32·7
Llama	2·75	1·75	0·875	66·7	33·3
Elephant (Indian)	35·75	20·5	14·75	58·2	41·8
Horse	8·25	4·75	3·5	57·6	42·4
Horse (large Clydesdale)	15·5	8·5	7·0	54·8	45·2

It will be observed that in all these animals the load upon the fore-feet preponderates considerably over that upon the hind, the preponderance being rather greater in the elephant than in the horse, and markedly greater in the camel and the llama than in the other two. But while these weights are helpful and suggestive, it is obvious that they do not go nearly far enough to give us a full insight into the constructional diagram to which the animals are conformed. For such a purpose we should require to weigh the total load, not in two portions but in many; and we should also have to take close account of the general form of the animal, of the relation between that form and the distribution of the load, and of the actual directions of each bone and ligament by which the forces of compression and tension were transmitted. All this lies beyond us for the present; but never-

[1] I owe the first four of these determinations to the kindness of Sir P. Chalmers Mitchell, who had them made for me at the Zoological Society's Gardens; while the great Clydesdale carthorse was weighed for me by a friend in Dundee.

theless we may consider, very briefly, the principal cases involved in our enquiry, of which the above animals form a partial and preliminary illustration.

(1) Wherever we have a heavily loaded anterior cantilever arm, that is to say whenever the head and neck represent a considerable fraction of the whole weight of the body, we tend to have large bending-moments over the fore-legs, and correspondingly high spines over the vertebrae of the withers. This is the case in the great majority of four-footed terrestrial animals, the chief exceptions being found in animals with comparatively small heads but large and heavy tails, such as the anteaters or the Dinosaurian reptiles, and also (very

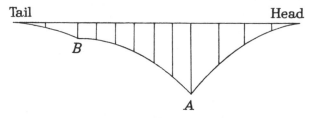

Fig. 116. Stress-diagram of Titanotherium.

naturally) in animals such as the crocodile, where the 'bridge' can scarcely be said to be developed, for the long heavy body sags down to rest upon the ground. The case is sufficiently exemplified by the horse, and still more notably by the stag, the ox, or the pig. It is illustrated in the skeleton of a bison (Fig. 104), or in the accompanying diagram of the conditions in the great extinct Titanotherium.

(2) In the elephant and the camel we have similar conditions, but slightly modified. In both cases, and especially in the latter, the weight on the fore-quarters is relatively large; and in both cases the bending-moments are all the larger, by reason of the length and forward extension of the camel's neck and the forward position of the heavy tusks of the elephant. In both cases the dorsal spines are large, but they do not strike us as exceptionally so; but in both cases, and especially in the elephant, they slope backwards in a marked degree. Each spine, as already explained, must in all cases assume the position of the diagonal in the parallelogram of forces defined by the tensions acting on it at its extremity; for it constitutes a 'hinged lever', by which the bending-moments on either side are automatically balanced; and it is plain that the more the spine slopes backwards the more it indicates a relatively large strain thrown upon the great ligament of the neck, and a relief of strain upon the more directly acting, but

255

weaker, ligaments of the back and loins. In both cases, the bending-moments would seem to be more evenly distributed over the region of the back than, for instance, in the stag, with its light hind-quarters and heavy load of antlers: and in both cases the high 'girder' is considerably prolonged, by an extension of the tall spines backwards in the direction of the loins. When we come to such a case as the mammoth, with its immensely heavy and immensely elongated tusks, we perceive at once that the bending-moments over the fore-legs are now very severe; and we see also that the dorsal spines in this region are much more conspicuously elevated than in the ordinary elephant.

(3) In the case of the giraffe we have, without doubt, a very heavy load upon the fore-legs, though no weighings are at hand to define the ratio: but as far as possible this disproportionate load would seem to be relieved by help of a downward as well as backward thrust, through the sloping back to the unusually low hind-quarters. The dorsal spines of the vertebrae are very high and strong, and the whole girder-system very perfectly formed. The elevated rather than protruding position of the head lessens the anterior bending-moment as far as possible, but it leads to a strong compressional stress transmitted almost directly downwards through the neck: in correlation with which we observe that the bodies of the cervical vertebrae are exceptionally large and strong, and steadily increase in size and strength from the head downwards.

(4) In the kangaroo, the fore-limbs are entirely relieved of their load, and accordingly the tall spines over the withers, which were so conspicuous in all heavy-headed *quadrupeds*, have now completely vanished. The creature has become bipedal, and body and tail form the extremities of *a single* balanced cantilever, whose maximal bending-moments are marked by strong, high lumbar and sacral vertebrae, and by iliac bones of peculiar form, of exceptional strength and nearly upright position.

Precisely the same condition is illustrated in the Iguanodon, and better still by reason of the great bulk of the creature and of the heavy load which falls to be supported by the great cantilever and by the hind-legs which form its piers. The long heavy body and neck require a balance-weight (as in the kangaroo) in the form of a long heavy tail; and the double-armed cantilever, so constituted, shows a beautiful parabolic curvature in the graded heights of the whole series of vertebral spines, which rise to a maximum over the haunches and die away slowly towards the neck and towards the tip of the tail.

256

(5) In the case of some of the great American fossil reptiles such as Diplodocus, it has always been a more or less disputed question whether or not they assumed, like Iguanodon, an erect, bipedal attitude. In all of them we see an elongated pelvis, and, in still more marked degree, we see elevated spinous processes of the vertebrae over the hind-limbs; in all of them we have a long heavy tail, and in most of them we have a marked reduction in size and weight both of the fore-limb and of the head itself. The great size of these animals is not of itself a proof against the erect attitude; because it might well have been accompanied by an aquatic or partially submerged

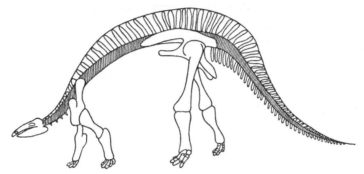

Fig. 117. Diagram of Stegosaurus.

habitat, and the crushing stress of the creature's huge bulk proportionately relieved. But we must consider each such case in the whole light of its own evidence; and it is easy to see that, just as the quadrupedal mammal may carry the greater part but not all of its weight upon its fore-limbs, so a heavy-tailed reptile may carry the greater part upon its hind-limbs, without this process going so far as to relieve its fore-limbs of all weight whatsoever. This would seem to be the case in such a form as Diplodocus, and also in Stegosaurus, whose restoration by Marsh is doubtless substantially correct.[1] The fore-limbs, though comparatively small, are obviously fashioned for support, but the weight which they have to carry is far less than that which the hind-limbs bear. The head is small and the neck short, while on the other hand the hind-quarters and the tail are big and massive. The backbone bends into a great double-armed cantilever,

[1] This pose of Diplodocus, and of other Sauropodous reptiles, has been much discussed. Cf. (*int. al.*) O. Abel, *Abh. zool. bot. Ges. Wien*, 5, 1909–10 (60 pp.); Tornier, *S.B. Ges. Naturf. Fr. Berl.* (1909), pp. 193–209; O. P. Hay, *Amer. Nat.* (October 1908); *Trans. Wash. Acad. Sci.* 42 (1910), 1–25; Holland, *Amer. Nat.* (May 1910), pp. 259–83; Matthew, *ibid.* pp. 547–60; C. W. Gilmore (*Restoration of Stegosaurus*), *Proc. U.S. Nat. Mus.* (1915).

culminating over the pelvis and the hind-limbs, and here furnished with its highest and strongest spines to separate the tension-member from the compression-member of the girder. The fore-legs form a secondary supporting pier to this great continuous cantilever, the greater part of whose weight is poised upon the hind-limbs alone.

(6) In the slender body of a weasel, neither head nor tail is such as to form an efficient cantilever; and though the lithe body is arched in active exercise, our parallel of the bridge no longer works well. What else to compare it with is far from clear; but the mechanism has some resemblance (perhaps) to an elastic spring. Animals of this habit of body are all small; their bodily weight is a light burden, and gravity becomes an ineffectual force.

(7) An abnormal and very curious case is that of the sloth, which hangs by hooked hands and feet, head downwards, from high branches in the Brazilian forest. The vertebrae are unusually numerous, they are all much alike one to another, and (as we might well suppose) the whole pensile chain of vertebrae hangs in what closely approximates to a catenary curve.[1]

Aquatic Animals

We find a highly important corollary in the case of aquatic animals. For here the effect of gravity is neutralised; we have neither piers nor cantilevers; and we find accordingly in all aquatic mammals of whatsoever group—whales, seals or sea-cows—that the high arched vertebral spines over the withers, or corresponding structures over the hind-limbs, have both entirely disappeared.

But in the whale or dolphin (and not less so in the aquatic bird), *stiffness* must be ensured in order to enable the muscles to act against the resistance of the water in the act of swimming; and accordingly Nature must provide against bending-moments irrespective of gravity. In the dolphin, at any rate as regards its tail-end, the conditions will be not very different from those of a column or beam with fixed ends, in which,

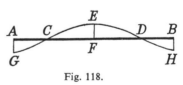

Fig. 118.

under deflection, there will be two points of contrary flexure, as at C, D, in Fig. 118.

---

[1] A *heavy* cord, or a cord carrying equal weights for equal distances along its line, hangs in a catenary: imagine it frozen and inverted, and we have an arch, carrying the same sort of load, and under compression only. On the other hand, a flexible cable (itself of negligible weight), carrying a uniform load along the line of its horizontal projection, hangs in the form of a parabola.

Here, between C and D we have a varying bending-moment, represented by a continuous curve with its maximal elevation midway between the points of inflection. And correspondingly, in our dolphin, we have a continuous series of high dorsal spines, rising to a maximum about the middle of the animal's body, and falling to nil at some distance from the end of the tail. It is their business (as usual) to keep the tension-member, represented by the strong supraspinous ligaments, wide apart from the compression-member, which is as usual represented by the backbone itself. But in our diagram we see that on the farther side of C and D we have a *negative* curve of bending-moments, or bending-moments in a contrary direction. Without enquiring how these stresses are precisely met towards the dolphin's head (where the coalesced cervical vertebrae suggest themselves as a partial explanation), we see at once that towards the tail they are met by the strong series of chevron-bones, which in the caudal region, where tall *dorsal* spines are no longer needed, take their place *below* the vertebrae, in precise correspondence with the bending-moment diagram. In many cases other than these aquatic ones, when we have to deal with animals with long and heavy tails (like the Iguanodon and the kangaroo of which we have already spoken), we are apt to meet with similar, though usually shorter chevron-bones; and in all these cases we may see without difficulty that a negative bending-moment in the vertical direction has to be resisted or controlled.

In the dolphin we may find an illustration of the fact that not only is it necessary to provide for rigidity in the vertical direction but often also in the horizontal, where a tendency to bending must be resisted on either side. This function is effected in part by the ribs with their associated muscles, but they extend but a little way and their efficacy for this purpose can be but small. We have, however, behind the region of the ribs and on either side of the backbone a strong series of elongated and flattened transverse processes, forming a web for the support of a tension-member in the usual form of ligament, and so playing a part precisely analogous to that performed by the dorsal spines in the same animal. In an ordinary fish, such as a cod or a haddock, we see precisely the same thing: the backbone is stiffened by the indispensable help of its *three series* of ligament-connected processes, the dorsal and the two transverse series; but there are no such stiffeners in the eel. When we come to the region of the tail, where rigidity gives place to lateral flexibility, the three stiffeners give place to two—the dorsal and haemal spines of the caudal vertebrae. And here we see that the three series of processes, or struts, tend (when all three are present) to be arranged wellnigh at equal angles, of 120°, with one another, giving the greatest and most uniform strength of which such a system is capable. On the other hand, in a flat fish, such as a plaice, where from the natural mode of progression it is necessary that the backbone should be flexible in one direction while stiffened in another, we find the whole outline of the fish comparable to that of a double bowstring girder, the compression-member being (as usual) the backbone itself, the tension-member

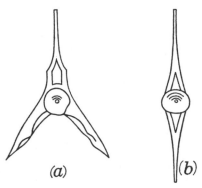

Fig. 119. (a) Dorsal and (b) caudal vertebrae of haddock.

on either side being constituted by the interspinous ligaments and muscles, while the web or filling is very beautifully represented by the long and evenly graded neural and haemal spines, which spring symmetrically up and down from each individual vertebra.

In the skeleton of the flat fishes, the web of the otherwise perfect parabolic girder has to be cut away and encroached on to make room for the viscera. When the body is long and the vertebrae many, as in the sole, the space required is small compared with the length of the girder, and the strength of the latter is not much impaired. In the shorter, rounder kinds with fewer vertebrae, like the turbot, the visceral cavity is large compared with the length of the fish, and its presence would seem to weaken the girder very seriously. But Nature repairs the breach by framing in the hinder part of the space with a strong curved bracket or angle-iron, which takes the place very efficiently of the bony struts which have been cut away.

The main result at which we have now arrived, in regard to the construction of the vertebral column and its associated parts, is

260

that we may look upon it as a certain type of *girder*, whose depth is everywhere very nearly proportional to the height of the corresponding ordinate in the diagram of moments: just as it is in a girder designed by a modern engineer. In short, after the nineteenth or twentieth century engineer has done his best in framing the design of a big cantilever, he may find that some of his best ideas had, so to speak, been anticipated ages ago in the fabric of the great saurians and the larger mammals.

But it is possible that the modern engineer might be disposed to criticise the skeleton girder at two or three points; and in particular he might think the girder, as we see it for instance in Diplodocus or Stegosaurus, not deep enough for carrying the animal's enormous weight of some twenty tons. If we adopt a much greater depth (or ratio of depth to length) as in the modern cantilever, we shall greatly increase the *strength* of the structure; but at the same time we should greatly increase its *rigidity*, and this is precisely what, in the circumstances of the case, it would seem that Nature is bound to avoid. We need not suppose that the great saurian was by any means active and limber; but a certain amount of activity and flexibility he was bound to have, and in a thousand ways he would find the need of a backbone that should be *flexible* as well as *strong*. Now this opens up a new aspect of the matter and is the beginning of a long, long story, for in every direction this double requirement of strength and flexibility imposes new conditions upon the design. To represent all the correlated quantities we should have to construct not only a diagram of moments but also a diagram of elastic deflection and its so-called 'curvature'; and the engineer would want to know something more about the *material* of the ligamentous tension-member—its flexibility, its modulus of elasticity in direct tension, its elastic limit, and its safe working stress.

In various ways our structural problem is beset by 'limiting conditions'. Not only must rigidity be associated with flexibility, but also stability must be ensured in various positions and attitudes; and the primary function of support or weight-carrying must be combined with the provision of *points d'appui* for the muscles concerned in locomotion. We cannot hope to arrive at a numerical or quantitative solution of this complicated problem, but we have found it possible to trace it out in part towards a qualitative solution. And speaking broadly we may certainly say that in each case the problem has been solved by Nature herself, very much as she solves the difficult problems of minimal areas in a system of soap-bubbles; so that each animal is fitted with a backbone adapted to his own individual needs, or (in other words) corresponding to the mean resultant of the many stresses to which as a mechanical system it is exposed.

Throughout this short discussion of the principles of construction, we see the same general principles at work in the skeleton as a whole as we recognised in the plan and construction of an individual bone. That is to say, we see a tendency for material to be laid down just in the lines of *stress*, and so to evade thereby the distortions and disruptions due to *shear*. In these phenomena there lies a definite law of growth, whatever its ultimate expression or explanation may come to be. Let us not press either argument or hypothesis too far: but be content to see that skeletal form, as brought about by growth, is to a very large extent determined by mechanical considerations, and tends to manifest itself as a diagram, or reflected image, of mechanical stress. If we fail, owing to the immense complexity of the case, to unravel all the mathematical principles involved in the construction of the skeleton, we yet gain something, and not a little, by applying this method to the familiar objects of anatomical study: *obvia conspicimus, nubem pellente mathesi.*[1]

The Whole Animal

Before we leave this subject of mechanical adaptation, let us dwell once more for a moment upon the considerations which arise from our conception of a field of force, or field of stress, in which tension and compression (for instance) are inevitably combined, and are met by the materials naturally fitted to resist them. It has been remarked over and over again how harmoniously the whole organism hangs together, and how throughout its fabric one part is related and fitted to another in strictly functional correlation. But this conception, though never denied, is sometimes apt to be forgotten in the course of that process of more and more minute analysis by which, for simplicity's sake, we seek to unravel the intricacies of a complex organism.

As we analyse a thing into its parts or into its properties, we tend to magnify these, to exaggerate their apparent independence, and to hide from ourselves (at least for a time) the essential integrity and individuality of the composite whole. We divide the body into its organs, the skeleton into its bones, as in very much the same fashion

[1] The motto was Macquorn Rankine's, in 1857; cf. *Trans. Roy. Soc. Edinb.* 26 (1872), 715.

we make a subjective analysis of the mind, according to the teachings of psychology, into component factors: but we know very well that judgment and knowledge, courage or gentleness, love or fear, have no separate existence, but are somehow mere manifestations, or imaginary coefficients, of a most complex integral. And likewise, as biologists, we may go so far as to say that even the bones themselves are only in a limited and even a deceptive sense, separate and individual things. The skeleton begins as a *continuum*, and a *continuum* it remains all life long. The things that link bone with bone, cartilage, ligaments, membranes, are fashioned out of the same primordial tissue, and come into being *pari passu* with the bones themselves. The entire fabric has its soft parts and its hard, its rigid and its flexible parts; but until we disrupt and dismember its bony, gristly and fibrous parts one from another, it exists simply as a 'skeleton', as one integral and individual whole.

A bridge was once upon a time a loose heap of pillars and rods and rivets of steel. But the identity of these is lost, just as if they were fused into a solid mass, when once the bridge is built; their separate functions are only to be recognised and analysed in so far as we can analyse the stresses, the tensions and the pressures, which affect this part of the structure or that; and these forces are not themselves separate entities, but are the resultants of an analysis of the whole field of force. Moreover, when the bridge is broken it is no longer a bridge, and all its strength is gone. So is it precisely with the skeleton. In it is reflected a field of force: and keeping pace, as it were, in action and interaction with this field of force, the whole skeleton and every part thereof, down to the minute intrinsic structure of the bones themselves, is related in form and in position to the lines of force, to the resistances it has to encounter; for by one of the mysteries of biology, resistance begets resistance, and where pressure falls there growth springs up in strength to meet it. And, pursuing the same train of thought, we see that all this is true not of the skeleton alone but of the whole fabric of the body. Muscle and bone, for instance, are inseparably associated and connected; they are moulded one with another; they come into being together, and act and react together.[1] We may study them apart, but it is a concession to our weakness and to the narrow outlook of our minds. We see, dimly perhaps but yet with all the assurance of conviction, that

---

[1] John Hunter was seldom wrong; but I cannot believe that he was right when he said (*Scientific Works*, ed. Owen, I, 371), 'The bones, in a mechanical view, appear to be the first that are to be considered. We can study their shape, connections, number, uses, etc., *without considering any other part of the body*'

between muscle and bone there can be no change in the one but it is correlated with changes in the other; that through and through they are linked in indissoluble association; that they are only separate entities in this limited and subordinate sense, that they are *parts* of a whole which, when it loses its composite integrity, ceases to exist.

The biologist, as well as the philosopher, learns to recognise that the whole is not merely the sum of its parts. It is this, and much more than this. For it is not a bundle of parts but an organisation of parts, of parts in their mutual arrangement, fitting one with another, in what Aristotle calls 'a single and indivisible principle of unity'; and this is no merely metaphysical conception, but is in biology the fundamental truth which lies at the basis of Geoffroy's (or Goethe's) law of 'compensation', or 'balancement of growth'.

Nevertheless, Darwin found no difficulty in believing that 'natural selection will tend in the long run to reduce *any part* of the organisation, as soon as, through changed habits, it becomes superfluous: without by any means causing some other part to be largely developed in a corresponding degree. And conversely, that natural selection may perfectly well succeed in largely developing an organ without requiring as a necessary compensation the reduction of some adjoining part.'[1] This view has been developed into a doctrine of the 'independence of single characters' (not to be confused with the germinal 'unit characters' of Mendelism), especially by the palaeontologists. Thus Osborn asserts a 'principle of hereditary correlation', combined with a 'principle of *hereditary separability*, whereby the body is a colony, a mosaic, of single individual and separable characters'.[2] I cannot think that there is more than a very small element of truth in this doctrine. As Kant said, 'die Ursache der Art der Existenz bei jedem Theile eines lebenden Körpers *ist im Ganzen enthalten*'. And, according to the trend or aspect of our thought, we may look upon the co-ordinated parts, now as related and fitted *to the end or function of* the whole, and now as related to or resulting *from the physical causes* inherent in the entire system of forces to which the whole has been exposed, and under whose influence it has come into being.[3]

---

[1] *Origin of Species* (6th ed.), p. 118.
[2] *Amer. Nat.* (April, 1915), p. 198, etc. Cf. *infra*, p. 275.
[3] Driesch saw in 'Entelechy' that something which differentiates the whole from the sum of its parts in the case of the organism: 'The organism, we know, is a system the single constituents of which are inorganic in themselves; only the whole constituted by them in their typical order or arrangement owes its specificity to "Entelechy"' (*Gifford Lectures*, 1908, p. 229): and I think it could be shown that many other philosophers have said precisely the same thing. So far as the argument goes, I fail to see

The Problem of Phylogeny

It would, I dare say, be an exaggeration to see in every bone nothing more than a resultant of immediate and direct physical or mechanical conditions; for to do so would be to deny the existence, in this connection, of a principle of heredity. And though I have tried throughout this book to lay emphasis on the direct action of causes other than heredity, in short to circumscribe the employment of the latter as a working hypothesis in morphology, there can still be no question whatsoever but that heredity is a vastly important as well as a mysterious thing; it is *one* of the great factors in biology, however we may attempt to figure to ourselves, or howsoever we may fail even to imagine, its underlying physical explanation. But I maintain that it is no less an exaggeration if we tend to neglect these direct physical and mechanical modes of causation altogether, and to see in the characters of a bone merely the results of variation and of heredity, and to trust, in consequence, to those characters as a sure and certain and unquestioned guide to affinity and phylogeny. Comparative anatomy has its physiological side, which filled men's minds in John Hunter's day, and in Owen's day; it has its classificatory and phylogenetic aspect, which all but filled men's minds in the early days of Darwinism; and we can lose sight of neither aspect without risk of error and misconception.

It is certain that the question of phylogeny, always difficult, becomes especially so in cases where a great change of physical or mechanical conditions has come about, and where accordingly the former physical and physiological constraints are altered or removed. The great depths of the sea differ from other habitations of the living, not least in their eternal quietude. The fishes which dwell therein are quaint and strange; their huge heads, prodigious jaws, and long tails and tentacles are, as it were, gross exaggerations of the common and conventional forms. We look in vain for any purposeful cause or

how *this* Entelechy is shown to be peculiarly or specifically related to the *living* organism. The conception (at the bottom of General Smuts's '*Holism*') that the whole is *always* something very different from its parts is a very ancient doctrine. The reader will perhaps remember how, in another vein, the theme is treated by Martinus Scriblerus (Huxley quoted it once, for his own ends): 'In every Jack there is a *meat-roasting* Quality, which neither resides in the fly, nor in the weight, nor in any particular wheel of the Jack, but is the result of the whole composition; etc., etc.' Indeed it was at that very time, in the early eighteenth century, that the terms *organism* and *organisation* were coming into use, to connote that harmonious combination of parts 'qui conspirent toutes ensembles à produire cet effet général que nous nommons la vie' (Buffon). Cf. Ch. Robin, 'Recherches sur l'origine et le sens des termes organisme et organisation', *J. Anat.*, Paris (1880), 1–55.

physiological explanation of these enormities; and are left under a vague impression that life has been going on in the security of all but perfect equilibrium, and that the resulting forms, liberated from many ordinary constraints, have grown with unusual freedom.

To discuss these questions at length would be to enter on a discussion of Lamarck's philosophy of biology, and of many other things besides. But let us take one single illustration. The affinities of the whales constitute, as will be readily admitted, a very hard problem in phylogenetic classification. We know now that the extinct Zeuglodons are related to the old Creodont carnivores, and thereby (though distantly) to the seals;[1] and it is supposed, but it is by no means so certain, that in turn they are to be considered as representing, or as allied to, the ancestors of the modern toothed whales.[2] The proof of any such a contention becomes, to my mind, extraordinarily difficult and complicated; and the arguments commonly used in such cases may be said (in Bacon's phrase) to allure, rather than to extort assent. Though the Zeuglodons were aquatic animals, we do not know, and we have no right to suppose or to assume, that they swam after the fashion of a whale (any more than the seal does), that they dived like a whale, or leaped like a whale. But the fact that the whale does these things, and the way in which he does them, is reflected in many parts of his skeleton—perhaps more or less in all: so much so that the lines of stress which these actions impose are the very plan and working-diagram of a great part of his structure. That the Zeuglodon has a scapula like that of a whale is to my mind no necessary argument that he is akin by blood-relationship to a whale: that his dorsal vertebrae are very different from a whale's is no conclusive argument that such blood-relationship is lacking. The former fact goes a long way to prove that he used his flippers very much as a whale does; the latter goes still farther to prove that his general movements and equilibrium in the water were totally different. The whale may be descended from the Carnivora, or might for that matter, as an older school of naturalists believed, be descended from the Ungulates; but whether or no, we need not expect to find in him the scapula, the pelvis or the vertebral column of the lion or of the cow, for it would be physically impossible that he could live the

[1] See (*int. al.*) my paper 'On the affinities of *Zeuglodon*' in *Studies from the Museum of University College, Dundee* (1889).

[2] 'There can be no doubt that Fraas is correct in regarding this type (*Procetus*) as an annectant form between the Zeuglodonts and the Creodonta, but, although the origin of the Zeuglodonts is thus made clear, it still seems to be by no means so certain as that author believes, that they may not themselves be the ancestral forms of the *Odontoceti*' (Andrews, *Tertiary Vertebrata of the Fayum*, 1906, p. 235).

life he does with any one of them. In short, when we hope to find the missing links between a whale and his terrestrial ancestors, it must be not by means of conclusions drawn from a scapula, an axis, or even from a tooth, but by the discovery of forms so intermediate in their general structure as to indicate an organisation and, *ipso facto*, a mode of life, intermediate between the terrestrial and the Cetacean form. There is no valid syllogism to the effect that *A* has a flat curved scapula like a seal's, and *B* has a flat curved scapula like a seal's: and therefore *A* and *B* are related to the seals and to each other; it is merely a flagrant case of an 'undistributed middle'. But there is validity in an argument that *B* shows in its general structure, extending over this bone and that bone, resemblances both to *A* and to the seals: and that therefore he may be presumed to be related to both, in his hereditary habits of life and in actual kinship by blood. It is cognate to this argument that (as every palaeontologist knows) we find clues to affinity more easily, that is to say with less confusion and perplexity, in certain structures than in others. The deep-seated rhythms of growth which, as I venture to think, are the chief basis of morphological heredity, bring about similarities of form which endure in the absence of conflicting forces; but a new system of forces, introduced by altered environment and habits, impinging on those particular parts of the fabric which lie within this particular field of force, will assuredly not be long of manifesting itself in notable and inevitable modifications of form. And if this be really so, it will further imply that modifications of form will tend to manifest themselves, not so much in small and *isolated* phenomena, in this part of the fabric or in that, in a scapula for instance or a humerus: but rather in some slow, *general*, and more or less uniform or graded modification, spread over a number of correlated parts, and at times extending over the whole, or over great portions, of the body. Whether any such general tendency to widespread and correlated transformation exists, we shall attempt to discuss in the following chapter.

CHAPTER IX

ON THE THEORY OF TRANSFORMATIONS, OR THE COMPARISON OF RELATED FORMS

This is the most celebrated chapter of the book and it has been widely commented upon in biological literature. I have not made any careful survey, but I suspect that the well-known diagrams of transformations have been reproduced in sundry scientific writings a large number of times.

The comments almost invariably have a few points in common, which I shall briefly summarise here. In the first place it is surprising that despite their fame, the Cartesian transformations have been used very little. This is because, to use Medawar's term, they are 'analytically unwieldy'. The few times the method has been applied is in the development or change of form in a single system, as for instance, Richards and Riley's* study of developing amphibians under different conditions, and Medawar's† analysis of tissue culture growth.

A far more significant result in terms of practical application is that the system of transformations of D'Arcy Thompson stimulated and contributed to the much simpler method of analysis of allometric growth, which has found widespread use, mainly through the work of J. S. Huxley.‡ Here instead of attempting to analyse a whole structure in two (or three) divisions, two factors are isolated and compared on a logarithmic scale. In this way it is possible to discover the ratio of the growth-rates of different structures, a method which has found application in embryology, taxonomy, palaeontology and even ecology.

Therefore despite the fact that they are 'analytically unwieldy', Cartesian transformations have been influential, and very likely they will continue to be so and stimulate new thoughts and new methods in the future. It is characteristic of D'Arcy Thompson's contributions in general, and this one in particular, that they are original, and their originality has a remarkable staying power. It is perhaps that he continually raises new problems that stretch the imagination, but in leaving them to varying degrees unsolved he livens the curiosity.

Mathematics and Form

In the foregoing chapters of this book we have attempted to study the inter-relations of growth and form, and the part which the physical forces play in this complex interaction; and, as part of

---

* *J. Exp. Zool.* 77 (1937), 159.
† In *Essays on Growth and Form* (Oxford, 1945), p. 157.
‡ *Problems of Relative Growth* (Methuen, London, 1932).

the same enquiry, we have tried in comparatively simple cases to use mathematical methods and mathematical terminology to describe and define the forms of organisms. We have learned in so doing that our own study of organic form, which we call by Goethe's name of Morphology, is but a portion of that wider Science of Form which deals with the forms assumed by matter under all aspects and conditions, and, in a still wider sense, with forms which are theoretically imaginable.

The study of form may be descriptive merely, or it may become analytical. We begin by describing the shape of an object in the simple words of common speech: we end by defining it in the precise language of mathematics; and the one method tends to follow the other in strict scientific order and historical continuity. Thus, for instance, the form of the earth, of a raindrop or a rainbow, the shape of the hanging chain, or the path of a stone thrown up into the air, may all be described, however inadequately, in common words; but when we have learned to comprehend and to define the sphere, the catenary, or the parabola, we have made a wonderful and perhaps a manifold advance. The mathematical definition of a 'form' has a quality of precision which was quite lacking in our earlier stage of mere description; it is expressed in few words or in still briefer symbols, and these words or symbols are so pregnant with meaning that thought itself is economised; we are brought by means of it in touch with Galileo's aphorism (as old as Plato, as old as Pythagoras, as old perhaps as the wisdom of the Egyptians), that 'the Book of Nature is written in characters of Geometry'.[1]

We are apt to think of mathematical definitions as too strict and rigid for common use, but their rigour is combined with all but endless freedom. The precise definition of an ellipse introduces us to all the ellipses in the world; the definition of a 'conic section' enlarges our concept, and a 'curve of higher order' all the more extends our range of freedom.[2] By means of these large limitations, by this con-

[1] Cf. Plutarch, *Symp.* viii, 2, on the meaning of Plato's aphorism ('if it actually was Plato's'): πῶς Πλάτων ἔλεγε τὸν θεὸν ἀεὶ γεωμετρεῖν.

[2] So said Gustav Theodor Fechner, the author of Fechner's Law, a hundred years ago. ('Ueber die mathematische Behandlung organischer Gestalten und Processe', *Berichte sächs. Gesellsch., Math.-phys. Cl.*, Leipzig, 1849, pp. 50–64.) Fechner's treatment is more purely mathematical and less physical in its scope and bearing than ours, and his paper is but a short one, but the conclusions to which he is led differ little from our own. Let me quote a single sentence which, together with its context, runs precisely on the lines which we have followed in this book: 'So ist also die mathematische Bestimmbarkeit im Gebiete des Organischen ganz eben so gut vorhanden als in dem des Unorganischen, und in letzterem eben solchen oder äquivalenten Beschränkungen unterworfen als in ersterem; und nur sofern die unorganischen Formen und

trolled and regulated freedom, we reach through mathematical analysis to mathematical synthesis. We discover homologies or identities which were not obvious before, and which our descriptions obscured rather than revealed: as for instance, when we learn that, however we hold our chain, or however we fire our bullet, the contour of the one or the path of the other is always mathematically homologous.

Once more, and this is the greatest gain of all, we pass quickly and easily from the mathematical concept of form in its statical aspect to form in its dynamical relations: we rise from the conception of form to an understanding of the forces which gave rise to it; and in the representation of form and in the comparison of kindred forms, we see in the one case a diagram of forces in equilibrium, and in the other case we discern the magnitude and the direction of the forces which have sufficed to convert the one form into the other. Here, since 'a change of material form is only effected by the movement of matter,[1] we have once again the support of the Schoolman's and the philosopher's axiom, *Ignorato motu, ignoratur Natura*.

There is yet another way—we learn it of Henri Poincaré—to regard the function of mathematics, and to realise why its laws and its methods *are bound* to underlie all parts of physical science. Every natural phenomenon, however simple, is really composite, and every visible action and effect is a summation of countless subordinate actions. Here mathematics shows her peculiar power, to combine and to generalise. The concept of an average, the equation to a curve, the description of a froth or cellular tissue, all come within the scope of mathematics for no other reason than that they are summations of more elementary principles or phenomena. Growth and Form are throughout of this composite nature; therefore the laws of mathematics are bound to underlie them, and her methods to be peculiarly fitted to interpret them.

For one reason or another there are very many organic forms which we cannot describe, still less define, in mathematical terms: just as there are problems even in physical science beyond the mathematics of our age. We never even seek for a formula to define this fish or that, or this or that vertebrate skull. But we may already use mathematical language to describe, even to define in general terms, the shape of a snail-shell, the twist of a horn, the outline of a leaf, the

das unorganische Geschehen sich einer einfacheren Gesetzlichkeit mehr nähern als die organischen, kann die Approximation im unorganischen Gebiet leichter und weiter getrieben werden als im organischen. Dies wäre der ganze, sonach rein relative, Unterschied.' Here, in a nutshell, is the gist of the whole matter.

[1] 'We can *move* matter, that is all we can do to it' (Oliver Lodge).

texture of a bone, the fabric of a skeleton, the stream-lines of fish or bird, the fairy lace-work of an insect's wing. Even to do this we must learn from the mathematician to eliminate and to discard; to keep the type in mind and leave the single case, with all its accidents, alone; and to find in this sacrifice of what matters little and conservation of what matters much one of the peculiar excellences of the method of mathematics.[1]

Method of Co-ordinates

In a very large part of morphology, our essential task lies in the comparison of related forms rather than in the precise definition of each; and the *deformation* of a complicated figure may be a phenomenon easy of comprehension, though the figure itself have to be left unanalysed and undefined. This process of comparison, of recognising in one form a definite permutation or *deformation* of another, apart altogether from a precise and adequate understanding of the original 'type' or standard of comparison, lies within the immediate province of mathematics, and finds its solution in the elementary use of a certain method of the mathematician. This method is the Method of Co-ordinates, on which is based the Theory of Transformations.[2]

I imagine that when Descartes conceived the method of co-ordinates, as a generalisation from the proportional diagrams of the artist and the architect, and long before the immense possibilities of this analysis could be foreseen, he had in mind a very simple purpose; it was perhaps no more than to find a way of translating the *form* of a curve (as well as the position of a point) into *numbers* and into *words*. This is precisely what we do, by the method of co-ordinates, every time we study a statistical curve; and conversely translate numbers into form whenever we 'plot a curve', to illustrate a table of mortality, a rate of growth, or the daily variation of temperature or barometric pressure. In precisely the same way it is possible to inscribe in a net of rectangular co-ordinates the outline, for instance, of a fish, and so to translate it into a table of numbers, from which again we may at pleasure reconstruct the curve.

[1] Cf. W. H. Young, 'The Mathematical Method and its Limitations', *Congresso dei Matematici* (Bologna, 1928).

[2] The mathematical Theory of Transformations is part of the Theory of Groups, of great importance in modern mathematics. A distinction is drawn between Substitution-groups and Transformation-groups, the former being discontinuous, the latter continuous—in such a way that within one and the same group each transformation is infinitely little different from another. The distinction among biologists between a mutation and a variation is curiously analogous.

But it is the next step in the employment of co-ordinates which is of special interest and use to the morphologist; and this step consists in the alteration, or deformation, of our system of co-ordinates, and in the study of the corresponding transformation of the curve or figure inscribed in the co-ordinate network.

Let us inscribe in a system of Cartesian co-ordinates the outline of an organism, however complicated, or a part thereof: such as a fish, a crab, or a mammalian skull. We may now treat this complicated figure, in general terms, as a function of x, y. If we submit our rectangular system to deformation on simple and recognised lines, altering, for instance, the direction of the axes, the ratio of x/y, or substituting for x and y some more complicated expressions, then we obtain a new system of co-ordinates, whose deformation from the original type the inscribed figure will precisely follow. In other words, we obtain a new figure which represents the old figure under a more or less homogeneous *strain*, and is a function of the new co-ordinates in precisely the same way as the old figure was of the original co-ordinates x and y.

The problem is closely akin to that of the cartographer who transfers identical data to one projection or another;[1] and whose object is to secure (if it be possible) a complete correspondence, *in each small unit of area*, between the one representation and the other. The morphologist will not seek to draw his organic forms in a new and artificial projection; but, in the converse aspect of the problem, he will enquire whether two different but more or less obviously related forms can be so analysed and interpreted that each may be shown to be a transformed representation of the other. This once demonstrated, it will be a comparatively easy task (in all probability) to postulate the direction and magnitude of the force capable of effecting the required transformation. Again, if such a simple alteration of the system of forces can be proved adequate to meet the case, we may find ourselves able to dispense with many widely current and more complicated hypotheses of biological causation. For it is a maxim in physics that an effect ought not to be ascribed to the joint operation of many causes if few are adequate to the production of it: *Frustra fit per plura, quod fieri potest per pauciora.*

---

[1] Cf. (e.g.) Tissot, *Mémoire sur la représentation des surfaces, et les projections des cartes géographiques* (Paris, 1881).

Related Forms

We might suppose that by the combined action of appropriate forces any material form could be transformed into any other: just as out of a 'shapeless' mass of clay the potter or the sculptor models his artistic product; or just as we attribute to Nature herself the power to effect the gradual and successive transformation of the simple germ into the complex organism. But we need not let these considerations deter us from our method of comparison of *related* forms. We shall strictly limit ourselves to cases where the transformation necessary to effect a comparison shall be of a simple kind, and where the transformed, as well as the original, co-ordinates shall constitute an harmonious and more or less symmetrical system. We should fall into deserved and inevitable confusion if, whether by the mathematical or any other method, we attempted to compare organisms separated far apart in Nature and in zoological classification. We are limited, both by our method and by the whole nature of the case, to the comparison of organisms such as are manifestly related to one another and belong to the same zoological class. For it is a grave sophism, in natural history as in logic, to make a transition into another kind.[1]

Our enquiry lies, in short, just within the limits which Aristotle himself laid down when, in defining a 'genus', he showed that (apart from those superficial characters, such as colour, which he called 'accidents') the essential differences between one 'species' and another are merely differences of proportion, of relative magnitude, or (as he phrased it) of 'excess and defect'. 'Save only for a difference in the way of excess or defect, the parts are identical in the case of such animals as are of one and the same genus; and by "genus" I mean, for instance, Bird or Fish.' And again: 'Within the limits of the same genus, as a general rule, most of the parts exhibit differences...in the way of multitude or fewness, magnitude or parvitude, in short, in the way of excess or defect. For "the more" and "the less" may be represented as "excess" and "defect".'[2] It is precisely this difference of relative magnitudes, this Aristotelian 'excess and defect' in the case of form, which our co-ordinate method is especially adapted to analyse, and to reveal and demonstrate as the main cause of what (again in the Aristotelian sense) we term 'specific' differences.

The applicability of our method to particular cases will depend upon, or be further limited by, certain practical considerations or

---

[1] The saying *heterogena comparari non possunt* is discussed by Coleridge in his *Aids to Reflexion*.   [2] *Historia Animalium* I, 1.

qualifications. Of these the chief, and indeed the essential, condition is, that the form of the entire structure under investigation should be found to vary in a more or less uniform manner, after the fashion of an approximately homogeneous and isotropic body. But an imperfect isotropy, provided always that some 'principle of continuity' run through its variations, will not seriously interfere with our method; it will only cause our transformed co-ordinates to be somewhat less regular and harmonious than are those, for instance, by which the physicist depicts the motions of a perfect fluid, or a theoretic field of force in a uniform medium.

Again, it is essential that our structure vary in its entirety, or at least that 'independent variants' should be relatively few. That independent variations occur, that localised centres of diminished or exaggerated growth will now and then be found, is not only probable but manifest; and they may even be so pronounced as to appear to constitute new formations altogether. Such independent variants as these Aristotle himself clearly recognised: 'It happens further that some have parts which others have not; for instance, some [birds] have spurs and others not, some have crests, or combs, and others not; but, as a general rule, most parts and those that go to make up the bulk of the body are either identical with one another, or differ from one another in the way of contrast and of excess and defect. For "the more" and "the less" may be represented as "excess" or "defect".'[1]

If, in the evolution of a fish, for instance, it be the case that its several and constituent parts—head, body and tail, or this fin and that fin—represent so many independent variants, then our co-ordinate system will at once become too complex to be intelligible; we shall be making not one comparison but several separate comparisons, and our general method will be found inapplicable. Now precisely this independent variability of parts and organs—here, there, and everywhere within the organism—would appear to be implicit in our ordinary accepted notions regarding variation; and, unless I am greatly mistaken, it is precisely on such a conception of the easy, frequent, and normally independent variability of parts that our conception of the process of natural selection is fundamentally based. For the morphologist, when comparing one organism with another, describes the differences between them point by point, and

---

[1] Aristotle's argument is even more subtle and far-reaching; for the differences of which he speaks are not merely those between one bird and another, but between them all and the very type itself, or Platonic 'idea' of a bird.

'character' by 'character'.[1] If he is from time to time constrained to admit the existence of 'correlation' between characters (as a hundred years ago Cuvier first showed the way), yet all the while he recognises this fact of correlation somewhat vaguely, as a phenomenon due to causes which, except in rare instances, he can hardly hope to trace; and he falls readily into the habit of thinking and talking of evolution as though it had proceeded on the lines of his own descriptions, point by point, and character by character.[2]

With the 'characters' of Mendelian genetics there is no fault to be found; tall and short, rough and smooth, plain or coloured are opposite tendencies or contrasting qualities, in plain logical contradistinction. But when the morphologist compares one animal with another, point by point or character by character, these are too often the mere outcome of artificial dissection and analysis. Rather is the living body one integral and indivisible whole, in which we cannot find, when we come to look for it, any strict dividing line even between the head and the body, the muscle and the tendon, the sinew and the bone. Characters which we have differentiated insist on integrating themselves again; and aspects of the organism are seen to be conjoined which only our mental analysis had put asunder. The co-ordinate diagram throws into relief the integral solidarity of the organism, and enables us to see how simple a certain kind of *correlation* is which had been apt to seem a subtle and a complex thing.

But if, on the other hand, diverse and dissimilar fishes can be referred as a whole to identical functions of very different co-ordinate systems, this fact will of itself constitute a proof that variation has proceeded on definite and orderly lines, that a comprehensive 'law of growth' has pervaded the whole structure in its integrity, and that some more or less simple and recognisable system of forces has been in control. It will not only show how real and deep-seated is the phenomenon of 'correlation', in regard to form, but it will also demonstrate the fact that a correlation which had seemed too complex for analysis or comprehension is, in many cases, capable of

---

[1] Cf. *supra*, p. 264.

[2] Cf. H. F. Osborn, 'On the Origin of Single Characters, as Observed in Fossil and Living Animals and Plants', *Amer. Nat.* **49** (1915), 193–239 (and other papers); *ibid.* p. 194, 'Each individual is composed of a vast number of somewhat similar new or old characters, each character has its independent and separate history, each character is in a certain stage of evolution, each character is correlated with the other characters of the individual.... The real problem has always been that of the origin and development of characters. Since the *Origin of Species* appeared, the terms variation and variability have always referred to single characters; if a species is said to be variable, we mean that a considerable number of the single characters or groups of characters of which it is composed are variable' etc.

very simple graphic expression. This, after many trials, I believe to be in general the case, bearing always in mind that the occurrence of independent or localised variations must sometimes be considered.

Cartesian Transformations

If we begin by drawing a net of rectangular equidistant co-ordinates (about the axes x and y), we may alter or *deform* this network in various ways, several of which are very simple indeed. Thus (1) we may alter the dimensions of our system, extending it along one or other axis, and so converting each little square into a corresponding and proportionate oblong (Figs. 120, 121). It follows that any figure

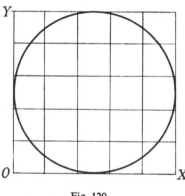

Fig. 120.

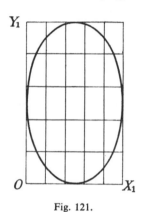

Fig. 121.

which we may have inscribed in the original net, and which we transfer to the new, will thereby be *deformed* in strict proportion to the deformation of the entire configuration, being still defined by corresponding points in the network and being throughout in conformity with the original figure. For instance, a circle inscribed in the original 'Cartesian' net will now, after extension in the y-direction, be found elongated into an ellipse. In elementary mathematical language, for the original x and y we have substituted x_1 and cy_1, and the equation to our original circle, $x^2+y^2 = a^2$, becomes that of the ellipse, $x_1^2+c^2y_1^2 = a^2$.

If I draw the cannon-bone of an ox (Fig. 122, a), for instance, within a system of rectangular co-ordinates, and then transfer the same drawing, point for point, to a system in which for the x of the original diagram we substitute $x' = 2x/3$, we obtain a drawing (b) which is a very close approximation to the cannon-bone of the

276

sheep. In other words, the main (and perhaps the only) difference betwcen the two bones is simply that that of the sheep is elongated along the vertical axis as compared with that of the ox, in the proportion of 3/2. And similarly, the long slender cannon-bone of the giraffe (c) is referable to the same identical type, subject to a reduction of breadth, or increase of length, corresponding to $x'' = x/3$.

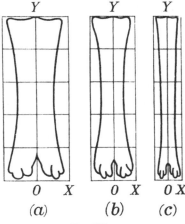

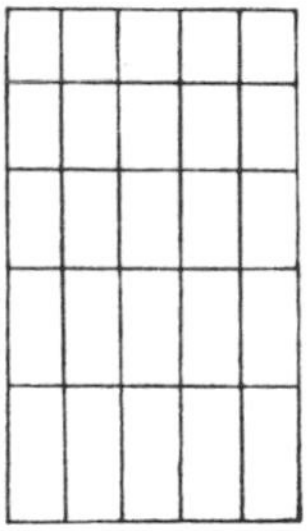

Fig. 123.

Fig. 122.

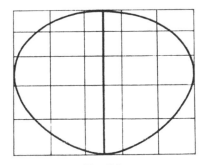

Fig. 124.

(2) The second type is that where extension is not equal or uniform at all distances from the origin: but grows greater or less, as, for instance, when we stretch a *tapering* elastic band. In such cases, as I have represented it in Fig. 123, the ordinate increases logarithmically, and for y we substitute ϵ^y. It is obvious that this logarithmic extension may involve both abscissae and ordinates, x becoming ϵ^x while y becomes ϵ^y. The circle in our original figure is now deformed into some such shape as that of Fig. 124. This method of deformation

277

is a common one, and will often be of use to us in our comparison of organic forms.

(3) Our third type is the 'simple shear', where the rectangular co-ordinates become 'oblique', their axes being inclined to one another at a certain angle ω. Our original rectangle now becomes such a figure as that of Fig. 125. The system may now be described in terms of the oblique axes X, Y; or may be directly referred to new rectangular co-ordinates ξ, η by the simple transposition $x = \xi - \eta$ cot ω, $y = \eta$ cosec ω.

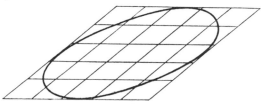

Fig. 125.

Radial Co-ordinates

(4) Yet another important class of deformations may be represented by the use of radial co-ordinates, in which one set of lines are represented as radiating from a point or 'focus', while the other set are transformed into circular arcs cutting the radii orthogonally.

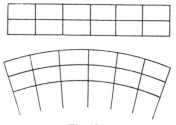

Fig. 126.

These radial co-ordinates are especially applicable to cases where there exists (either within or without the figure) some part which is supposed to suffer no deformation; a simple illustration is afforded by the diagrams which illustrate the flexure of a beam (Fig. 126). In biology these co-ordinates will be especially applicable in cases where the growing structure includes a 'node', or point where growth is absent or at a minimum; and about which node the rate of growth may be assumed to increase symmetrically. Precisely such a case is furnished us in a leaf of an ordinary dicotyledon. The leaf of a typical

278

monocotyledon—such as a grass or a hyacinth, for instance—grows continuously from its base, and exhibits no node or 'point of arrest'. Its sides taper off gradually from its broad base to its slender tip, according to some law of decrement specific to the plant; and any alteration in the relative velocities of longitudinal and transverse growth will merely make the leaf a little broader or narrower, and will effect no other conspicuous alteration in its contour. But if there once come into existence a node, or 'locus of no growth', about which we may assume growth—which in the hyacinth leaf was longitudinal

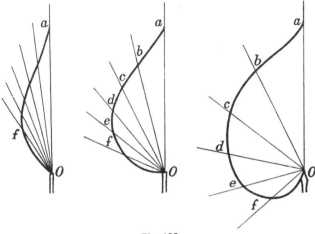

Fig. 127.

and transverse—to take place radially and transversely to the radii, then we shall soon see the sloping sides of the hyacinth leaf give place to a more typical and 'leaf-like' shape. If we alter the ratio between the radial and tangential velocities of growth—in other words, if we increase the angles between corresponding radii—we pass successively through the various configurations which the botanist describes as the lanceolate, the ovate, and the cordiform leaf. These successive changes may to some extent, and in appropriate cases, be traced as the individual leaf grows to maturity; but as a much more general rule, the balance of forces, the ratio between radial and tangential velocities of growth, remains so nicely and constantly balanced that the leaf increases in size without conspicuous modification of form. It is rather what we may call a long-period variation, a tendency for the relative velocities to alter from one generation to another, whose result is brought into view by this method of illustration.

There are various corollaries to this method of describing the form of a leaf which may be here alluded to. For instance, the so-called asymmetrical leaf[1] of a begonia, in which one side of the leaf may be merely ovate while the other has a cordate outline, is seen to be really a case of *unequal*, and not truly asymmetrical, growth on either side of the midrib. There is nothing more mysterious in its conformation than, for instance, in that of a forked twig in which one limb of the fork has grown longer than the other. The case of the begonia leaf is

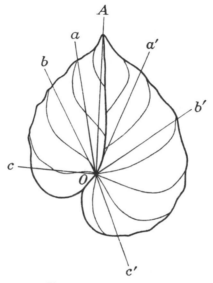

Fig. 128. *Begonia daedalea.*

of sufficient interest to deserve illustration, and in Fig. 128 I have outlined a leaf of the large *Begonia daedalea*. On the smaller left-hand side of the leaf I have taken at random three points *a*, *b*, *c*, and have measured the angles, *AOa*, etc., which the radii from the hilus of the leaf to these points make with the median axis. On the other side of the leaf I have marked the points *a'*, *b'*, *c'*, such that the radii drawn to this margin of the leaf are equal to the former, *Oa'* to *Oa*, etc. Now if the two sides of the leaf are mathematically similar to one another, it is obvious that the respective angles should be in

[1] Cf. Sir Thomas Browne, in *The Garden of Cyrus*: 'But why ofttimes one side of the leaf is unequall unto the other, as in Hazell and Oaks, why on either side the master vein the lesser and derivative channels stand not directly opposite, not at equall angles, respectively unto the adverse side, but those of one side do often exceed the other, as the Wallnut and many more, deserves another enquiry.'

continued proportion, i.e. as AOa is to AOa', so should AOb be to AOb'. This proves to be very nearly the case. For I have measured the three angles on one side, and one on the other, and have then compared, as follows, the calculated with the observed values of the other two:

	AOa	AOb	AOc	AOa'	AOb'	AOc'
Observed values	12°	28·5°	88°	—	—	157°
Calculated values	—	—	—	21·5°	51·1°	—
Observed values	—	—	—	20	52	—

The agreement is very close, and what discrepancy there is may be amply accounted for, first, by the slight irregularity of the sinuous margin of the leaf; and secondly, by the fact that the true axis or mid-rib of the leaf is not straight but slightly curved, and therefore that it is curvilinear and not rectilinear triangles which we ought to have measured. When we understand these few points regarding the peripheral curvature of the leaf, it is easy to see that its principal veins approximate closely to a beautiful system of isogonal co-ordinates. It is also obvious that we can easily pass, by a process of shearing, from those cases where the principal veins start from the base of the leaf to those where they arise successively from the midrib, as they do in most dicotyledons.

It may sometimes happen that the node,[1] or 'point of arrest', is at the upper instead of the lower end of the leaf-blade; and occasionally there is a node at both ends. In the former case, as we have it in the daisy, the form of the leaf will be, as it were, inverted, the broad, more or less heart-shaped, outline appearing at the upper end, while below the leaf tapers gradually downwards to an ill-defined base. In the latter case, as in *Dionaea*, we obtain a leaf equally expanded, and similarly ovate or cordate, at both ends. We may notice, lastly, that the shape of a solid fruit, such as an apple or a cherry, is a solid of revolution, developed from similar curves and to be explained on the same principle. In the cherry we have a 'point of arrest' at the base of the berry, where it joins its peduncle, and about this point the fruit (in imaginary section) swells out into a cordate outline; while in the apple we have two such well-marked points of arrest, above and below, and about both of them the same conformation tends to arise. The bean and the human kidney owe their 'reniform' shape to precisely the same phenomenon, namely, to the existence of a node or 'hilus', about which the forces of growth are radially and symmetrically arranged. When the seed is small and the pod roomy, the

---

[1] 'Node', in the botanical, not the mathematical, sense.

seed may grow round, or nearly so, like a pea; but it is flattened and bean-shaped, or elliptical like a kidney-bean, when compressed within a narrow and elongated pod. If the original seed have any simple pattern, of the nature for instance of meridians or parallels of latitude, it is easy to see how these will suffer a conformal transformation, corresponding to the deformation of the sphere.

We might go farther, and farther than we have room for here, to illustrate the shapes of leaves by means of radial co-ordinates, and even to attempt to define them by polar equations. We may look upon the curve of sines as an easy, gradual and natural transition— perhaps the simplest and most natural of all—from minimum to

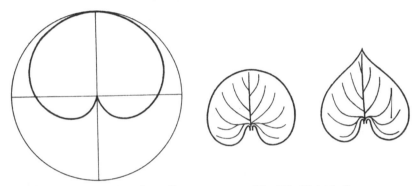

Fig. 129. Curve resembling the outline of a reniform leaf: $r = \sin \theta/2$.

Fig. 130. Violet leaf.

corresponding maximum, and so on alternately and continuously; and we found the same curve going round like the hands of a clock, when plotted on radial co-ordinates and (so to speak) prevented from leaving its place. Either way it represents a 'simple harmonic motion'. Now we have just seen an ordinary dicotyledonous leaf to have a 'point of arrest', or zero-growth in a certain direction, while in the opposite direction towards the tip it has grown with a maximum velocity. This progress from zero to maximum suggests one-half of the sine-curve; in other words, if we look on the outline of the leaf as a vector-diagram of its own growth, at rates varying from zero to zero in a complete circuit of 360°, this suggests, as a possible and very simple case, the plotting of $r = \sin \theta/2$. Doing so, we obtain a curve (Fig. 129) closely resembling what the botanists call a *reniform* (or kidney-shaped) leaf, that is to say, with a cordate outline at the base formed of two 'auricles', one on either side, and then rounded

282

off with no projecting apex.[1] The ground-ivy and the dog-violet (Fig. 130) illustrate such a leaf; and sometimes, as in the violet, the veins of the leaf show similar curves congruent with the outer edge. Moreover, the violet is a good example of how the reniform leaf may be drawn out more and more into an acute and ovate form.

From $\sin\theta/2$ we may proceed to any other given fraction of θ, and plot, for instance, $r = \sin 5\theta/3$, as in Fig. 131; which now no longer represents a single leaf but has become a diagram of the five petals of a pentamerous flower. Abbot Guido Grandi, a Pisan mathematician of the early eighteenth century, drew such a curve and pointed out its botanical analogies; and we still call the curves of this family 'Grandi's curves'.[2]

Fig. 131. Grandi's curves based on $r = \sin \frac{5}{3}\theta$, and illustrating the five petals of a simple flower.

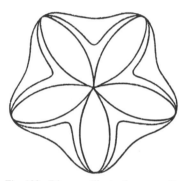

Fig. 132. Diagram illustrating a corolla of five petals, or of five lobes, based on the equation $r = a + b \cos\theta$.

The gamopetalous corolla is easily transferred to polar co-ordinates, in which the radius vector now consists of two parts, the one a constant, the other expressing the amplitude (or half-amplitude) of the sine-curve; we may write the formula $r = a + b \cos n\theta$. In Fig. 132 $n = 5$; in this figure, if the radius of the outermost circle be taken as unity, the outer of the two sinuous curves has $a:b$ as 9:1, and the inner curve as 3:1; while the five petals become separate when $a = b$, and the formula reduces to $r = \cos^2 5\theta/2$.

In Fig. 133 we have what looks like a first approximation to a horse-chestnut leaf. It consists of so many separate leaflets, akin

[1] Fig. 129 illustrates the whole leaf, but only shows one-half of the sine-curve. The rest is got by reflecting the moiety already drawn in the horizontal axis ($\theta = \pi/2$).
[2] Dom Guido Grandus, *Flores geometrici ex rhodonearum et cloeliarum curvarum descriptione resultantes*...(Florentiae, 1728). Cf. Alfred Lartigues, *Biodynamique générale* (Paris, 1930)—a curious but eccentric book.

to the five petals in Fig. 132; but these are now inscribed in (or have a *locus* in) the cordate or reniform outline of Fig. 129. The new curve is, in short, a composite one; and its general formula is $r = \sin\theta/2.\sin n\theta$. The small size of the two leaflets adjacent to the petiole is characteristic of the curve, and helps to explain the development of 'stipules'.

In this last case we have combined one curve with another, and the doing so opens out a new range of possibilities. On the outline of the simple leaf, whether ovate, lanceolate or cordate, we may superpose secondary sine-curves of lesser period and varying amplitude, after the fashion of a Fourier series; and the results will vary from a mere crenate outline to the digitate lobes of an ivy-leaf, or

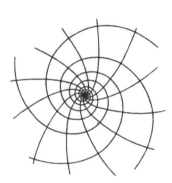

Fig. 133. Outline of a compound leaf, like a horse-chestnut, based on a composite sine-curve, of the form $r = \sin\theta/2.\sin n\theta$.

Fig. 134.

to separate leaflets such as we have just studied in the horse-chestnut. Or again, we may inscribe the separate petals of Fig. 134 within a spiral curve, equable or equiangular as the case may be; and then, continuing the series on and on, we shall obtain a figure resembling the clustered leaves of a stonecrop, or the petals of a water-lily or other polypetalous flower.

Most of the transformations which we have hitherto considered (other than that of the simple shear) are particular cases of a general transformation, obtainable by the method of conjugate functions and equivalent to the projection of the original figure on a new plane. Appropriate transformations, on these general lines, provide for the cases of a coaxial system where the Cartesian co-ordinates are replaced by coaxial circles, or a confocal system in which they are replaced by confocal ellipses and hyperbolas.

Yet another curious and important transformation, belonging to the same class, is that by which a system of straight lines becomes transformed into a conformal system of logarithmic spirals: the straight line $Y - AX = c$ corresponding to the logarithmic spiral $\theta - A \log r = c$ (Fig. 134). This beautiful and simple transformation lets us at once convert, for instance, the straight conical shell of the Pteropod or the *Orthoceras* into the logarithmic spiral of the Nautiloid; it involves a mathematical symbolism which is but a slight extension of that which we have employed in our elementary treatment of the logarithmic spiral.

These various systems of co-ordinates, which we have now briefly considered, are sometimes called 'isothermal co-ordinates', from the fact that, when employed in this particular branch of physics, they perfectly represent the phenomena of the conduction of heat, the contour lines of equal temperature appearing, under appropriate conditions, as the orthogonal lines of the co-ordinate system. And it follows that the 'law of growth' which our biological analysis by means of orthogonal co-ordinate systems presupposes, or at least foreshadows, is one according to which the organism grows or develops along *stream-lines*, which may be defined by a suitable mathematical transformation.

When the system becomes no longer orthogonal, as in many of the following illustrations—for instance, that of *Orthagoriscus* (Fig. 155) —then the transformation is no longer within the reach of comparatively simple mathematical analysis. Such departure from the typical symmetry of a 'stream-line' system is, in the first instance, sufficiently accounted for by the simple fact that the developing organism is very far from being homogeneous and isotropic, or, in other words, does not behave like a perfect fluid. But though under such circumstances our co-ordinate systems may be no longer capable of strict mathematical analysis, they will still indicate *graphically* the relation of the new co-ordinate system to the old, and conversely will furnish us with some guidance as to the 'law of growth', or play of forces, by which the transformation has been effected.

Some Special Cases

Before we pass from this brief discussion of transformations in general, let us glance at one or two cases in which the forces applied are more or less intelligible, but the resulting transformations are, from the mathematical point of view, exceedingly complicated.

The 'marbled papers' of the bookbinder are a beautiful illustration of visible 'stream-lines'. On a dishful of a sort of semi-liquid gum the workman dusts a few simple lines or patches of colouring matter; and then, by passing a comb through the liquid, he draws the colour-bands into the streaks, waves, and spirals which constitute the marbled pattern, and which he then transfers to sheets of paper laid down upon the gum. By some such system of shears, by the effect of unequal traction or unequal growth in various directions and superposed on an originally simple pattern, we may account for the not dissimilar marbled patterns which we recognise, for instance, on a large serpent's skin. But it must be remarked, in the case of the marbled paper, that though the method of application of the forces is simple, yet in the aggregate the system of forces set up by the many teeth of the comb is exceedingly complex, and its complexity is revealed in the complicated 'diagram of forces' which constitutes the pattern.

To take another and still more instructive illustration. To turn one circle (or sphere) into two circles (or spheres) would be, from the point of view of the mathematician, an extraordinarily difficult transformation; but, physically speaking, its achievement may be extremely simple. The little round gourd grows naturally, by its symmetrical forces of expansive growth, into a big, round, or somewhat oval pumpkin or melon.[1]* But the Moorish husbandman ties a rag round its middle, and the same forces of growth, unaltered save for the presence of this trammel, now expand the globular structure into two superposed and connected globes. And again, by varying the position of the encircling band, or by applying several such ligatures instead of one, a great variety of artificial forms of 'gourd'

[1] Analogous structural differences, especially in the fibrovascular bundles, help to explain the differences between (e.g.) a smooth melon and a cantelupe, or between various elongate, flattened and globular varieties. These breed true to type, and obey, when crossed, the laws of Mendelian inheritance. Cf. E. W. Sinnott, 'Inheritance of Fruit-shape in Cucurbita', *Bot. Gaz.* **74** (1922), 95–103, and other papers.

* Sinnott has illustrated changes in cucurbit shape using D'Arcy Thompson's method of transformations, *Ann. N.Y. Acad. Sci.* **71** (1958), p. 1223, fig. 7.

may be, and actually are, produced. It is clear, I think, that we may account for many ordinary biological processes of development or transformation of form by the existence of trammels or lines of constraint, which limit and determine the action of the expansive forces of growth that would otherwise be uniform and symmetrical. This case has a close parallel in the operations of the glass-blower, to which we have already, more than once, referred in passing.[1] The glass-blower starts his operations with a *tube*, which he first closes at one end so as to form a hollow vesicle, within which his blast of air exercises a uniform pressure on all sides; but the spherical conformation which this uniform expansive force would naturally tend to produce is modified into all kinds of forms by the trammels or resistances set up as the workman lets one part or another of his bubble be unequally heated or cooled. It was Oliver Wendell Holmes who first showed this curious parallel between the operations of the glass-blower and those of Nature, when she starts, as she so often does, with a simple tube.[2] The alimentary canal, the arterial system including the heart, the central nervous system of the vertebrate, including the brain itself, all begin as simple tubular structures. And with them Nature does just what the glass-blower does, and, we might even say, no more than he. For she can expand the tube here and narrow it there; thicken its walls or thin them; blow off a lateral off-shoot or caecal diverticulum; bend the tube, or twist and coil it; and infold or crimp its walls as, so to speak, she pleases. Such a form as that of the human stomach is easily explained when it is regarded from this point of view; it is simply an ill-blown bubble, a bubble that has been rendered lopsided by a trammel or restraint along one side, such as to prevent its symmetrical expansion—such a trammel as is produced if the glass-blower lets one side of his bubble get cold, and such as is actually present in the stomach itself in the form of a muscular band.

The Florence flask, or any other handiwork of the glass-blower, is always beautiful, because its graded contours are, as in its living analogues, a picture of the graded forces by which it was conformed. It is an example of mathematical beauty, of which the machine-made, moulded bottle has no trace at all. An alabaster bottle is different again. It is no longer an unduloid figure of equilibrium. Turned on

---

[1] Where gourds are common, the glass-blower is still apt to take them for a proto-type, as the prehistoric potter also did. For instance, a tall, annulated Florence oil-flask is an exact but no longer a conscious imitation of a gourd which has been converted into a bottle in the manner described.

[2] Cf. *Elsie Venner*, ch. II.

a lathe, it is a solid of revolution, and not without beauty; but it is not near so beautiful as the blown flask or bubble.

The gravitational field is part of the complex field of force by which the form of the organism is influenced and determined. Its share is seldom easy to define, but there is a resultant due to gravity in hanging breasts and tired eyelids and all the sagging wrinkles of the old. Now and then we see gravity at work in the normal construction of the body, and can describe its effect on form in a general, or qualitative, way. Each pair of ribs in man forms a hoop which droops of its own weight in front, so flattening the chest, and at the same time twisting the rib on either hand near its point of suspension.[1] But in the dog each costal hoop is dragged straight downwards, into a vertical instead of a transverse ellipse, and is even narrowed to a point at the sternal border.

Rectangular Co-ordinates

We may now proceed to consider and illustrate a few permutations or transformations of organic form, out of the vast multitude which are equally open to this method of enquiry.

We have already compared in a preliminary fashion the metacarpal or cannon-bone of the ox, the sheep, and the giraffe (Fig. 122); and we have seen that the essential difference in form between these three bones is a matter of relative length and breadth, such that, if we reduce the figures to an identical standard of length (or identical values of y), the breadth (or value of x) will be approximately two-thirds that of the ox in the case of the sheep and one-third that of the ox in the case of the giraffe. We may easily, for the sake of closer comparison, determine these ratios more accurately, for instance, if it be our purpose to compare the different racial varieties within the limits of a single species. And in such cases, by the way, as when we compare with one another various breeds or races of cattle or of horses, the ratios of length and breadth in this particular bone are extremely significant.[2]

[1] See T. P. Anderson Stuart, 'How the Form of the Thorax is Partly Determined by Gravitation', *Proc. Roy. Soc.* **49** (1891), 143.

[2] This significance is particularly remarkable in connection with the development of speed, for the metacarpal region is the seat of very important leverage in the propulsion of the body. In a certain Scottish Museum there stand side by side the skeleton of an immense carthorse (celebrated for having drawn all the stones of the Bell Rock Lighthouse to the shore), and a beautiful skeleton of a racehorse, long supposed to be the actual skeleton of Eclipse. When I was a boy my grandfather used to point out to me that the cannon-bone of the little racer is not only relatively, but actually, longer than that of the great Clydesdale.

288

If, instead of limiting ourselves to the cannon-bone, we inscribe the entire foot of our several Ungulates in a co-ordinate system, the same ratios of x that served us for the cannon-bones still give us a first approximation to the required comparison; but even in the case of such closely allied forms as the ox and the sheep there is evidently something wanting in the comparison. The reason is that the relative elongation of the several parts, or individual bones, has not proceeded equally or proportionately in all cases; in other words, that the equations for x will not suffice without some simultaneous modification of the values of y (Fig. 135). In such a case it may be found possible

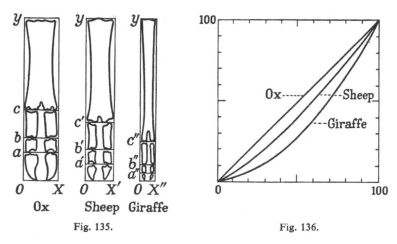

Fig. 135.   Fig. 136.

to satisfy the varying values of y by some logarithmic or other formula; but, even if that be possible, it will probably be somewhat difficult of discovery or verification in such a case as the present, owing to the fact that we have too few well-marked points of correspondence between the one object and the other, and that especially along the shaft of such long bones as the cannon-bone of the ox, the deer, the llama, or the giraffe there is a complete lack of easily recognisable corresponding points. In such a case a brief tabular statement of apparently corresponding values of y, or of those obviously corresponding values which coincide with the boundaries of the several bones of the foot, will, as in the following example, enable us to dispense with a fresh equation.

		a	b	c	d
y (ox)	0	18	27	42	100
y' (sheep)	0	10	19	36	100
y'' (giraffe)	0	5	10	24	100

289

This summary of values of y', coupled with the equations for the value of x, will enable us, from any drawing of the ox's foot, to construct a figure of that of the sheep or of the giraffe with remarkable accuracy.

That underlying the varying amounts of extension to which the parts or segments of the limb have been subject there is a law, or principle of continuity, may be discerned from such a diagram as the above (Fig. 136), where the values of y in the case of the ox are plotted as a straight line, and the corresponding values for the sheep (extracted from the above table) are seen to form a more or less regular and even curve. This simple graphic result implies the existence of a comparatively simple equation between y and y'.

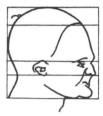 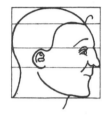 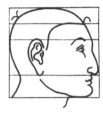

Fig. 137. (After Albrecht Dürer.)

An elementary application of the principle of co-ordinates to the study of proportion, as we have here used it to illustrate the varying proportions of a bone, was in common use in the sixteenth and seventeenth centuries by artists in their study of the human form. The method is probably much more ancient, and may even be classical;[1] it is fully described and put in practice by Albrecht Dürer in his *Geometry*, and especially in his *Treatise on Proportion*.[2] In this latter work, the manner in which the human figure, features, and facial expression are all transformed and modified by slight variations in the relative magnitude of the parts is admirably and copiously illustrated (Fig. 137).

In a tapir's foot there is a striking difference, and yet at the same time there is an obvious underlying resemblance, between the middle toe and either of its unsymmetrical lateral neighbours. Let us take

[1] Cf. Vitruvius, III, 1.
[2] *Les quatres livres d'Albert Dürer de la proportion des parties et pourtraicts des corps humains* (Arnheim, 1613), folio (and earlier editions). Cf. also Lavater, *Essays on Physiognomy* (1799), III, 271; also H. Meige, 'La géométrie des visages d'après Albert Dürer', *La Nature* (December 1927). On Dürer as mathematician, cf. Cantor, II, 459; S. Günther, *Die geometrische Näherungsconstructione Albrecht Dürers* (Ansbach, 1866); H. Staigmuller, *Dürer als Mathematiker* (Stüttgart, 1891).

the median terminal phalanx and inscribe its outline in a net of rectangular equidistant co-ordinates (Fig. 138*a*). Let us then make a similar network about axes which are no longer at right angles, but inclined to one another at an angle of about 50° (*b*). If into this new network we fill in, point for point, an outline precisely corresponding to our original drawing of the middle toe, we shall find that

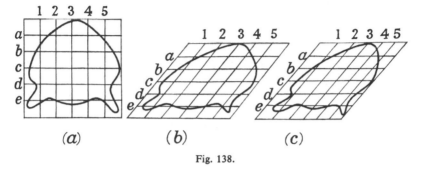

Fig. 138.

we have already represented the main features of the adjacent lateral one. We shall, however, perceive that our new diagram looks a little too bulky on one side, the inner side, of the lateral toe. If now we substitute for our equidistant ordinates, ordinates which get gradually closer and closer together as we pass towards the median side of the toe, then we shall obtain a diagram which differs in no essential respect

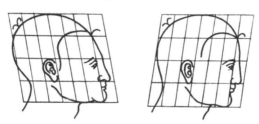

Fig. 139. (After Albrecht Dürer.)

from an actual outline copy of the lateral toe (*c*). In short, the difference between the outline of the middle toe of the tapir and the next lateral toe may be almost completely expressed by saying that if the one be represented by rectangular equidistant co-ordinates, the other will be represented by oblique co-ordinates, whose axes make an angle of 50°, and in which the abscissal interspaces decrease in a certain logarithmic ratio. We treated our original complex curve or projection of the tapir's toe as a function of the form $F(x, y) = 0$. The

291

figure of the tapir's lateral toe is a precisely identical function of the form $F(e^x, y_1) = 0$, where x_1, y_1 are oblique co-ordinate axes inclined to one another at an angle of 50°.

Dürer was acquainted with these oblique co-ordinates also, and I have copied two illustrative figures from his book.[1]

Crustacea

In Fig. 140 I have sketched the common Copepod *Oithona nana*, and have inscribed it in a rectangular net, with abscissae three-fifths the length of the ordinates. Side by side (Fig 141) is drawn a very different Copepod, of the genus *Sapphirina*; and about it is drawn

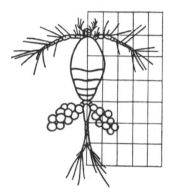

Fig. 140. *Oithona nana.*

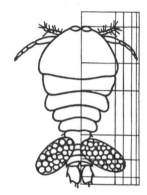

Fig. 141. *Sapphirina.*

a network such that each co-ordinate passes (as nearly as possible) through points corresponding to those of the former figure. It will be seen that two differences are apparent. (1) The values of *y* in Fig. 141 are large in the upper part of the figure, and diminish rapidly towards its base. (2) The values of *x* are very large in the

[1] It was these very drawings of Dürer's that gave to Peter Camper his notion of the 'facial angle'. Camper's method of comparison was the very same as ours, save that he only drew the axes, without filling in the network, of his co-ordinate system; he saw clearly the essential fact, that the skull *varies as a whole*, and that the 'facial angle' is the index to a general deformation. 'The great object was to show that natural differences might be reduced to rules, of which the direction of the facial line forms the *norma* or canon; and that these directions and inclinations are always accompanied by correspondent form, size and position of the other parts of the cranium', etc.; from Dr T. Cogan's preface to Camper's work *On the Connexion between the Science of Anatomy and the Arts of Drawing, Painting and Sculpture* (1768?), quoted in Dr R. Hamilton's Memoir of Camper, in *Lives of Eminent Naturalists* (*Nat. Libr.*), Edinburgh, 1840. See also P. Camper, *Dissertation sur les différences réelles que présentent les Traits du Visage chez les hommes de différents pays et de différents âges,* Paris, 1791 (*op. posth.*); cf. P. Topinard, 'Études sur Pierre Camper, et sur l'angle facial dit de Camper', *Rev. Anthropol.* **2** (1874).

neighbourhood of the origin, but diminish rapidly as we pass towards either side, away from the median vertical axis; and it is probable that they do so according to a definite, but somewhat complicated, ratio. If, instead of seeking for an actual equation, we simply tabulate our values of x and y in the second figure as compared with the first (just as we did in comparing the feet of the Ungulates), we get the dimensions of a net in which, by simply projecting the figure of *Oithona*, we obtain that of *Sapphirina* without further trouble, e.g.:

x (*Oithona*)	0	3	6	9	12	15	—
x' (*Sapphirina*)	0	8	10	12	13	14	—
y (*Oithona*)	0	5	10	15	20	25	30
y' (*Sapphirina*)	0	2	7	3	23	32	40

In this manner, with a single model or type to copy from, we may record in very brief space the data requisite for the production of approximate outlines of a great number of forms. For instance, the difference, at first sight immense, between the attenuated body of a *Caprella* and the thick-set body of a *Cyamus* is obviously little, and is probably nothing more than a difference of relative magnitudes, capable of tabulation by numbers and of complete expression by means of rectilinear co-ordinates.

The Crustacea afford innumerable instances of more complex deformations. Thus we may compare various higher Crustacea with one another, even in the case of such dissimilar forms as a lobster and a crab. It is obvious that the whole body of the former is elongated as compared with the latter, and that the crab is relatively broad in the region of the carapace, while it tapers off rapidly towards its attenuated and abbreviated tail. In a general way, the elongated rectangular system of co-ordinates in which we may inscribe the outline of the lobster becomes a shortened triangle in the case of the crab. In a little more detail we may compare the outline of the carapace in various crabs one with another: and the comparison will be found easy and significant, even, in many cases, down to minute details, such as the number and situation of the marginal spines, though these are in other cases subject to independent variability.

If we choose, to begin with, such a crab as *Geryon* (Fig. 142, *a*) and inscribe it in our equidistant rectangular co-ordinates, we shall see that we pass easily to forms more elongated in a transverse direction, such as *Matuta* or *Lupa* (*e*), and conversely, by transverse compression, to such a form as *Corystes* (*b*). In certain other cases the carapace conforms to a triangular diagram, more or less curvilinear, as in Fig. 142, *d*, which represents the genus *Paralomis*. Here we can easily see that the posterior border is transversely elongated as compared with that of *Geryon*, while at the same time the anterior part is longitudinally extended as compared with the posterior. A system of slightly curved and converging ordinates, with ortho-

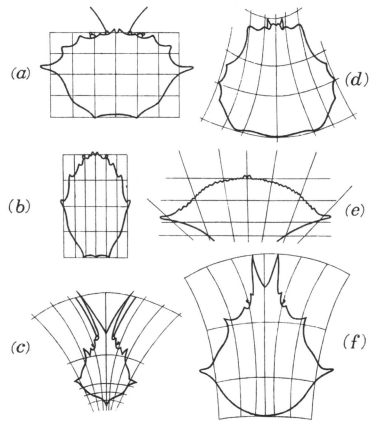

Fig. 142. Carapaces of various crabs. (*a*) *Geryon*; (*b*) *Corystes*; (*c*) *Scyramathia*; (*d*) *Paralomis*; (*e*) *Lupa*; (*f*) *Chorinus*.

gonal and logarithmically interspaced abscissal lines, as shown in the figure, appears to satisfy the conditions.

In an interesting series of cases, such as the genus *Chorinus*, or *Scyramathia*, and in the spider-crabs generally, we appear to have just the converse of this. While the carapace of these crabs presents a somewhat triangular form, which seems at first sight more or less similar to those just described, we soon see that the actual posterior border is now narrow instead of broad, the broadest part of the carapace corresponding precisely, not to that which is broadest in *Paralomis*, but to that which was broadest in *Geryon*; while the most striking difference from the latter lies in an antero-posterior lengthening of the forepart of the carapace, culminating in a great elongation of the frontal region, with its two spines or 'horns'. The curved ordinates here converge posteriorly and diverge widely in front (Fig. 142, *c* and *f*), while the decremental interspacing of the abscissae is very marked indeed.

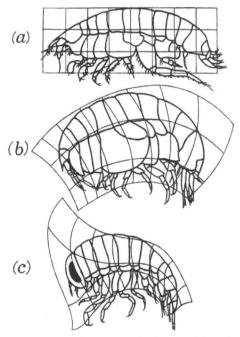

Fig. 143. (*a*) *Harpinia plumosa* Kr.; (*b*) *Stegocephalus inflatus* Kr.;
(*c*) *Hyperia galba*.

We put our method to a severer test when we attempt to sketch an entire and complicated animal than when we simply compare corresponding parts such as the carapaces of various Malacostraca, or related bones as in the case of the tapir's toes. Nevertheless, up to a certain point, the method stands the test very well. In other words, one particular mode and direction of variation is often (or even usually) so prominent and so paramount throughout the entire organism, that one comprehensive system of co-ordinates suffices to give a fair picture of the actual phenomenon. To take another

295

illustration from the Crustacea, I have drawn roughly in Fig. 143 *a* a little amphipod of the family Phoxocephalidae (*Harpinia* sp.). Deforming the co-ordinates of the figure into the curved orthogonal[1] system in Fig. 143 *b*, we at once obtain a very fair representation of an allied genus, belonging to a different family of amphipods, namely *Stegocephalus*. As we proceed further from our type our co-ordinates will require greater deformation, and the resultant figure will usually be somewhat less accurate. In Fig. 143 *c* I show a network, to which, if we transfer our diagram of *Harpinia* or of *Stegocephalus*, we shall obtain a tolerable representation of the aberrant genus *Hyperia*,[2] with its narrow abdomen, its reduced pleural lappets, its great eyes, and its inflated head.

Hydroids

The hydroid zoophytes constitute a 'polymorphic' group, within which a vast number of species have already been distinguished; and the labours of the systematic naturalist are constantly adding to the number. The specific distinctions are for the most part based, not upon characters directly presented by the living animal, but upon the form, size and arrangement of the little cups, or 'calycles', secreted and inhabited by the little individual polyps which compose the compound organism. The variations, which are apparently infinite, of these conformations are easily seen to be a question of relative magnitudes, and are capable of complete expression, sometimes by very simple, sometimes by somewhat more complex, co-ordinate networks.

For instance, the varying shapes of the simple wineglass-shaped cups of the Campanularidae are at once sufficiently represented and compared by means of simple Cartesian co-ordinates (Fig. 144). In the two allied familes of Plumulariidae and Aglaopheniidae the calycles are set unilaterally upon a jointed stem, and small cup-like structures (holding rudimentary polyps) are associated with the large calycles in definite number and position. These small calyculi are variable in number, but in the great majority of cases they accompany the large calycle in groups of three—two standing by its upper border, and one, which is especially variable in form and magnitude, lying at its base. The stem is liable to flexure and, in a high

---

[1] Similar co-ordinates are treated of by Lamé, *Leçons sur les coordonnées curvilignes* (Paris, 1859).

[2] For an analogous, but more detailed comparison, see H. Mogk, 'Versuch einer Formanalyse bei Hyperiden', *Int. Rev. ges. Hydrobiol.*, etc., **14** (1923), 276–311; **17** (1926), 1–98.

degree, to extension or compression; and these variations extend, often on an exaggerated scale, to the related calycles. As a result we find that we can draw various systems of curved or sinuous co-ordinates, which express, all but completely, the configuration of the

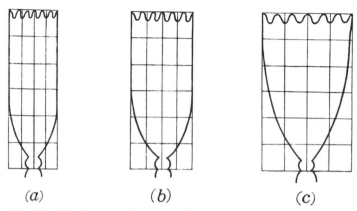

(a)   (b)   (c)

Fig. 144. (a) *Campanularia macroscyphus* Allm.; (b) *Gonothyraea hyalina* Hincks; (c) *Clytia Johnstoni* Alder.

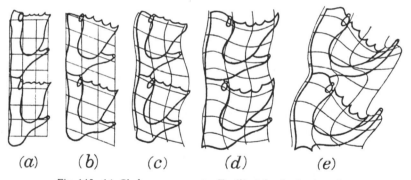

(a)   (b)   (c)   (d)   (e)

Fig. 145. (a) *Cladocarpus crenatus* F.; (b) *Aglaophenia pluma* L.; (c) *A. rhynchocarpa* A.; (d) *A. cornuta* K.; (e) *A. ramulosa* K.

various hydroids which we inscribe therein (Fig. 145). The comparative smoothness of denticulation of the margin of the calycle, and the number of its denticles, constitutes an independent variation, and requires separate description; we have already seen (p. 68) that this denticulation is in all probability due to a particular physical cause.

Among countless other invertebrate animals which we might illustrate, did space and time permit, we should find the bivalve molluscs showing certain things extremely well. If we start with a

297

more or less oblong shell, such as *Anodon* or *Mya* or *Psammobia*, we can see how easily it may be transformed into a more circular or orbicular, but still closely related form; while on the other hand a simple shear is wellnigh all that is needed to transform the oblong *Anodon* into the triangular, pointed *Mytilus*, *Avicula* or *Pinna*. Now suppose we draw the shell of *Anodon* in the usual rectangular co-ordinates, and deform this network into the corresponding oblique co-ordinates of *Mytilus*, we may then proceed to draw within the same two nets the anatomy of the same two molluscs. Then of the two adductor muscles, co-equal in *Anodon*, one becomes small, the other large, when transferred to the oblique network of *Mytilus*; at the same time the foot becomes stunted and the siphonal aperture enlarged. In short, having 'transformed' one shell into the other we may perform an identical transformation on their contained anatomy: and so (provided the two are not too distantly related) deduce the bodily structure of the one from our knowledge of the other, to a first but by no means negligible approximation.

Fish

Among the fishes we discover a great variety of deformations, some of them of a very simple kind, while others are more striking and more unexpected. A comparatively simple case, involving a simple shear, is illustrated by Figs. 146 and 147. The one represents, within Cartesian co-ordinates, a certain little oceanic fish known as *Argyropelecus olfersi*. The other represents precisely the same outline, transferred to a system of oblique co-ordinates whose axes are inclined at an angle of 70°; but this is now (as far as can be seen on the scale of the drawing) a very good figure of an allied fish, assigned to a different genus, under the name of *Sternoptyx diaphana*. The deformation illustrated by this case of *Argyropelecus* is precisely analogous to the simplest and commonest kind of deformation to which fossils are subject as the result of shearing-stresses in the solid rock.

Fig. 148 is an outline diagram of a typical Scaroid fish. Let us deform its rectilinear co-ordinates into a system of (approximately) coaxial circles, as in Fig. 149, and then filling into the new system, space by space and point by point, our former diagram of *Scarus*, we obtain a very good outline of an allied fish, belonging to a neighbouring family, of the genus *Pomacanthus*. This case is all the more interesting, because upon the body of our *Pomacanthus* there are striking colour bands, which correspond in direction very closely

298

to the lines of our new curved ordinates. In like manner, the still more bizarre outlines of other fishes of the same family of Chaetodonts will be found to correspond to very slight modifications of

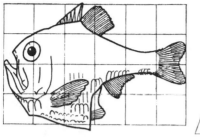

Fig. 146. *Argyropelecus olfersi.*

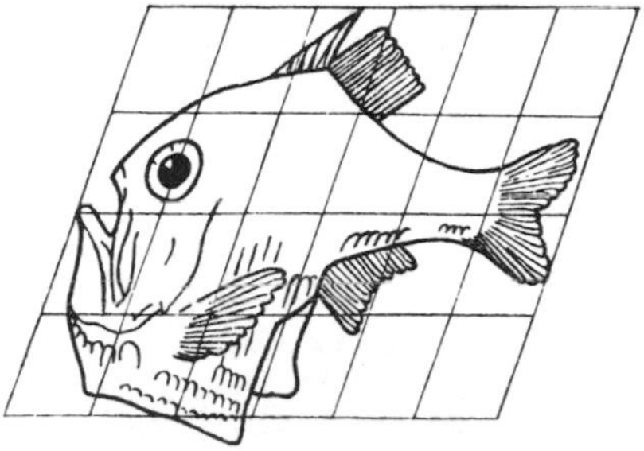

Fig. 147. *Sternoptyx diaphana.*

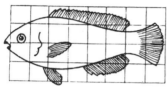

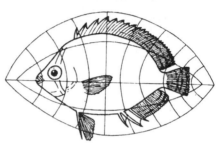

Fig. 148. *Scarus* sp.

Fig. 149. *Pomacanthus.*

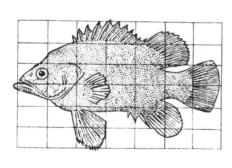

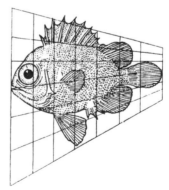

Fig. 150. *Polyprion.*

Fig. 151. *Pseudopriacanthus altus.*

similar co-ordinates; in other words, to small variations in the values of the constants of the coaxial curves.

In Figs. 150–153 I have represented another series of Acanthopterygian fishes, not very distantly related to the foregoing. If we

299

start this series with the figure of *Polyprion*, in Fig. 150, we see that the outlines of *Pseudopriacanthus* (Fig. 151) and of *Sebastes* or *Scorpaena* (Fig. 152) are easily derived by substituting a system of triangular, or radial, co-ordinates for the rectangular ones in which we had inscribed *Polyprion*. The very curious fish *Antigonia capros*, an oceanic relative of our own boar-fish, conforms closely to the peculiar deformation represented in Fig. 153.

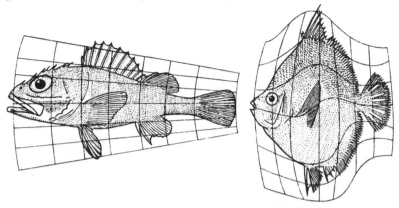

Fig. 152. *Scorpaena* sp.   Fig. 153. *Antigonia capros.*

Fig. 154 is a common, typical *Diodon* or porcupine-fish, and in Fig. 155 I have deformed its vertical co-ordinates into a system of concentric circles, and its horizontal co-ordinates into a system of curves which, approximately and provisionally, are made to resemble a system of hyperbolas. The old outline, transferred in its integrity to the new network, appears as a manifest representation of the closely allied, but very different looking, sunfish, *Orthagoriscus mola*. This is a particularly instructive case of deformation or transformation. It is true that, in a mathematical sense, it is not a perfectly satisfactory or perfectly regular deformation, for the system is no longer isogonal; but nevertheless, it is symmetrical to the eye, and obviously approaches to an isogonal system under certain conditions of friction or constraint. And as such it accounts, by one single integral transformation, for all the apparently separate and distinct external differences between the two fishes. It leaves the parts near to the origin of the system, the whole region of the head, the opercular orifice and the pectoral fin, practically unchanged in form, size and position; and it shows a greater and greater apparent modification of size and form as we pass from the origin towards the periphery of the system.

300

In a word, it is sufficient to account for the new and striking contour in all its essential details, of rounded body, exaggerated dorsal and ventral fins, and truncated tail. In like manner, and using precisely the same co-ordinate networks, it appears to me possible to

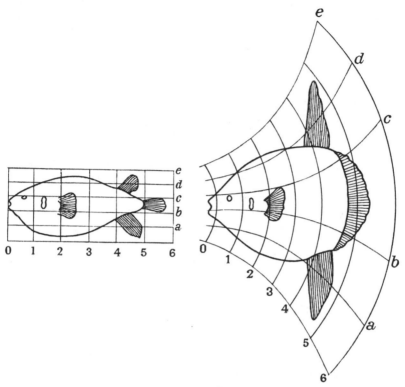

Fig. 154.   Fig. 155. *Orthagoriscus.*

show the relations, almost bone for bone, of the skeletons of the two fishes; in other words, to reconstruct the skeleton of the one from our knowledge of the skeleton of the other, under the guidance of the same correspondence as is indicated in their external configuration.

Reptiles

The family of the crocodiles has had a special interest for the evolutionist ever since Huxley pointed out that, in a degree only second to the horse and its ancestors, it furnishes us with a close and almost unbroken series of transitional forms, running down in continuous succession from one geological formation to another. I should be inclined to transpose this general statement into other terms, and to say that the Crocodilia constitute a case in which, with unusually little complication from the presence of independent variants, the trend of one particular mode of transformation is visibly manifested. If we exclude meanwhile from our comparison a few of the oldest of the crocodiles, such as *Belodon*, which differ more fundamentally from the rest, we shall find a long series of genera in which we can refer not only the changing contours of the skull, but even the shape and size of the many constituent bones and their intervening spaces or 'vacuities', to one and the same simple system of transformed co-ordinates. The manner in which the skulls of various Crocodilians differ from one another may be sufficiently illustrated by three or four examples.

Let us take one of the typical modern crocodiles as our standard of form, e.g. *Crocodilus porosus*, and inscribe it, as in Fig. 156 *a*, in the usual Cartesian co-ordinates. By deforming the rectangular network into a triangular system, with the apex of the triangle a little way in front of the snout, as in (*b*), we pass to such a form as *C. americanus*. By an exaggeration of the same process we at once get an approximation to the form of one of the sharp-snouted, or longirostrine, crocodiles, such as the genus *Tomistoma*; and, in the species figured, the oblique position of the orbits, the arched contour of the occipital border, and certain other characters suggest a certain amount of curvature, such as I have represented in the diagram (Fig. 156*b*), on the part of the horizontal co-ordinates. In the still more elongated skull of such a form as the Indian Gavial, the whole skull has undergone a great longitudinal extension, or, in other words, the ratio of x/y is greatly diminished; and this extension is not uniform, but is at a maximum in the region of the nasal and maxillary bones. This especially elongated region is at the same time narrowed in an exceptional degree, and its excessive narrowing is represented by a curvature, convex towards the median axis, on the part of the vertical ordinates. Let us take as a last illustration one of the Mesozoic crocodiles, the little *Notosuchus*, from the Cretaceous formation. This little crocodile is very different from our type in the proportions of its skull. The region of the snout, in front of and including the frontal bones, is greatly shortened; from constituting fully two-thirds of the whole length of the skull in *Crocodilus*, it now constitutes less than half, or, say, three-sevenths of the whole; and the whole skull, and especially its posterior part, is curiously

compact, broad, and squat. The orbit is unusually large. If in
the diagram of this skull we select a number of points obviously
corresponding to points where our rectangular co-ordinates intersect
particular bones or other recognisable features in our typical crocodile, we shall easily discover that the lines joining these points in
Notosuchus fall into such a co-ordinate network as that which is
represented in Fig. 156 c. To all intents and purposes, then, this not

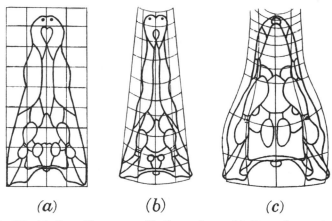

(a)   (b)   (c)

Fig. 156. (a) *Crocodilus porosus*; (b) *C. americanus*; (c) *Notosuchus terrestris*.

very complex system, representing one harmonious 'deformation',
accounts for *all* the differences between the two figures, and is sufficient to enable one at any time to reconstruct a detailed drawing,
bone for bone, of the skull of *Notosuchus* from the model furnished
by the common crocodile.

The many diverse forms of Dinosaurian reptiles, all of which manifest a strong family likeness underlying much superficial diversity,
furnish us with plentiful material for comparison by the method of
transformations. As an instance, I have figured the pelvic bones of
Stegosaurus and of *Camptosaurus* (Fig. 157a, b) to show that, when
the former is taken as our Cartesian type, a slight curvature and an
approximately logarithmic extension of the *x*-axis brings us easily to
the configuration of the other. In the original specimen of *Camptosaurus* described by Marsh,[1] the anterior portion of the iliac bone is
missing; and in Marsh's restoration this part of the bone is drawn as
though it came somewhat abruptly to a sharp point. In my figure I
have completed this missing part of the bone in harmony with the

---

[1] *Dinosaurs of North America* (1896), pl. LXXXI, etc.

general co-ordinate network which is suggested by our comparison of the two entire pelves; and I venture to think that the result is more natural in appearance, and more likely to be correct than was Marsh's conjectural restoration. It would seem, in fact, that there is an obvious field for the employment of the method of co-ordinates in this task of reproducing missing portions of a structure to the proper scale and in harmony with related types. To this subject we shall presently return.

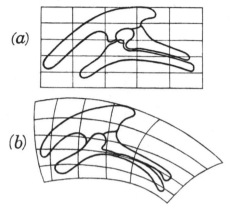

Fig. 157. Pelvis of (*a*) *Stegosaurus*; (*b*) *Camptosaurus*.

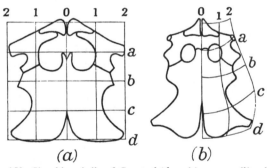

Fig. 158. Shoulder-girdle of *Cryptocleidus*. (*a*) young; (*b*) adult.

In Fig. 158 *a*, *b*, I have drawn the shoulder-girdle of *Cryptocleidus*, a Plesiosaurian reptile, half-grown in the one case and full-grown in the other. The change of form during growth in this region of the body is very considerable, and its nature is well brought out by the two co-ordinate systems. In Fig. 159 I have drawn the shoulder-girdle of an Ichthyosaur, referring it to *Cryptocleidus* as a standard of comparison. The interclavicle, which is present in *Ichthyosaurus*, is minute and hidden in *Cryptocleidus*; but the numerous other differences between the two forms, chief among which is the great elongation in *Ichthyosaurus* of the two clavicles, are all seen by our diagrams to be part and parcel of one general and systematic deformation.

Before we leave the group of reptiles we may glance at the very strangely modified skull of *Pteranodon*, one of the extinct flying reptiles, or Pterosauria. In this very curious skull the region of

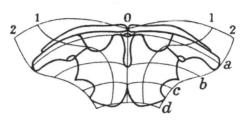

Fig. 159. Shoulder-girdle of *Ichthyosaurus*.

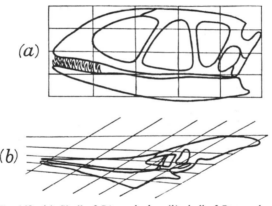

Fig. 160. (*a*) Skull of *Dimorphodon*; (*b*) skull of *Pteranodon*.

the jaws, or beak, is greatly elongated and pointed; the occipital bone is drawn out into an enormous backwardly directed crest; the posterior part of the lower jaw is similarly produced backwards; the orbit is small; and the quadrate bone is strongly inclined downwards and forwards. The whole skull has a configuration which stands, apparently, in the strongest possible contrast to that of a more normal Ornithosaurian such as *Dimorphodon*. But if we inscribe the latter in Cartesian co-ordinates (Fig. 160*a*), and refer our *Pteranodon* to a system of oblique co-ordinates (*b*), in which the

305

two co-ordinate systems of parallel lines become each a pencil of diverging rays, we make manifest a correspondence which extends uniformly throughout all parts of these very different-looking skulls.

The Pelvis of Birds

We have dealt so far, and for the most part we shall continue to deal, with our co-ordinate method as a means of comparing one known structure with another. But it is obvious, as I have said, that it may

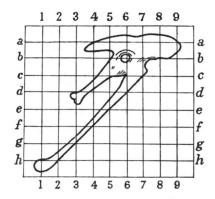

Fig. 161. Pelvis of *Archaeopteryx*.

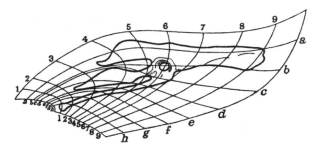

Fig. 162. Pelvis of *Apatornis*.

also be employed for drawing hypothetical structures, on the assumption that they have varied from a known form in some definite way. And this process may be especially useful, and will be most obviously legitimate, when we apply it to the particular case of representing intermediate stages between two forms which are actually known to exist, in other words, of reconstructing the transitional stages through which the course of evolution must have successively travelled if it

306

has brought about the change from some ancestral type to its presumed descendant. Some years ago I sent my friend, Mr Gerhard Heilmann of Copenhagen, a few of my own rough co-ordinate diagrams, including some in which the pelves of certain ancient and primitive

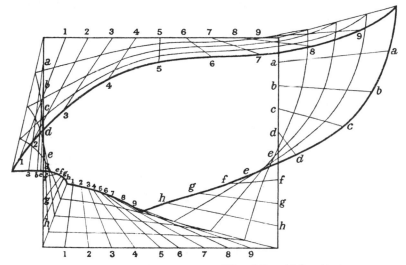

Fig. 163. The co-ordinate systems of Figs. 161 and 162, with three intermediate systems interpolated.

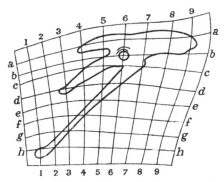

Fig. 164. The first intermediate co-ordinate network, with its corresponding inscribed pelvis.

birds were compared one with another. Mr Heilmann, who is both a skilled draughtsman and an able morphologist, returned me a set of diagrams which are a vast improvement on my own, and which are reproduced in Figs. 161–166. Here we have, as extreme cases, the pelvis of *Archaeopteryx*, the most ancient of known birds, and

307

that of *Apatornis*, one of the fossil 'toothed' birds from the North American Cretaceous formations—a bird showing some resemblance to the modern terns. The pelvis of *Archaeopteryx* is taken as our type, and referred accordingly to Cartesian co-ordinates (Fig. 161); while the corresponding co-ordinates of the very different pelvis of *Apatornis* are represented in Fig. 162. In Fig. 163 the outlines of these two co-ordinate systems are superposed upon one another, and those of

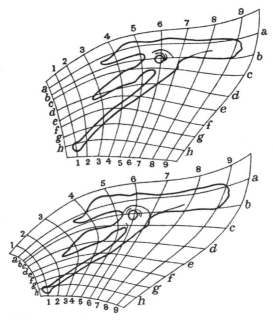

Fig. 165. The second and third intermediate co-ordinate networks, with their corresponding inscribed pelves.

three intermediate and equidistant co-ordinate systems are interpolated between them. From each of these latter systems, so determined by direct interpolation, a complete co-ordinate diagram is drawn, and the corresponding outline of a pelvis is found from each of these systems of co-ordinates, as in Figs. 164, 165. Finally, in Fig. 166 the complete series is represented, beginning with the known pelvis of *Archaeopteryx*, and leading up by our three intermediate hypothetical types to the known pelvis of *Apatornis*.

Among mammalian skulls I will take two illustrations only, one drawn from a comparison of the human skull with that of the higher apes, and another from the group of Perissodactyle Ungulates, the group which includes the rhinoceros, the tapir, and the horse.

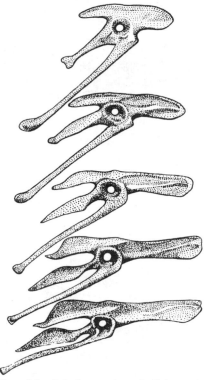

Fig. 166. The pelvis of *Archaeopteryx* and of *Apatornis*, with three transitional types interpolated between them.

Let us begin by choosing as our type the skull of *Hyrachyus agrarius* Cope, from the Middle Eocene of North America, as figured by Osborn in his Monograph of the Extinct Rhinoceroses (Fig. 167).[1]

The many other forms of primitive rhinoceros described in the monograph differ from *Hyrachyus* in various details—in the characters of the teeth, sometimes in the number of the toes, and so forth;

[1] *Mem. Amer. Mus. Nat. Hist.* **1**, pt. 3 (1898).

and they also differ very considerably in the general appearance of the skull. But these differences in the conformation of the skull, conspicuous as they are at first sight, will be found easy to bring under the conception of a simple and homogeneous transformation, such as would result from the application of some not very compli-

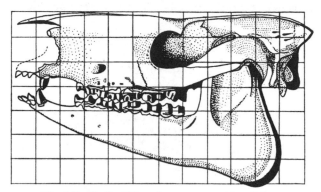

Fig. 167. Skull of *Hyrachyus agrarius*. After Osborn.

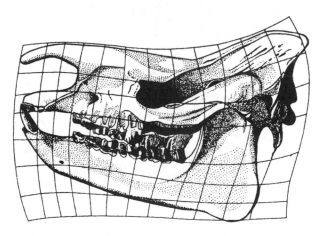

Fig. 168. Skull of *Aceratherium tridactylum*. After Osborn.

cated stress. For instance, the corresponding co-ordinates of *Aceratherium tridactylum*, as shown in Fig. 168, indicate that the essential difference between this skull and the former one may be summed up by saying that the long axis of the skull of *Aceratherium* has undergone a slight double curvature, while the upper parts of the skull have at the same time been subject to a vertical expansion, or to growth in somewhat greater proportion than the lower parts. Precisely the

310

same changes, on a somewhat greater scale, give us the skull of an existing rhinoceros.

Among the species of *Aceratherium*, the posterior, or occipital, view of the skull presents specific differences which are perhaps more conspicuous than those furnished by the side view; and these differences are very strikingly brought out by the series of conformal transformations which I have represented in Fig. 169. In this case it will perhaps be noticed that the correspondence is not always quite accurate in small details. It could easily have been made

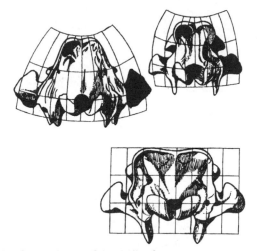

Fig. 169. Occipital view of the skulls of various extinct rhinoceroses
(*Aceratherium* spp.). After Osborn.

much more accurate by giving a slightly sinuous curvature to certain of the co-ordinates. But as they stand, the correspondence indicated is very close, and the simplicity of the figures illustrates all the better the general character of the transformation.

By similar and not more violent changes we pass easily to such allied forms as the Titanotheres (Fig. 170); and the well-known series of species of *Titanotherium*, by which Professor Osborn has illustrated the evolution of this genus, constitutes a simple and suitable case for the application of our method.

But our method enables us to pass over greater gaps than these, and to discern the general, and to a very large extent even the detailed, resemblances between the skull of the rhinoceros and those of the tapir or the horse. From the Cartesian co-ordinates in which we have begun by inscribing the skull of a primitive rhinoceros,

311

we pass to the tapir's skull (Fig. 171), firstly, by converting the rectangular into a triangular network, by which we represent the depression of the anterior and the progressively increasing elevation of the posterior part of the skull; and secondly, by giving to the vertical ordinates a curvature such as to bring about a certain longitudinal compression, or condensation, in the forepart of the skull, especially in the nasal and orbital regions.

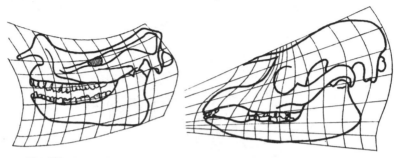

Fig. 170. *Titanotherium robustum.*   Fig. 171. Tapir's skull.

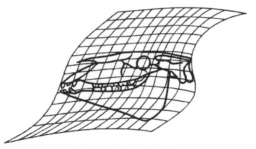

Fig. 172. Horse's skull.

The conformation of the horse's skull departs from that of our primitive Perissodactyle (that is to say our early type of rhinoceros, *Hyrachyus*) in a direction that is nearly the opposite of that taken by *Titanotherium* and by the recent species of rhinoceros. For we perceive, by Fig. 172, that the horizontal co-ordinates, which in these latter cases become transformed into curves with the concavity upwards, are curved, in the case of the horse, in the opposite direction. And the vertical ordinates, which are also curved, somewhat in the same fashion as in the tapir, are very nearly equidistant, instead of being, as in that animal, crowded together anteriorly. Ordinates and abscissae form an obiique system, as is shown in the figure. In this case I have attempted to produce the network beyond the region

312

which is actually required to include the diagram of the horse's skull, in order to show better the form of the general transformation, with a part only of which we have actually to deal.

It is at first sight not a little surprising to find that we can pass, by a cognate and even simpler transformation, from our Perissodactyle skulls to that of the rabbit; but the fact that we can easily do so is a simple illustration of the undoubted affinity which exists between the Rodentia, especially the family of the Leporidae, and the more primitive Ungulates. For my part, I would go further; for I think there is strong reason to believe that the Perissodactyles are more closely related to the Leporidae than the former are to the other Ungulates, or than the Leporidae are to the rest of the Rodentia. Be that as it may, it is obvious from Fig. 173 that the rabbit's skull

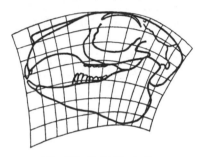

Fig. 173. Rabbit's skull.

conforms to a system of co-ordinates corresponding to the Cartesian co-ordinates in which we have inscribed the skull of *Hyrachyus*, with the difference, first, that the horizontal ordinates of the latter are transformed into equidistant curved lines, approximately arcs of circles, with their concavity directed downwards; and secondly, that the vertical ordinates are transformed into a pencil of rays approximately orthogonal to the circular arcs. In short, the configuration of the rabbit's skull is derived from that of our primitive rhinoceros by the unexpectedly simple process of submitting the latter to a strong and uniform flexure in the downward direction (cf. Fig. 167). In the case of the rabbit the configuration of the individual bones does not conform quite so well to the general transformation as it does when we are comparing the several Perissodactyles one with another; and the chief departures from conformity will be found in the size of the orbit and in the outline of the immediately surrounding bones. The simple fact is that the relatively enormous eye of the rabbit con-

313

stitutes an independent variation, which cannot be brought into the general and fundamental transformation, but must be dealt with separately. The enlargement of the eye, like the modification in form and number of the teeth, is a separate phenomenon, which supplements but in no way contradicts our general comparison of the skulls taken in their entirety.

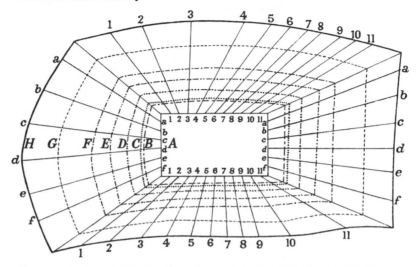

Fig. 174. A, outline diagram of the Cartesian co-ordinates of the skull of *Hyracotherium* or *Eohippus*, as shown in Fig. 175. *A–H*, outline of the corresponding projection of the horse's skull. *B–G*, intermediate, or interpolated, outlines.

Before we leave the Perissodactyla and their allies, let us look a little more closely into the case of the horse and its immediate relations or ancestors, doing so with the help of a set of diagrams which I again owe to Mr Gerhard Heilmann.[1] Here we start afresh, with the skull (Fig. 175 *A*) of *Hyracotherium* (or *Eohippus*), inscribed in a simple Cartesian network. At the other end of the series (*H*) is a skull of *Equus*, in its own corresponding network; and the intermediate stages (*B–G*) are all drawn by direct and simple interpolation, as in Mr Heilmann's former series of drawings of *Archaeopteryx* and *Apatornis*. In this present case, the relative magnitudes are shown, as well as the forms, of the several skulls. Alongside of these reconstructed diagrams are set figures of certain extinct 'horses' (Equidae or Palaeotheriidae), and in two cases, viz. *Mesohippus* and

[1] These and also other co-ordinate diagrams will be found in Mr G. Heilmann's beautiful and original book *Fuglenes Afstamning* (Copenhagen, 1916), 398 pp.; see especially pp. 368–80.

Protohippus (*M*, *P*), it will be seen that the actual fossil skull coincides in the most perfect fashion with one of the hypothetical forms or stages which our method shows to be implicitly involved in the transition from *Hyracotherium* to *Equus*.[1] In a third case, that of *Parahippus* (*Pa*), the correspondence (as Mr Heilmann points out) is by no means exact. The outline of this skull comes nearest to that of the hypothetical transition stage *D*, but the 'fit' is now a bad one; for the skull of *Parahippus* is evidently a longer, straighter and narrower skull, and differs in other minor characters besides. In short, though some writers have placed *Parahippus* in the direct line of descent between *Equus* and *Eohippus*, we see at once that there is no place for it there, and that it must, accordingly, represent a somewhat divergent branch or offshoot of the Equidae.[2] It may be noticed, especially in the case of *Protohippus* (*P*), that the configuration of the angle of the jaw does not tally quite so accurately with that of our hypothetical diagrams as do other parts of the skull. As a matter of fact, this region is somewhat variable, in different species of a genus, and even in different individuals of the same species; in the small figure (*Pp*) of *Protohippus placidus* the correspondence is more exact.

In considering this series of figures we cannot but be struck, not only with the regularity of the succession of 'transformations', but also with the slight and inconsiderable differences which separate each recorded stage from the next, and even the two extremes of the whole series from one another. These differences are no greater (save in regard to actual magnitude) than those between one human skull and another, at least if we take into account the older or remoter races; and they are again no greater, but if anything less, than the range of variation, racial and individual, in certain other human bones, for instance the scapula.[3]

The variability of this latter bone is great, but it is neither

---

[1] Cf. Zittel, *Grundzüge d. Palaeontologie* (1911), p. 463.

[2] Cf. W. B. Scott (*Amer. J. Sci.* **48**, 1894, 335–74), 'We find that any mammalian series at all complete, such as that of the horses, is remarkably continuous, and that the progress of discovery is steadily filling up what few gaps remain. So closely do successive stages follow upon one another that it is sometimes extremely difficult to arrange them all in order, and to distinguish clearly those members which belong in the main line of descent, and those which represent incipient branches. Some phylogenies actually suffer from an embarrassment of riches.'

[3] Cf. T. Dwight, 'The Range of Variation of the Human Scapula', *Amer. Nat.* **21** (1887), 627–38. Cf. also Turner, *Chall. Rep.* XLVII, on Human Skeletons, p. 86, 1886: 'I gather both from my own measurements, and those of other observers, that the range of variation in the relative length and breadth of the scapula is very considerable in the same race, so that it needs a large number of bones to enable one to obtain an accurate idea of the mean of the race.'

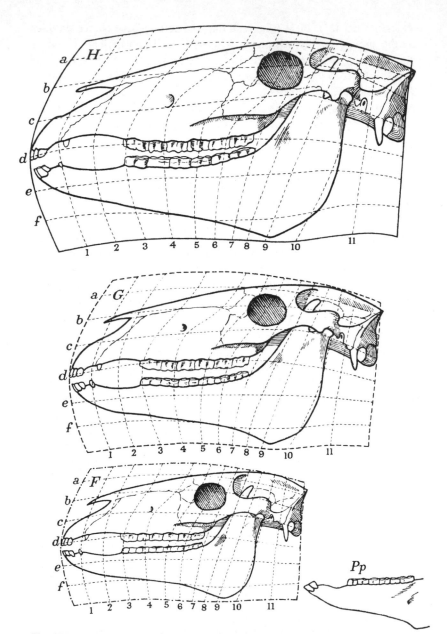

Fig. 175. *A*, skull of *Hyracotherium*, from the Eocene, after W. B. Scott; *H*, skull of horse, represented as a co-ordinate transformation of that of *Hyracotherium*, and to the same scale of magnitude; *B–G*, various artificial or imaginary types, reconstructed as intermediate stages between *A* and *H*; *M*, skull of *Mesohippus*, from the Oligocene,

surprising nor peculiar; for it is linked with all the considerations of mechanical efficiency and functional modification which we dealt with in our last chapter. The scapula occupies, as it were, a focus in a very important field of force; and the lines of force converging on it will be very greatly modified by the varying development of the muscles over a large area of the body and of the uses to which they are habitually put.

Let us now inscribe in our Cartesian co-ordinates the outline of

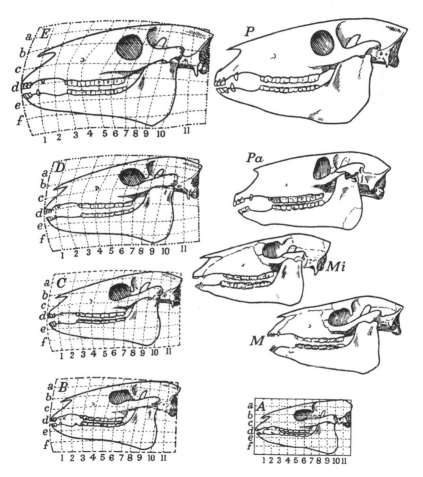

after Scott, for comparison with *C*; *P*, skull of *Protohippus*, from the Miocene, after Cope, for comparison with *E*; *Pp*, lower jaw of *Protohippus placidus* (after Matthew and Gidley), for comparison with *F*; *Mi*, *Miohippus* (after Osborn), *Pa*, *Parahippus* (after Peterson), showing resemblance, but less perfect agreement, with *C* and *D*.

317

a human skull (Fig. 177), for the purpose of comparing it with the skulls of some of the higher apes. We know beforehand that the main differences between the human and the simian types depend upon the enlargement or expansion of the brain and braincase in man, and the relative diminution or enfeeblement of his jaws. Together with these changes, the 'facial angle' increases from an oblique

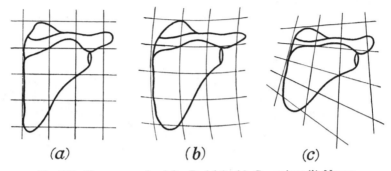

Fig. 176. Human scapulae (after Dwight). (a) Caucasian; (b) Negro; (c) North American Indian (from Kentucky Mountains).

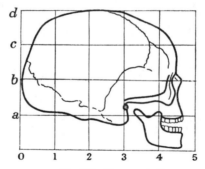

Fig. 177. Human skull.

angle to nearly a right angle in man, and the configuration of every constituent bone of the face and skull undergoes an alteration. We do not know to begin with, and we are not shown by the ordinary methods of comparison, how far these various changes form part of one harmonious and congruent transformation, or whether we are to look, for instance, upon the changes undergone by the frontal, the occipital, the maxillary, and the mandibular regions as a congeries of separate modifications or independent variants. But as soon as we have marked out a number of points in the gorilla's or chimpanzee's skull, corresponding with those which our co-ordinate network inter-

318

sected in the human skull, we find that these corresponding points may be at once linked up by smoothly curved lines of intersection, which form a new system of co-ordinates and constitute a simple 'projection' of our human skull. The network represented in Fig. 178 constitutes such a projection of the human skull on what we may call, figuratively speaking, the 'plane' of the chimpanzee; and the full

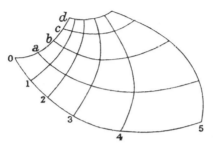

Fig. 178. Co-ordinates of chimpanzee's skull, as a projection of the Cartesian co-ordinates of Fig. 177.

Fig. 179. Skull of chimpanzee.   Fig. 180. Skull of baboon.

diagram in Fig. 179 demonstrates the correspondence. In Fig. 180 I have shown the similar deformation in the case of a baboon, and it is obvious that the transformation is of precisely the same order, and differs only in an increased intensity or degree of deformation.[1] These anthropoid skulls, then, which we can transform one into another by a 'continuous transformation', are admirable examples of what Listing called 'topological similitude'.

In both dimensions, as we pass from above downwards and from behind forwards, the corresponding areas of the network are seen

[1] The empirical co-ordinates which I have sketched in for the chimpanzee as a conformal transformation of the Cartesian co-ordinates of the human skull look as if they might find their place in an equipotential elliptic field. They are indeed closely analogous to some already figured by MM. Y. Ikada and M. Kuwaori, 'Some Conformal Representations by means of the Elliptic Integrals', *Sci. Pap. Inst. Phys. Res., Tokyo*, **26** (1936), 208–15: c.g. pl. xxxib.

to increase in a gradual and approximate logarithmic order in the lower as compared with the higher type of skull; and, in short, it becomes at once manifest that the modifications of jaws, brain-case, and the regions between are all portions of one continuous and integral process. It is of course easy to draw the inverse diagrams, by which the Cartesian co-ordinates of the ape are transformed into curvilinear and non-equidistant co-ordinates in man.[1]

From this comparison of the gorilla's or chimpanzee's with the human skull we realise that an inherent weakness underlies the anthropologist's method of comparing skulls by reference to a small number of axes. The most important of these are the 'facial' and 'basicranial' axes, which include between them the 'facial angle'. But it is, in the first place, evident that these axes are merely the principal axes of a system of co-ordinates, and that their restricted and isolated use neglects all that can be learned from the filling in of the rest of the co-ordinate network. And, in the second place, the 'facial axis', for instance, as ordinarily used in the anthropological comparison of one human skull with another, or of the human skull with the gorilla's, is in all cases treated as a straight line; but our investigation has shown that rectilinear axes only meet the case in the simplest and most closely related transformations; and that, for instance, in the anthropoid skull no rectilinear axis is homologous with a rectilinear axis in a man's skull, but what is a straight line in the one has become a certain definite curve in the other.

Mr Heilmann tells me that he has tried, but without success, to obtain a transitional series between the human skull and some pre-human, anthropoid type, which series (as in the case of the Equidae) should be found to contain other known types in direct linear sequence. It appears impossible, however, to obtain such a series, or to pass by successive and continuous gradations through such forms as Mesopithecus, Pithecanthropus, *Homo neanderthalensis*, and the lower or higher races of modern man. The failure is not the fault of our method. It merely indicates that no one straight line of descent, or of consecutive transformation, exists; but on the contrary, that among human and anthropoid types, recent and extinct, we have to do with a complex problem of divergent, rather than of continuous,

---

[1] Speaking of 'diagrams in pairs', and doubtless thinking of his own 'reciprocal diagrams', Clerk Maxwell says (in his article *Diagrams* in the *Encyclopaedia Britannica*): 'The method in which we simultaneously contemplate two figures, and recognise a correspondence between certain points in the one figure and certain points in the other, is one of the most powerful and fertile methods hitherto known in science....It is sometimes spoken of as the method or principle of duality.'

variation.* And in like manner, easy as it is to correlate the baboon's and chimpanzee's skulls severally with that of man, and easy as it is to see that the chimpanzee's skull is much nearer to the human type than is the baboon's, it is also not difficult to perceive that the series is not, strictly speaking, continuous, and that neither of our two apes lies *precisely* on the same direct line or sequence of deformation by which we may hypothetically connect the other with man.

After easily transforming our co-ordinate diagram of the human skull into a corresponding diagram of ape or of baboon, we may effect a further transformation of man or monkey into dog no less easily; and we are thereby encouraged to believe that any two mammalian skulls may be compared with, or transformed into, one another by this method. There is something, an essential and indispensable something, which is common to them all, something which is the subject of all our transformations, and remains *invariant* (as the mathematicians say) under them all. In these transformations of ours every point may change its place, every line its curvature, every area its magnitude; but on the other hand every point and every line continues to exist, and keeps its relative order and position throughout all distortions and transformations. A series of points, *a, b, c*, along a certain line persist as corresponding points *a', b', c'*, however the line connecting them may lengthen or bend; and as with points, so with lines, and so also with areas. Ear, eye and nostril, and all the other great landmarks of cranial anatomy, not only continue to exist but retain their relative order and position throughout all our transformations.

We can discover a certain *invariance*, somewhat more restricted than before, between the mammalian skull and that of fowl, frog or even herring. We have still something common to them all; and using another mathematical term (somewhat loosely perhaps) we may speak of the *discriminant characters* which persist unchanged, and continue to form the subject of our transformation. But the method, far as it goes, has its limitations. We cannot fit both beetle and cuttlefish into the same framework, however we distort it; nor by any co-ordinate transformation can we turn either of them into one another or into the vertebrate type. They are essentially different; there is nothing about them which can be legitimately compared. Eyes they all have, and mouth and jaws; but what we call by these

* This turns out to be quite correct. As G. E. Hutchinson, *Amer. Sci.* **36**, 600, points out and illustrates with new information on man's ancestry, it is now possible to trace a sequence of early man by co-ordinate transformations.

names are no longer in the same order or relative position; they are no longer the same thing, there is no *invariant* basis for transformation. The cuttlefish eye seems as perfect, optically, as our own; but the lack of an invariant relation of position between them, or lack of true homology between them (as we naturalists say), is enough to show that they are unrelated things, and have come into existence independently of one another.

As a final illustration I have drawn the outline of a dog's skull (Fig. 181), and inscribed it in a network comparable with the Cartesian network of the human skull in Fig. 177. Here we attempt to

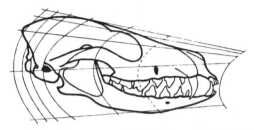

Fig. 181. Skull of dog, compared with the human skull of Fig. 177.

bridge over a wider gulf than we have crossed in any of our former comparisons. But, nevertheless, it is obvious that our method still holds good, in spite of the fact that there are various specific differences, such as the open or closed orbit, etc., which have to be separately described and accounted for. We see that the chief essential differences in plan between the dog's skull and the man's lie in the fact that, relatively speaking, the former tapers away in front, a triangular taking the place of a rectangular conformation; secondly, that, coincident with the tapering off, there is a progressive elongation, or pulling out, of the whole forepart of the skull; and lastly, as a minor difference, that the straight vertical ordinates of the human skull become curved, with their convexity directed forwards, in the dog. While the net result is that in the dog, just as in the chimpanzee, the brain-pan is smaller and the jaws are larger than in man, it is now conspicuously evident that the co-ordinate network of the ape is by no means intermediate between those which fit the other two. The mode of deformation is on different lines; and, while it may be correct to say that the chimpanzee and the baboon are more brute-like, it would be by no means accurate to assert that they are more dog-like, than man.

Three-dimensional Co-ordinate Systems

In this brief account of co-ordinate transformations and of their morphological utility I have dealt with plane co-ordinates only, and have made no mention of the less elementary subject of co-ordinates in three-dimensional space. In theory there is no difficulty whatsoever in such an extension of our method; it is just as easy to refer the form of our fish or of our skull to the rectangular co-ordinates x, y, z, or to the polar co-ordinates ξ, η, ζ, as it is to refer their plane projections to the two axes to which our investigation has been confined. And that it would be advantageous to do so goes without saying, for it is the shape of the solid object, not that of the mere drawing of the object, that we want to understand; and already we have found some of our easy problems in solid geometry leading us (as in the case of the form of the bivalve and even of the univalve shell) quickly in the direction of co-ordinate analysis and the theory of conformal transformations. But this extended theme I have not attempted to pursue, and it must be left to other times, and to other hands. Nevertheless, let us glance for a moment at the sort of simple cases, the simplest possible cases, with which such an investigation might begin; and we have found our plane co-ordinate systems so easily and effectively applicable to certain fishes that we may seek among them for our first and tentative introduction to the three-dimensional field.

It is obvious enough that the same method of description and analysis which we have applied to one plane, we may apply to another; drawing by observation, and by a process of trial and error, our various cross-sections and the co-ordinate systems which seem best to correspond. But the new and important problem which now emerges is to *correlate* the deformation or transformation which we discover in one plane with that which we have observed in another: and at length, perhaps, after grasping the general principles of such correlation, to forecast approximately what is likely to take place in the third dimension when we are acquainted with two, that is to say, to determine the values along one axis in terms of the other two.

Let us imagine a common 'round' fish, and a common 'flat' fish, such as a haddock and a plaice. These two fishes are not as nicely adapted for comparison by means of plane co-ordinates as some which we have studied, owing to the presence of essentially unimportant, but yet conspicuous differences in the position of the eyes, or in the number of the fins—that is to say in the manner in which the con-

323

tinuous dorsal fin of the plaice appears in the haddock to be cut or scolloped into a number of separate fins. But speaking broadly, and apart from such minor differences as these, it is manifest that the chief factor in the case (so far as we at present see) is simply the broadening out of the plaice's body, as compared with the haddock's, in the dorso-ventral direction, that is to say, along the y axis; in other words, the ratio x/y is much less (and indeed little more than half as great) in the haddock than in the plaice. But we also recognise at once that while the plaice (as compared with the haddock) is expanded in one direction, it is also flattened, or thinned out, in the other: y increases, but z diminishes, relatively to x. And furthermore, we soon see that this is a common or even a general phenomenon. The high, expanded body in our Antigonia or in our sun-fish or in a John Dory is at the same time flattened or *compressed* from side to side, in comparison with the related fishes which we have chosen as standards of reference or comparison; and conversely, such a fish as the skate, while it is expanded from side to side in comparison with a shark or dogfish, is at the same time flattened or *depressed* in its vertical section. We hasten to enquire whether there be any simple relation of *magnitude* discernible between these twin factors of expansion and compression; and the very fact that the two dimensions of breadth and depth tend to vary inversely assures us that, in the general process of deformation, the volume and the area of cross-section are less affected than are those two linear dimensions. Some years ago, when I was studying the weight-length coefficient in fishes (of which we have already spoken in chapter II), that is to say the coefficient k in the formula $W = kL^3$, I was not a little surprised to find that k (let us call it in this case k_1) was all but identical in two such different-looking fishes as the haddock and the plaice: thus indicating that these two fishes have approximately the same *volume* when they are equal in *length*; or, in other words, that the extent to which the plaice has broadened is *just about compensated for* by the extent to which it has also got flattened or thinned. In short, if we might conceive of a haddock being transformed directly into a plaice, a very large part of the change would be accounted for by supposing the round fish to be 'rolled out' into the flat one, as a baker rolls a piece of dough. This is, as it were, an extreme case of the *balancement des organes*, or 'compensation of parts'.

We must not forget, while we consider the 'deformation' of a fish, that the fish, like the bird, is subject to certain strict limitations of form. What we happen to have found in a particular case was

324

observed fifty years ago, and brought under a general rule, by a naval engineer studying fishes from the shipbuilder's point of view. Mr Parsons compared the contours and the sectional areas of a number of fishes and of several whales;[1] and he found the sectional areas to be always very much the same at the same proportional distances from the front end of the body.[2] Increase in depth was balanced (as we also have found) by diminution of breadth; and the magnitude of the 'entering angle' presented to the water by the advancing fish was fairly constant. Moreover, according to Parsons, the position of the greatest cross-section is fixed for all species, being situated at 36 per cent of the length behind the snout. We need not stop to consider such extreme cases as the eel or the globefish (*Diodon*), whose ways of propulsion and locomotion are materially modified. But it is certainly curious that no sooner do we try to correlate deformation in one direction with deformation in another, than we are led towards a broad generalisation, touching on hydrodynamical conditions and the limitations of form and structure which are imposed thereby.

Our simple, or simplified, illustrations carry us but a little way, and only half prepare us for much harder things. But interesting as the whole subject is we must meanwhile leave it alone; recognising, however, that if the difficulties of description and representation could be overcome, it is by means of such co-ordinates in space that we should at last obtain an adequate and satisfying picture of the processes of deformation and the directions of growth.

[1] H. de B. Parsons, 'Displacements and Area-curves of Fish', *Trans. Amer. Soc. of Mechan. Engrs*, **9** (1888), 679–95.

[2] That is to say, if the areas of cross-section be plotted against their distances from the front end of the body, the results are very much alike, for all the species examined. See also Selig Hecht, 'Form and Growth in Fishes', *J. Morph.* **27** (1916), 379–400.

X. EPILOGUE

The fact that I set little store by certain postulates (often deemed to be fundamental) of our present-day biology the reader will have discovered and I have not endeavoured to conceal. But it is not for the sake of polemical argument that I have written, and the doctrines which I do not subscribe to I have only spoken of by the way. My task is finished if I have been able to show that a certain mathematical aspect of morphology, to which as yet the morphologist gives little heed, is interwoven with his problems, complementary to his descriptive task, and helpful, nay essential, to his proper study and comprehension of Growth and Form. *Hic artem remumque repono.*

And while I have sought to show the naturalist how a few mathematical concepts and dynamical principles may help and guide him, I have tried to show the mathematician a field for his labour—a field which few have entered and no man has explored. Here may be found homely problems, such as often tax the highest skill of the mathematician, and reward his ingenuity all the more for their trivial associations and outward semblance of simplicity. *Haec utinam excolant, utinam exhauriant, utinam aperiant nobis Viri mathematice docti.*[1]

That I am no skilled mathematician I have had little need to confess. I am 'advanced in these enquiries no farther than the threshold'; but something of the use and beauty of mathematics I think I am able to understand. I know that in the study of material things, number, order and position are the threefold clue to exact knowledge; that these three, in the mathematician's hands, furnish the 'first outlines for a sketch of the Universe'; that by square and circle we are helped, like Emile Verhaeren's carpenter, to conceive 'Les lois indubitables et fécondes Qui sont la règle et la clarté du monde'.

For the harmony of the world is made manifest in Form and Number, and the heart and soul and all the poetry of Natural Philosophy are embodied in the concept of mathematical beauty. A greater than Verhaeren had this in mind when he told of 'the golden compasses prepared In God's eternal store'. A greater than Milton had magnified the theme and glorified Him 'that sitteth upon the circle of the earth', saying: He hath measured the waters in the

[1] So Boerhaave, in his *Oratio de Usu Ratiocinii Mechanici in Medicina* (1703).

hollow of his hand, and meted out heaven with the span, and comprehended the dust of the earth in a measure.

Moreover, the perfection of mathematical beauty is such (as Colin Maclaurin learned of the bee), that whatsoever is most beautiful and regular is also found to be most useful and excellent.

Not only the movements of the heavenly host must be determined by observation and elucidated by mathematics, but whatsoever else can be expressed by number and defined by natural law. This is the teaching of Plato and Pythagoras, and the message of Greek wisdom to mankind. So the living and the dead, things animate and inanimate, we dwellers in the world and this world wherein we dwell— πάντα γα μὰν τὰ γιγνωσκόμενα—are bound alike by physical and mathematical law. 'Conterminous with space and coeval with time is the kingdom of Mathematics; within this range her dominion is supreme; otherwise than according to her order nothing can exist, and nothing takes place in contradiction to her laws.' So said, some sixty years ago, a certain mathematician;[1] and Philolaus the Pythagorean had said much the same.

But with no less love and insight has the science of Form and Number been appraised in our own day and generation by a very great Naturalist indeed[2]—by that old man eloquent, that wise student and pupil of the ant and the bee, who died while this book was being written; who in his all but saecular life had tasted of the firstfruits of immortality; who curiously conjoined the wisdom of antiquity with the learning of today; whose Provençal verse seems set to Dorian music; in whose plainest words is a sound as of bees' industrious murmur; and who, being of the same blood and marrow with Plato and Pythagoras, saw in Number *le comment et le pourquoi des choses*, and found in it *la clef de voûte de l'Univers*.

[1] William Spottiswoode, in his presidential address to the British Association at Dublin in 1878.
[2] Henri Fabre.

327

INDEX

Abel, O., 257 n.
Acanthaceae, cystoliths of, 134
Acantharia, strontium sulphate skeletons of, 154, 167
Aceratherium tridactylum, transformation of skull of, 310–11
Acervularidae, shell of, 196
Actinomma arcadophorum, skeleton of, 157, 158
Actinophrys, vacuoles in, 92
Actinosphaerium, vacuoles in, 92, 156
adaptation of organisms by operation of mechanical forces, 221–67
Addison, J., 4
adsorption at surfaces of cells, 101, 139, 141–5, 156
Aglaophenidae, shapes of calycles of, 296–7
Agyropelecus olfersi, transformation of, 298–9
albatross, hollow wing-bones of, 226
Alcyonaria, spicules of, 134
Allman, W., 71, 73
Amelung, E., 39 n.
ammonites, spiral shells of, 189, 190, 191
Amoeba, physical properties of cell of, 11–12
Amoeba spp., inter-relations of, 196
amphibia, size of blood corpuscles of, 39
Anabaena, chains of cells in, 98
Anaxagoras, 8
Andrews, C. W., 266 n.
angle
    of enveloping cone (angle β), in shells, 191, 192; in horns, 207
    of lag between rotations of tusk and body in narwhal, 219
    of Maraldi, 107, 110–14, 118 n., 123, 141, 160, 161, 163, 164
    of maximum value of shear, 236
    of retardation of spiral (angle γ), in shells, 191, 193
    of spiral (angle α), in beaks, 215; in shells, 189, 191, 192; in tusks, 216
angles, facial, of man and apes, 318, 320
angles
    of branching of blood vessels, 128–9
    of cell-walls of honeycomb, 97, 100, 107–19; of segmenting eggs, 100
    of radii of leaves, 280–1
    of triradiate spicules of sponges, 138–40
    of tusks of mastodons, 216
    of vertebral processes, 260
anisotropy
    and asymmetry, 76–80, 212
    in treacle, 240

Anodon, transformation of shell of, 298
anteater, stresses in backbone of, 255
antelopes, horns of, 202, 204–8, 210, 212
Anthony, R., 35 n.
Antigonia capros, transformation of, 300, 324
Apatornis, transformation of pelvis of, 300, 324
apes, skulls of, 318–22
apple, shape of, 281
aquatic animals
    motion of, 32, 218, 318
    size of, 22, 43
    speed of, 22–3
    structure of, 235, 258–60, 266
Arabian Nights, 108
arch, comparison of quadruped structure to, 244
Archaeopteryx, transformation of pelvis of, 306–9
Archibald, R. C., 178
Archimedean bodies, 121
    spiral, 175–7
Archimedes, 15
Archiscenium, skeleton of, 163
areas, Law of Minimal, 50, 51, 53, 61, 65, 99, 104
Argonauta, shell of, 193
Aristotle, viii, 3, 4, 5, 6, 8, 9 n., 181, 182, 198, 264, 273, 274
Aristotelean 'excess and defect' in differences between species, 273
arrest, point of (or node), in growth of leaf or fruit, 278–82
arthropods, bristles, hairs, and spines of, 81
    hollow structure of, 227
Asparagus sprengeri, 14-hedral cells of, 122
Astrorhiza, amoeboid body of, 198, 199, 201
asymmetry, molecular, 136–8, 240
asymmetry
    and anisotropy, 76–80, 212
    and ciliated organisms, 81
    in leaves, 280
    of skull, in some Cetacea, 218, 220
Auerbach, F., 9
Aulastrum triceros, skeleton of, 160
Aulonia hexagona, siliceous skeleton of, 156, 157, 158
Ausonius, 108
Avicula, transformation of shell of, 298

Babak, E., 35 n.
Babirussa (wart-hog), tusks of, 215, 216

329

baboon, transformation of skull of, 319, 321, 322
Backhouse, T. W., 50 n.
Bacon, Lord, 60, 266
Bacon, Roger, 1, 6
bacteria
    absence of nucleus from, 41
    rate of fall of, 43, 44
    size of, 42
Baer, K. E. von, 3
Ball, O. M., 230 n., 239 n.
bamboo
    size of, 21 n.
    stiffening rings in, 230
Barclay, J., 119 n.
Bartholin, E., 109, 116, 153 n.
Basil, St, 108 n.
bat, hearing of, and sounds made by, 34
Bateson, W., 217 n.
Batsch, A. J. G. C., 197 n.
Baylis, H. A., 214 n.
beading
    of edge of splash, 67
    of filaments, 65, 66, 75, 82
beaks, shapes of, 214–16
bean, shape of, 281
bear, bones of foot of, 235
beaver, curvature of incisors of, 175, 215, 216
Becher, S., 146 n.
bees
    food of, 25
    honeycomb of, 2, 97, 100, 101, 107–19, 139
    motion of swarm of, 46 n.
    rate of wing-beat of, 34 n.
beetles, minimum length of, 42
Begonia, crystals of oxalate in, 133
Begonia daedalea, shape of leaf of, 280
Beijerinck, M. W., 42
Bell, Sir C., 32, 231, 240 n.
Belodon, skull of, 302
Bénard, H., 104, 105, 106, 107 n.
bending moments, 20, 179, 213, 226, 227, 248–52, 256, 258, 259
Bennett, G. T., 64, 184 n.
Benninghoff, A., 234 n.
Benton, J. R., 65 n.
Berezowski, A., 41 n.
Bergmann, C., 25 n.
Bergmann's Law, relating size of animals to temperature of surroundings, 26
Bergson, H., 201 n.
Berkeley, G., 16 n.
Bernard, Claude, 1, 3
Bernhardt, H., 234 n.
Bernoulli, J., 28 n., 178, 180
Berthold, G., 8, 66, 93
Bezold, W., 70 n.
birds
    flight of, 30–2

minimum weight of, 26
    transformations of pelves of, 306–9
    wing-bones of, 226, 235
Bishop, J., 29 n.
bison, skeletal structure of, 243, 255
Bjerknes, V., 38
blackcock, flight of, 31
Blackwall, J., 65
Blair, G. W. S., 239 n.
Blake, J. F., 191
blood corpuscles, size of, in different animals, 39 n.
blood vessels, branching of, 125–31
Blum, E., 128 n.
Blum, H. F., 14 n.
boar, canine teeth of, 215
Boerhaave, H., 326 n.
Boltzmann, L., 14 n., 62
Bonanni, A. P. P., 103, 119
Bonaventure, St, 14
bone
    deformation of, 210
    growth of, 238–9
bones, corresponding to lines of compression in body, 224, 242, 250
    mechanical aspects of structure of, 223–67
Bonnet, C., 118 n.
bony core of horns, 205–6, 209–11
Borelli, J. A., 8, 26–8, 241
Borelli's Law, of work of muscle, 27
Boscovich, R. J., 112, 113, 114
Bourgery, J. M., 234 n.
bourrelet of Plateau (Gibbs's ring), at boundaries of surfaces and cells, 92–4, 114, 140, 159 n., 163
Boveri, T., 41
Bower, F. O., 35 n.
Boyle, R., 55 n.
Bradfield, J. R. G., xiv
Bramanti, 19 n.
branching of blood vessels, 125–31
Brandt, A., 155 n.
Bravais, L. and A., 187, 188
Brazier, L. G., 227 n.
Breton, E. le, 39 n.
Bridgeman, P. W., 18 n.
bridges, comparison of quadruped structure to, 242–58, 260–3
Brobdingnag, 16, 17, 29 n.
Brody, S., xiii, 26 n.
Broglie, L. de, 45 n.
Brooke, Sir V., 204 n., 206, 213 n.
Brougham, Lord, 112, 114, 119 n.
Brown, R., 45, 46
Browne, Sir T., 103, 169 n., 280
Brownian movement, 44–8
Brownlee, J., 46 n.
bubbles
    bursting of, 66 n., 70 n.
    faces between, 76, 91, 94–100

bubbles (*cont.*)
  gravity and, 76
  shapes of, 55, 90, 160–2; in a froth, 121, 125
  surface tension and, 43, 51–2, 94–7, 142
  volume and surface area of, 38
Buch, L. von, 188
Buddenbroek, W., 178 n.
buffalo, Indian, balance of head of, 214
buffaloes, horns of, 205
Buffon, G. L. L., 107, 117, 122, 123, 124, 125, 226 n.
Bulimi, shells of, 191
bull, neck of, 19
Buller, A. H. R., 44 n.
bustard, flight of, 31
Butler, S., 4
Bütschli, E., 80, 154, 155
Büttel-Reepen, H. von, 116 n.

calcium salts in organisms, 132–5, 138, 145
Callimitra agnesae, siliceous skeleton of, 160–3
calorie output per kg. of mammals of different sizes, 25 n.
calycles of hydroids, shapes of, 67, 68, 296, 297
Calyptraea, shell of, 191
camel
  load on fore- and hind-feet of, 254
  stresses in backbone of, 255
Campanularidae, shapes of calycles of, 68, 296, 297
Camper, P., 276 n., 292 n.
Camptosaurus, transformation of pelvis of, 303, 304
canary, claws of, 175, 214
cancellous tissue, 231, 234 n.
Candolle, A. P. de, 21
cannon-bones
  of horses, 30
  of ox, sheep, and giraffe, 276–7, 288–90
cantilever structure
  compression and tension in, 227
  resemblance of quadruped to, 245–58, 260–2
capercailzie, flight of, 31
Capra falconeri, horns of, 208
Caprella, transformation of, 293
carapaces of crabs, shapes of, 293–5
Carnot, S., 6 n.
Carnoy, J. B., 156 n.
Carpenter, W. B., 45 n.
Carruccio, E., 112 n., 119 n.
Cartesian co-ordinates and transformations, 268, 271–325
Cassini, D., 153
Cassis, shell of, 193
Castellan, A. L., 114 n.
Castillon, G. F. M. M. S., 112 n., 113 n.

cat
  claws of, 193
  ganglion cell of, 40
  tail of, 20 n.
catenary curve, in backbone of sloth, 258
catenoid (surface of revolution)
  properties of, 53, 54, 57–9
  in ciliate *Trichodina pediculus*, 85
cattle
  cannon-bones of, 288
  rings in horns of, 203 n.
  see also ox
Caulerpa, size of cells of, 39
celestine (sulphate of strontium and barium), in capsules of *Collosphaera*, 155
cells
  aggregates of (tissues), 88–131
  division of, 13, 37
  forms of, 49–87
  partitions between, 94–101
  sizes of, 37–42
  walls of, 77, 92–7
Cenosphaera spp., skeletons of, 158, 159
centipede, rate of movement of, 29
Ceratorhinus, spiral curvature of horn of, 202
Cerithium, shell of, 193
Césaro, G., 107 n., 119 n.
Cetacea, asymmetry of skull in, 218, 220
Chabry, L., 28 n., 135 n.
Chambers, R., 66
chamois, horns of, 205
Chapman, A., 31
cherry, shape of, 281
chevron bones, 259
chimpanzee, transformation of skull of, 319–22
chitinous material of horn, akin to hair, 202, 203
chitinous pellicle in Foraminifera, 135
chitinous skeleton of *Clathrulina*, 159
Chiton, shell of, 187 n.
Chladni, E. F. F., 94
Chorinus, transformation of carapace of, 294
cilia
  associated with parts of low surface-tension, 81
  maintaining equilibrium of form, 81, 86–7
Circogonia icosahedra, skeleton of, 168
Circorrhegma dodecahedra, skeleton of, 168, 169
Circospathis novena, skeleton of, 168
Circoporus spp., skeletons of, 168
Cladocarpus crenatus, transformation of calycle of, 297
Cladonema, resemblance of, to breaking drop, 73, 74
Clathrulina, chitinous skeletons of, 159

331

332

Fontenelle, B. le B. de, 111, 242 n.
Foraminifera
    calcium in shells of, 135
    rate of fall of, in water, 44
    spiral shells of, 173, 193–6
    unduloid forms of, 82
    variation in shells of, 135, 196–201
Forbes, E., 155 n.
fore-feet of quadrupeds, load on, 254
form
    and correlation between characters in
        organisms, 275
    and mathematics, 268–85
    and mechanical efficiency, 221–67
    of an organism as diagram of forces, 11
    and phylogeny, 132, 149, 151, 267
    and size, 36
    and species, 149, 151, 166, 171, 197–9
    and strength, 18, 22, 225–325
forms, comparison of related, 268–325
forms
    of cells, 49–87
    of horns, 172, 174, 202–14
    of shells, 173, 174, 186, 188–96
    of tissues (cell aggregates), 88–131
Forth Bridge, 230, 245, 251
Fraas, E., 266 n.
fracture, Coulomb's theory of, 237
fracture of bone, effect of, on trabeculae,
    238
Fraenkel, G., 25 n.
friction, coefficients of, of salt and fresh
    water, 44
friction, and motion through water, 23,
    72 n.
Froude's Law of the Correspondence of
    Speeds, 23, 29, 31, 32
Froude's Law of Steamship Comparison,
    24
fruit-flies, effect of temperature shock on
    development of, 221–2
fruits
    shapes of, 19, 281–2
    strength and thickness of stalks of, 19
        239
Fucus, spherical oospheres of, 61 n.
Fulton, A. R., 230 n.

Galen, 3
Galiani, the Abbé, 14, 137
Galileo, 6 n., 8, 19, 20, 21, 226, 228, 249,
    250, 269
Galileo's Principle of Similitude, 19–26
Gamble, F. A., 152 n.
ganglion cells, size of, in different animals,
    40
Gans, R., 46
Gardiner, J. H., 173
gavial, transformation of skull of, 302
gazelle, size and shape of, 20
Gebhardt, W., 234 n.

gelatine, tension and compression in, **224**
gemsbok, horns of, 206
genetics, Mendelian, 264, 275, 286 n.
Geoffroy St Hilaire's Law of Compensa-
    tion, 264
Geomys, curvature of incisors of, 215
Geryon, transformation of carapace of,
    293
Ghetti, A., 115 n.
Gibbs, Willard, xiii
Gibbs's ring, at boundaries of surfaces and
    cells, 92–4, 114, 140, 159 n., 163
Gidley, J. W., 317
gills, in crustacea and worms, 35
Gilmore, C. W., 257 n.
giraffe
    cannon-bones of, 277, 288, 290
    stresses in backbone of, 256
girders, 18, 225–6, 244, 247, 248, 250–4,
    261
glacier, tension and compression in, 223–4
Glaisher, J. W. L., 112, 113, 114
glass-blowing, 51, 55, 63, 69, 75, 287
Globefish (Diodon), transformation of,
    300, 301, 325
Globigerina
    beaded protoplasm of, 65
    spiral shell of, 174, 176, 194, 195
gnat, size of limbs of, 22
gnomons, 181–7, 193
goats, horns of, 202, 203, 206–8, 210
Goethe, J. W. von, 2, 21, 41, 264, 269
Goliath beetle, and mouse, 42
Gonothyraea hyalina, transformation of
    calycle of, 297
Gontarsi, H., 115 n.
Goodsir, J., 155 n., 174 n.
goose
    flight of, 31
    proportion of bone in body of, 20
gorilla
    bones of foot of, 235
    transformation of skull of, 320
Gott, J. P., 68 n.
Gottlieb, H., 251 n.
Goupillière, H. de la, 174 n.
gourds, shapes of, 286
Grandi, G., 283
Grantia, spicules of, 140
graphic statics, 232, 241, 249 n.
grasshopper, jumping powers of, 27–8
gravity
    and Amoeba, 12
    and aquatic animals, 20
    and forms of organisms, 288
    and orbits of planets, 178
    and rate of travel of waves, 36
    as size-limiting factor, 32–3
    and size of stars, 38
    and small organisms, 36–7, 43, 48
    and swing of leg in walking, 28

340

Petersen, C., 180 n.
Peterson, O. A., 317
Pfeffer, W., 239
Pflüger, E., 230 n.
Phaeodaria, skeletons of, 167
Philolaus, 327
phylogeny
and forms of organisms, 132, 149–51, 200
the problem of, 265–7
physical sciences, and natural history, 1–14, 326–7
physiology, a problem in maxima and minima, 127
pigeon, size of eye of, 35
Pike, F. H., 127 n.
Pileopsis, shell of, 190
Pinna, transformation of shell of, 298
Pithecanthropus, skull of, 320
plaice
skeletal structure of, 260
transformation of, 323, 324
plane (surface of revolution), properties of, 53–5, 57–8
Planorbis, shell of, 189
plant fibres, strength of, 228–9, 239–40
plants
breaking-strength of shoots of, 239
growing shoots of, 102
spirals in, 187, 188, 229 n., 284
Plateau, F., 28, 53, 55, 58, 59, 60, 62, 63, 64 n., 65, 75, 76, 80, 92, 93, 94, 100, 159 n., 160, 161, 163
Plateau's bourrelet, at boundaries of surfaces and cells, 92–4, 140, 159 n.
Plateau's surfaces of revolution, 49, 53–61
biological examples of, 80–7
Plato, 2, 153, 169 n., 269, 327
Platonic bodies, 167
Pliny, 108, 172
Plochman, G. K., vii
Plumulariidae, transformations of calycles of, 296, 297
Plutarch, 269 n.
Pluteus larva, in calcium-free water, 135
Podocyrtis, skeleton of, 165
poecilothermic (variable-temperature) animals, 25
Poincaré, H., 270
Poiseuille, J. L. M., 130 n.
Poiseuille's Law, of work done in causing fluid to flow through a tube against resistance, 131
polar furrow, 100, 101, 145, 165
polarity, and liquid crystals, 169–71
Polistes, cells of, 108
Polyprion, transformation of, 299, 300
polyps, hydroid, resemblance of, to splashes, 70
Pomacanthus, transformation of, 298, 299
Popoff, M., 39 n.

porpoise, skeletal proportions of, 20
Porsild, M. P., 216 n.
Potamides, shell of, 189
potter, art of the, 69, 70
Pouchet, G., 135 n.
Powell, B., 119 n.
Price, A. T., 223 n.
Priestley, J. H., 94 n., 102 n.
Primula, giant race of, 42 n.
Prismatium tripodium, skeleton of, 164
Prochnow, O., 236
propulsion
importance of metacarpal region in, 288 n.
moment of, in the horse in motion, 246
Protococcus
cell wall of, 13
chains of cells in, 98
spherical form of, 80
Protohippus, transformation of skull of, 315, 317
protoplasm
beading of, 65–6
fluid properties of, 12, 78
overspreading of shell of Orbulina by, 195
protoplasm of Amoeba, 12
Przibram, K., 46 n., 47
Psammobia, transformation of shell of, 298
Pseudopriacanthus altus, transformation of, 299, 300
Pteranodon, transformation of skull of, 305
Pulvinulina, shell of, 194
Pupae, shells of, 191
Pütter, A., 35 n.
Pythagoras, 2, 181, 183, 269, 327

quadrupeds, resemblance of structure of, to cantilever structure, 245–58, 260–3
queen bees, cells of, 115
Quincke, H. H., 38, 106 n.
quincuncial symmetry, 103

rabbit
calorie output of, 25 n.
ganglion cell of, 40
growth of incisor teeth of, 214
transformation of skull of, 313
vertebral spines of, 251 n.
Rabelais, 4
Rabl, K., 100 n.
Radau, M., 28 n.
Radiolaria, skeletons of, 151–69
rainbow, sphericity of drops in, 50 n.
Rainey, G., 7 n.
ram, balance of head of, 213–14
Rameaux, J. F., 25 n.
Rankine, W. J. McQ., 249 n., 262 n.
Rasumowsky, V. I., 234 n.

341

stag-beetle, strength of, 26
Staigmüller, H., 290 n.
stalks of fruit, strength and thickness of, 19, 239
standing posture, 231, 241, 242
star, mass of, 38
statics, graphic, 232, 241, 249 n.
Stay, B., 113 n.
staying power, and size of animal, 24, 31
steady state, in diffusion models, 106
  (or pseudo-equilibrium) of living cell, 61
Stegocephalus inflatus, transformation of, 295–6
Stegosaurus
  skeletal structure of, 257, 261
  transformation of pelvis of, 303, 304
Steinbruck, C., 229 n.
Steiner, J., 60 n.
Stentor, nodoid form of cell of, 86
Sternoptyx diaphana, transformation of, 298, 299
Stillman, J. D., 246 n.
stilt, relation of weight to leg length in, 15 n.
Stirling, 'Old', 155 n.
Stokes, Sir G. G., 44
Stokes's Law, of rate of fall of small bodies, 43, 44, 48
Stomatella, shell of, 190
stonecrop, arrangement of leaves of, 284
Stoney, G. J., 47
stork, rate of wing-beat of, 31
strain
  as stimulus to growth, 238, 241
  and stress, 238–41
  in torsion of elastic body, 210–11
Strasbürger, E., 39 n.
Strauss-Dürckheim, H. E., 28 n.
straw, hollow structure of, 20 n., 226, 228
straws, shearing force of wind on, 237
Strehlke, F., 64 n.
strength
  and form, 18, 22, 225–30
  of materials, 225, 261
  of metal wires, 228
  of plant fibres, 228–9, 239–40
  and size, 23, 26
stress diagrams of bridges and quadrupeds, 247–55
stress and strain, 238–41
stress, *see* compression and tension
strontium sulphate, in skeletons of Acantharia, 154
Stuart, T. P. A., 288 n.
Succinea, shell of, 191
sunfish (*Orthagoriscus*), transformation of, 300, 301, 324
sunflower
  breaking-strength of shoots of, 239
  spiral in florets of, 172, 174, 175

surface
  of cell aggregates (tissues), 101–2
  and heat loss, 25
  of lung, 35
  and size, 15, 19
  and volume, 35–8
surface tension, 43–4
  in cell aggregates (tissues), 88–102
  and cell form, 49–50
  in cells, 37
  at curved elastic surfaces, 43, 51–2, 77–8, 125–6
  at edges of cells, 139
  in formation of spicules, 141, 142, 150
  and rate of travel of ripples, 36
  and small organisms, 33, 36, 48
surfaces of revolution
  Plateau's, 53–61
  biological examples of, 80–7
suture in shells, 190
Svedberg, T., 45 n.
Swammerdam, J., 8, 110, 113 n., 114, 188
swan, flight of, 31
Swann, M. M., xiii
Swift, J., 16 n.
Sylvester, J. J., 1
symmetry
  hexagonal, 103–19
  quincuncial, 103
Synapta spp., spicules of, 145–6
Syncoryne, resemblance of, to vorticoid drop, 73

Tait, P. G., 1, 38 n., 59 n.
Takahasi, K., 104 n.
tapir, transformation
  of skull of, 311, 312
  of toes of, 290–2
Tartar, V., 122 n.
Tate, J., 7 n.
Taylor, Sir G. I., 75 n.
Taylor, Jeremy, 108
teeth, shapes of, 214–16
Teissier, G., x
Teixeira, G. de, 178 n.
tension, lines of, followed by muscles and ligaments, 224, 242, 250
tension and compression, in engineering and in organisms, 223–62
Terebra, shell of, 219
Teredo, tube of, 135 n.
Terni, T., 39 n.
Terquem, O., 119 n.
Tetley, U., 102 n.
tetractinellid sponges, spicules of, 139, 143, 144
tetrahedron
  in skeleton of *Callimitra*, 160–3
  in skeletons of Radiolaria, 167
tetrakaidekahedron, as space-filling solid, 121, 122, 124, 125

344

Seven Clues to the Origin of Life

A. G. Cairns-Smith

This book addresses, in the spirit of an intriguing detective story, the question of how life may have arisen on earth. It relies on the methods of Sherlock Holmes, in particular his principle that one should use the most paradoxical features of a case to crack it. This approach to the essential biological problems is not merely light-hearted, but a fascinating scrutiny of some fundamental questions.

'It is a summary of the best evolutionary thinking as applied to the origins of life in which the important issues are addressed pertinently, economically and with a happy recourse to creative analogies.'

Nature

'... a splendid story ... here he [Cairns-Smith] sets it out in a way from which anyone – even those whose chemistry and biology stopped at 16 – can learn.'

New Statesman

1990   216 × 138 mm   143 pp
Paperback 0 521 39828 2